Connectivity and the Mobility Industry

Other SAE books of interest:

Green Technologies and the Mobility Industry
By Dr. Andrew Brown, Jr.
(Product Code: PT-146)

Active Safety and the Mobility Industry
By Dr. Andrew Brown, Jr.
(Product Code: PT-147)

Automotive 2030 – North America
By Bruce Morey
(Product Code: T-127)

Multiplexed Networks for Embedded Systems
By Dominique Paret
(Product Code: R-385)

Connectivity and the Mobility Industry

By Dr. Andrew Brown, Jr.

Warrendale, Pennsylvania, USA

400 Commonwealth Drive
Warrendale, PA 15096-0001 USA

E-mail: CustomerService@sae.org
Phone: 877-606-7323 (inside USA and Canada)
724-776-4970 (outside USA)
Fax: 724-776-0790

ISBN 978-0-7680-4767-7
Library of Congress Catalog Number 2011937693
SAE Order No. PT-148
DOI 10.4271/PT-148

To purchase bulk quantities, please contact

SAE Customer Service
e-mail: CustomerService@sae.org
phone: 877-606-7323 (inside USA and Canada)
724-776-4970 (outside USA)
fax: 724-776-0790

Visit the SAE International Bookstore at http://books.sae.org

Table of Contents

Introduction

Special Contributions

Challenges and Benefits

Communications – Vehicular Safety

Communications – Vehicle Networks

Communications – PHEV/EV Requirements

Technologies

Applications

Editor's and Special Contributors' Biographies

Introduction

Introduction

The nature of being connected while driving has changed dramatically over the years. It was once sufficient to have AM radio as base reception and to seek FM signals while driving long distances. There were regional lists of such stations that drivers could tune to for favorite programs. Now, with the emergence of portable nomadic devices (PNDs) not only does the vehicle occupant want AM/FM, but XM satellite radio, MP3 functionality, smart phone connectivity, and other applications. In addition, sophisticated networks such as DSRC, WIFI, WIMAX, Bluetooth, and 3G can enable driver-to-service, vehicle-to-vehicle, and vehicle-to-infrastructure functionality. The vehicle is regarded not only as a communication site, but also as a node on the internet.

The challenges of greenhouse gas (GHG), fuel economy improvement, emissions reductions, and even increased safety pose the need for the vehicle to connect to the electric power grid for battery charging. The electric power grid presents its own communication and connectivity issues augmented by the opportunity to perform wireless charging. The emergence of fourth generation (4G), long term evolution (LTE) represents an opportunity to provide seamless connectivity across silos in the process.

In this publication, we intend to explore these dimensions of *Connectivity and the Mobility Industry*, starting with three excellent articles newly written for this project, which lay the framework for our discussions.

The articles and authors are:

- "What to Expect Beyond 2015 - Fourth Generation (4G) Wireless and the Vehicle" by Steven Bayless and Scott Belcher
- "The Evolution of the Driving Experience and Associated Technologies" by Douglas Welk et al.
- "Wireless Charging of Electric Vehicle Converged with Communications Technology" by In-Soo Suh, PhD.

We augment these articles with 20 specially selected papers intended to elaborate on the dimensions of *Connectivity and the Mobility Industry*. For clarity, they are listed in the categories of: Challenges and Benefits, Communications - Vehicular Safety, Communication - Vehicle Networks, Communication – PHEV/EV Requirements, Technologies, and Applications.

Challenges and Benefits

- "Autonomous Driving - A Practical Roadmap" by Jeffrey D. Rupp & Anthony G. King, 2010.
- "The Line Within: Redrawing the Boundary of Connected Vehicle Systems Engineering" by Robert Gee, 2010.
- "Metrics for Evaluating Electronic Control System Architecture Alternatives" by Arkadeb Ghosal et al., 2010.
- "Connected Vehicle Accelerates Green Driving" by Tsuguo Nobe, 2010.

Communications – Vehicular Safety

- "Enabling Safety and Mobility through Connectivity" by Chris Domin, 2010.
- "Vehicle Safety Communications - Applications: System Design and Objective Testing Results" by Farid Ahmed-Zaid et al., 2010.
- "Prioritized CSMA Protocol for Roadside-to-Vehicle and Vehicle-to-Vehicle Communication Systems" by Jun Kosai et al., 2009.

Communications – Vehicle Networks

- "Vehicular Networks for Collision Avoidance at Intersections" by Seyed Reza Azimi et al., 2011.
- "Nomadic Device Connectivity Using the AMI-C HMI Architecture" by Frank Szczublewski et al., 2009.
- "Verify-on-Demand - A Practical and Scalable Approach for Broadcast Authentication in Vehicle-to-Vehicle Communication" by Hariharan Krishnan et al., 2011.

Communications – PHEV/EV Requirements

- "Communication Requirements for Plug-In Electric Vehicles" by Richard A. Scholer et al., 2011.
- "Communication between Plug-in Vehicles and the Utility Grid" by Richard A. Scholer et al., 2010.

Technologies

- "Intelligent Vehicle Technologies that Improve Safety, Congestion, and Efficiency: Overview and Public Policy Role" by Eric C. Sauck, 2009.
- "Eco Navigation with Vehicle Interaction" by Ricardo Takahira, 2010.
- "Comparative Analysis of Automatic Steering Technologies and Intelligent Transportation System Applied to BRT" by Leopoldo Yoshioka, 2010.
- "Development of HMI and Telematics Systems for a Reliable and Attractive Electric Vehicle" by Shoichi Yoshizawa et al., 2011.

Applications

- "Commercial Business Viability of IntelliDrive Safety Applications" by Robert White et al., 2010.
- "Performance of Aftermarket (DSRC) Antennas inside a Passenger Vehicle" by Radavan Miucic and Sue Bai, 2011.
- "Cybercars for Sustainable Urban Mobility - A European Collaborative Approach" by Michel Parent, 2010.
- "Merge Ahead: Integrating Heavy-duty Vehicle Networks with Wide Area Network Services" by Mark P. Zachos, 2010.

Personal mobility, individual communications, the internet, and the power grid are now merging in the transportation sector. These elements can help to improve the human condition or detract from it. It is the responsibility of the engineering profession to help provide the right solutions at the right price with the right quality at the right time. We expect this publication to aid in that endeavor.

Dr. Andrew Brown, Jr., P.E., FESD, NAE
2010 SAE President
Executive Director and Chief Technologist
Delphi Corporation

Special Contributions

What to Expect Beyond 2015 - Fourth Generation (4G) Wireless and the Vehicle

Steven H. Bayless
Scott Belcher
The Intelligent Transportation Society of America (ITS America)

What to Expect Beyond 2015 - Fourth Generation (4G) Wireless and the Vehicle.

Steven H. Bayless
Scott Belcher
The Intelligent Transportation Society of America (ITS America)

Abstract

Connected vehicle applications are set to explode in the next few years because of the development of mobile application platforms such as Google Android and others, and enterprise telematics/machine-to-machine support services. Supporting these platforms beyond 2015 is fourth generation (4G) Long Term Evolution (LTE) wireless and its progeny. This paper contends that automotive application developers will need to be cognizant of how application data is treated by 4G systems, and how innovations such as "traffic shaping" might improve quality of service for "off-board" (or "cloud-based") vehicular applications. Furthermore, "on-board" vehicular applications utilizing vehicle-to-vehicle, vehicle-to-infrastructure communications (such as femtocells, Wi-Fi, and for safety applications, Dedicated Short Range Communications/Wireless Access for Vehicular Environments (DSRC/WAVE) might also likely be integrated into 4G. Later versions of LTE (such as LTE Advanced) might establish and manage communication sessions that hop among many of the above-mentioned wireless technologies, a concept known as heterogeneous or "vertical" roaming.

Introduction

The Global System for Mobile Communications (GSM) Association predicts that there will be nearly 50 billion connected terminals by 2025, almost ten times the predicted human population of the world. Approximately three-quarters to one billion of these terminals will likely be automobiles. Automotive engineers designing connected systems will face a plethora of wireless technology options for vehicles that provide short range, long range, and regional and global connectivity. What the future holds for wireless technology, particularly for the communications infrastructure expected to be in place from 2015 and onward, is speculative. However, the broad outlines of what might be available are visible today.

Next generation wireless networks from 2015 on will have a repertoire of different techniques to prioritize communications based on application needs. Furthermore, not only will networks be smarter, but future mobile terminals will likely feature multiple standard air interfaces (such as WiFi, cellular, and even mobile satellite) and be able to choose the best network based on immediate needs (such as coverage, quality of service, or even cost). Vehicles will also likely have wireless systems devoted to safety, critical or highly mobile "spot" communications, to support such local applications as automatic collision notification, vehicle-to-vehicle cooperative collision avoidance, or vehicle/infrastructure applications (such as tolling or traffic signal preemption and intersection collision avoidance). Peer-to-peer communications systems, which seek opportunities to communicate directly to other nearby nodes without the need to sluggishly route data traffic to and from cell towers first, will likely expand as short range radio technologies improve.

With 4G cellular, we will see the completion of the extension of the internet protocols to the wireless environment. 4G likely represents the end of the traditional siloed telecommunications approach that has been the result of decades of investment in single-application "purpose built" wireless technologies (e.g., radio, TV, land mobile, and cellular) and regulatory practice. Current investment patterns in infrastructure provide a strong indication of technologies that will be available beyond 2015, as well as their quantity, quality, and cost.

This paper examines how next generation systems such as 4G (and to a lesser extent, 4G-interoperable Mobile Satellite Services) will be able to support vehicular applications. It suggests that over the long term, shorter range technologies such as 4G femto cells, WiFi, or DSRC/WAVE will likely be utilized in coordination or conjunction with "wide area" systems. Furthermore, the paper contends that automotive electronics engineers will need to be cognizant of how application data is treated by 4G systems, and how innovations such as "traffic shaping" may improve the performance of off-board (or cloud-based) vehicular applications.

Brief History of the Connected Vehicle

Motorola introduced the first commercial AM broadcast car radios in 1930. Large scale, two-way connected vehicle platforms, however, followed 65 years later when OnStar was formed as an alliance between General Motors (GM), Electronic Data Systems, and Hughes Electronics. PSA Citron, BMW, and a few others implemented telematics systems, while other companies such as Ford (under the brand name "WingCast") contemplated forays into telematics, but did not commit to major efforts until the late 2000s. In the meantime, aftermarket personal navigation devices (PND) took off as their prices dropped 50 to 60 percent over a 5-year period. Connected aftermarket PNDs began to find their way into vehicles. Ford joined the telematics field in collaboration with Microsoft to integrate smartphones into their telematics platform, Sync. Early telematics service providers used first and second generation (1G and 2G) cellular and some satellite communications systems in the 1990s and 2000s. In 2008, the FCC stopped requiring Mobile Network Operators (MNOs) to support first generation cellular in their infrastructure, requiring telematics pioneers such as OnStar to provide equipment upgrades to hundreds of thousands of existing customers.

In the meantime, the Intelligent Transportation Society of America (ITS), an industry association made up of auto, telecommunications and information technology/consumer electronics companies as well as public and private road operators, freight carriers, and transit operators, successfully petitioned the Federal Communications Commission (FCC) to allocate spectrum dedicated to ITS. By 1999, the FCC allocated 75 MHz of spectrum at 5.9GHz for DSRC to provide local-area vehicle-to-vehicle and vehicle-to-infrastructure communications that could support short range mobility and cooperative collision avoidance applications.

Since 2003, the U.S. Department of Transportation (USDOT) has sponsored research on applications utilizing DSRC/WAVE. It has also focused on application development using other communications systems such as cellular systems since 2007. The goal of the USDOT Connected Vehicle Research Program, known previously as the Vehicle Infrastructure Integration (VII)/IntelliDrive Program, was to establish concepts of operations, system requirements, and high-level architecture of a national scale, interoperable system that could utilize DSRC to support cooperative vehicle-to-vehicle and vehicle-to-infrastructure mobility and safety applications.

USDOT is currently determining research needs to conduct a potential National Highway Traffic Safety Administration (NHTSA) sponsored rulemaking in 2013. This rulemaking would likely explore a practical approach to wide-scale vehicle-to-vehicle collision avoidance application deployment. USDOT is on track to measure safety benefits and evaluate deployment strategies as a part of the rulemaking. In particular, USDOT is looking to the potential for aftermarket DSRC terminals to bridge the long equipment gap between old and new vehicles—DSRC as an embedded feature in new vehicles, with aftermarket devices supporting applications for older vehicles.

Finally, MNOs such as Verizon, AT&T, T-Mobile, and Sprint have aggressively marketed 4G technology as the "next generation of wireless" to retail-level consumers since 2009. MNO's are also attempting to convince enterprise customers (and in particular the auto industry), to upgrade to newer generations of equipment such as third generation (3G), which after a decade of provisioning is closing-in on near nationwide coverage. Once again, history may be repeating itself, as Telematics service providers, who committed to 1G technology in the 1990s only to have that system removed in the late 2000s, face the prospect of being stranded again with equipment in the 2010s that supports a technology that, in the MNO's view, is nearing obsolescence.

Navigating the Wireless Landscape

Digitization, and particularly convergence on internet protocol, is slowly transforming a number of legacy, single purpose-built wireless systems into more flexible, multi-purpose networks. The distribution of audio and video content and the provision of voice communications have for decades been synonymous with the wireless systems that supported them: radio, television, and telephone. Single purpose-built systems such as media broadcast or two-way communications—systems committed to a specific architecture and content transport function—are evolving on the margins to support generic data communications. The vibrant cellular wireless ecosystem, represented in large part by the appearance of application-based smartphones and other internet-enabled terminals, has been built off the back of a legacy network originally designed for wireless voice telephony.

There have been dramatic changes in wireless communications since the first light-vehicle telematics providers began operation in the late 1990s and the USDOT's connected vehicle and its precursor initiatives were first contemplated. The first trend, applicable to nearly all wireless systems, is digitization. The second trend, applicable to cellular systems, is the adoption of network services based on the suite of internet protocols. The structure of the wireless telecommunications industry—coarsely divided into cellular, land mobile, mobile satellite communications, terrestrial TV and broadcast radio, and satellite TV and broadcast radio—has been greatly influenced by past regulation, which has siloed technologies and services by application. Digitization and the adoption of the internet protocol in wireless networks has begun to break down the media and technology-oriented silos, revealing how different wireless technologies and network architectures might be best suited, or combined, to provide particular types of data-driven applications or content.

Because of the enormous growth in wireless internet demand and limited network capacity, since 2005 there has been huge investment on the part of MNOs to relieve bottlenecks. This has been done through new spectrum allocations, deployment of faster third and fourth generation radio access technologies (such as WiMax and LTE and their progeny), upgraded backhaul, and simpler, smarter, core networks that are able to route traffic from the radio access network faster, on through to the public internet and back.

For the most part, the cellular industry is the most dynamic part of the telecommunications sector in terms of coverage, the speed of integrating multiple generations of advanced technology in both networks and end-user devices, access to new spectrum, and the ability to adapt networks to the needs for internet connectivity. Mobile communication is now one of the most important growth sectors in both the telecommunications and information technology sectors. With the rollout of 3G and 4G in the early and late 2000s, MNOs have moved beyond mobile voice-only handsets to certify an enormous variety of mobile terminals including mobile computing platforms, nomadic Wi-Fi hotpots, and USB dongle modems.

New radio technologies and terminals, built to operate with cellular networks, are introducing more data-driven applications faster than satellite, TV, or radio with successive introductory deployments of 2G, 3G, and 4G over the period of a decade. In vehicles, the trend is toward at least one wide-area network terminal in the vehicle, either embedded or brought in as a consumer device. Because of the economies of scale already established by the large scale penetration of mobile voice services, cellular is able to bring together the activities of enormous numbers of innovative and competitive device and service suppliers within a single ecosystem.

Mobile Satellite Services (MSS) might follow closest behind cellular when measured by the standard of market potential for telematics and might begin to close the gap if it is able to adapt its architecture to support 4G cellular interoperability. MSS provides service where coverage may be unavailable from terrestrial-based networks. It plays a unique role in national communications infrastructure, especially for homeland security and public safety where a natural disaster such as an earthquake or hurricane can incapacitate terrestrial cellular or other fixed and mobile communications.

Two-way MSS has been an important part of commercial vehicle telematics because of its unique national coverage footprint. Satellite has been around for years, but has not achieved the mass adoption or success of cellular, primarily because of its higher cost of service and coverage limitations in built-up urban areas. This has meant that it has been adopted largely by enterprise, government, or rural subscribers who value either regional, global, or remote area coverage.

Currently, MSS providers are considering integrating cellular technology into their handsets through their network of ancillary base stations designed to provide coverage in urban areas. In the beginning of 2011, the FCC granted the major MSS provider LightSquared (formerly Skyterra) a request for modification of its authority for an ancillary terrestrial component (ATC). The grant allowed LightSquared and its wholesale customers to offer terrestrial-only 4G LTE compatible terminals rather than having to incorporate both satellite and terrestrial services. LightSquared has been hoping to launch commercially, in 2011, in four trial markets in Baltimore, Phoenix, Denver, and Las Vegas, but has been dogged by concerns by the GPS industry and several other federal government agencies about the potential of 4G ATC to interfere with GPS navigation services that are adjacent to the band.

Satellite and terrestrial broadcasting is less dynamic from a technological standpoint, partially because most broadcasters generally continue to operate as niche one-way distributors of non-data-oriented media content, although this could change over the long run. Satellite and terrestrial broadcast systems are supported by subscription-based or advertising-based news and entertainment content, but have been, and will likely continue to be, vital conduits for application data such as traffic information. Broadcast systems, however, might still be vital, as they are uniquely suited for classes of application data that have short shelf lives (information which loses value as time passes) and wide appeal for users such as traffic information and weather.

As outlined in its National Broadband Plan, the FCC is seeking congressional approval to initiate voluntary auctions of TV spectrum, allowing broadcasters to share in the auction proceeds, as a way to gain new spectrum for two-way mobile broadband services.[1] The argument for converting broadcast spectrum to cellular has focused on the fact that a large amount of TV content, and some radio content, is currently being provided via cable. Cellular systems such as 3G have begun to handle content multicasting or streaming of traditionally broadcast media content such as radio or TV. Future systems may be able to geo-cast, transmitting data based upon a user's declared location via GPS or determined through a network address. Geo-casting would be particularly useful for sending location-specific traffic or weather advisories to specific individual drivers that might be in the immediate path of a potentially disruptive event such as a storm or traffic jam.

Even if broadcast spectrum is not successfully converted to cellular use, broadcasters have digitized and will likely make their systems more data-oriented and media- and application-agnostic. Some purpose-built systems, such as broadcast TV or radio, have been adapted for data services such as traffic information. Work is underway to create platforms for Mobile TV such as Advanced Television Systems Committee Mobile/Handheld (ATSC-M/H) and of course, HD Radio. However, there are no large application ecosystems centered on these yet that are comparable to the ecosystem of cellular application-phones ("app-phones" also known as smartphones). Proponents of new mobile digital data broadcasting platforms might gain momentum if they can get mobile computing terminals such as app phones to adopt their technology to a meaningful degree.

Fourth Generation Wireless - Next Generation Vehicle Support

4G is likely the most relevant technology as far as the auto industry may be concerned. 4G is not just an industry quest for a faster radio technology. It is a remake of the entire cellular telecommunications system, with the objective of extending the internet suite of protocols beyond the wired environment. 4G represents the complete transition of cellular from a system designed for the unique requirements of voice to a general purpose system that can manage a number of applications. In 1983, for example, mobile subscribers experienced voice throughput of about 10 kilobits per second (kbps). A decade ago, end users could expect peak throughput of approximately 170 kbps with 2G technologies. 4G technologies are designed to meet, or at least approach, the International Telecommunications Union (ITU) requirement that targets peak data rates of up to approximately 100 Mbps for high mobility access (i.e., outdoor access with users moving at high speeds) and up to approximately 1 Gbps for low mobility or "nomadic" access (i.e., primarily indoor access with terminal movement at a minimum). [2]

The ITU originally contended that only two emerging technologies, LTE Advanced and WiMAX Release 2 (802.16m), qualify as 4G because they will be able to achieve peak data rates of 1 Gbps for a stationary user. Current systems being deployed (such as LTE or WiMax) are marketed as 4G, but do not meet this threshold. The ITU subsequently changed its definition to say that any technology offering a "meaningful improvement" over 3G can be classified as 4G.

No matter how 3G or 4G are defined, the success of these systems is not because they solved the problem of insufficient data rates. They have been successful because their deployment coincided with mobile terminals such as app-phones and other computing platforms such as tablets that provided portability, ease of use, and application flexibility and upgradability. These hardware and software platforms or ecosystems parallel wireline internet such as application/content/service development and distribution models using open application programming interfaces (APIs) and online distribution and maintenance. 3G also marked the point in time when the wireless telecommunications industry abandoned the "walled garden" approach of providing a fixed menu of applications proprietary to their networks, abandoning competition at the application layer and leaving it to the information technology (software) and consumer electronics (hardware) industry. Ecosystems thrive on economies of scale (number of users) and scope (the variety of applications, especially those niche sector applications in the "long tail" such as transportation). Ecosystems rely on the convergence and integration of complementary technologies from other parts of the technology chain such as wireline networks, hardware terminals, content, and applications.

Specifically for automotive consumers, automobile manufacturers and telematics service providers have contemplated replicating the same application ecosystem that is found on app phones, allowing third parties limited access to programming interfaces into the telematics/infotainment systems. Ford Sync or OnStar have interfaces with mobile phones and are contemplating the creation of app stores that can be used to vet, market, and distribute applications to consumers in the same way as applications are currently distributed by companies like Google, Apple, and RIM.

For enterprise-oriented applications, a new industry value chain is emerging as well. MNOs have also begun to establish, through partnerships with telematics providers, application- or platform-specific machine-to-machine (M2M) terminals, and shared network infrastructure. MNOs are fast-tracking certification of many new connected terminals beyond consumer handsets such as electric utility meters and telematics units. MNOs have relied upon Mobile Virtual Network Operators (MVNOs), who do not own their own network base stations, but buy wholesale airtime from multiple MNOs across multiple coverage areas. MVNOs sell wholesale airtime as a retail service to their M2M subscribers, and they bulk provision and activate M2M terminals in large, scalable batches for large enterprise clients.[3]

Mobile Virtual Network Enablers (MVNEs) have also come into existence. These companies provide infrastructure and services that enable M2M MVNOs to offer specialized shared infrastructure and value-added services such as verification/validation of message receipts and GPS data, remote M2M device diagnostics, and bearer testing of different carriers' services through which a device may be expected to roam.[4] There are numerous vertical M2M application specialties as well, which can be broadly divided into six sector categories: Consumer Home, Healthcare, Energy/Utilities, Security, Industrial/Building Control, and of course, Consumer and Commercial Fleet Telematics.[5]

Moving In for a Closer Look: Smaller "Smart" Cells and the Vehicle

They way in which future networks beyond 2015 will be designed and services provisioned will have a significant impact on machine-to-machine and automotive applications, and possibly vice-versa. A vehicle driving down a street in an urban area will enter and exit any number of coverage areas of a wide variety of wireless systems. Understanding which systems will be addressable, secure, and accessible will be of interest, whether they are short range systems such as Wi-Fi or DSRC/WAVE, wide-area networks such as 4G, or satellite. Furthermore, automakers will face the unique challenge of choosing a wireless communications technology that must remain useful and secure for the entire design/build/service lifecycle of an automobile. MNOs, on the other hand, will need to understand how the addition of millions of vehicles, and potentially billions of other M2M terminals, might impact their network operations and infrastructure provisioning strategy. In order to manage the diversity of infrastructure, technology, and applications, networks will need to get smarter in how they are provisioned, organized, and operated.

The number of wireless nodes with which a vehicle may potentially establish communications will likely multiply in the next decade. There will probably be a massive build-out of new wireless infrastructure, both cellular and short range "local area" systems such as WiFi, DSRC/WAVE, and others. The reason for the build-out is simple and is part of a century-long trend. Assuming no radical breakthroughs in wireless technology, the only way to create more network capacity is by splitting cells, re-using spectrum, multiplying access points, and allocating them to ever smaller coverage footprints. Spectrum re-use occurs when, instead of adding spectrum to increase capacity, existing spectrum frequencies are "re-used" and allocated to the same communications service in a regular pattern of smaller and smaller areas called "cells," each covered by one base station at reduced power to reduce the probability of inter-cell interference.[6] Accommodating the growing traffic that millions of mobile and machine devices will generate requires a completely new infrastructure provisioning strategy for MNOs, a strategy that possibly represents an accelerating trend toward much higher wireless access point density and accessibility.

Although the number of wireless data subscribers grew significantly in the last several years, traffic per subscriber has increased several orders of magnitude more. In the face of such traffic growth, a fixed line Internet Service Provider (ISP) would, for example, simply add additional link capacity, a strategy unavailable to an MNO that cannot provision beyond its fixed allocation of spectrum. Lack of new spectrum is a major constraint on MNOs. It is understood that we may be within a factor of ten of the maximum capacity that can be achieved within a given spectrum allocation, and that there are few options to increase capacity beyond more spectrum and aggressive cell splitting or spectrum re-use.[7] One of the keys to cellular's future success is the architecture's ability to "create" spectrum through the re-use of wireless channels over large geographical areas. The latest trend is to expand capacity by splitting cells into smaller and smaller units. This shrinking of cell sizes has been a consistent technology trend throughout several decades. For example, the first mobile telephones systems of the late 1970s had "macro" cells of nearly 700 square miles each, but current small (micro-, pico-, and femto-) cells range from one-half square mile to 50 square feet. Future networks will include combinations of macro cells and many smaller femtocells.

By creating and deploying faster radio access technologies, such as LTE or WiMAX, and deploying more cells, MNOs provide more capacity and speed in the wireless channel for users. However, this tends to move congestion upward into the backhaul networks that connect radio access network to the core MNO network and the rest of the internet. Expanding backhaul from cells is a high priority for MNOs, and the cost of backhaul influences whether splitting cells is cost effective. Whether smaller cells and cell splitting are a cost-effective strategy is hotly debated in wireless industry circles, as the marginal benefit/cost of cell splitting depends on the availability of cheap wireline backhaul (such as a customer's cable internet) and the cost of alternatives (such as the siting and construction of more "macro" cell towers). Femto cells will likely be built off the existing telecommunications plant supporting wireline broadband. Leveraging this backhaul will be critical for cost effectiveness and success of femtos.

New femto-cell infrastructure deployment is a potential MNO strategy to dislodge two capacity bottlenecks at once—spectrum and backhaul from the radio access network to the public internet. The spectrum bottleneck is being resolved by cell splitting into femtos, and the wireless backhaul problem is being overcome by taking advantage of existing home and business wireline internet access to connect these new mini- base stations. In the future, mobile phone or cable customers may not only purchase a phone with their data plan, but may also get a home femtocell base station that can be connected to their broadband cable internet to provide improved coverage indoors, but also to relieve congestion on MNOs' macro base-stations. The cost of these systems is dropping— the first sub-$100 femtocells can handle eight simultaneous calls and download speeds of up to 14.4Mbps. [8]

Femtocells have not taken off as wireless industry analysts have expected, as MNOs have focused more on installing and expanding macro base station infrastructure. However, despite the less than expected growth, Informa Telecoms & Media recently estimated that worldwide femtocells now outnumber macro-base stations, with 2.3 million 3G femtos deployed compared to 1.6 million 3G macrocells.[9] Furthermore, femtocells are also adding air interfaces to support multiple standards. Ubiquisys, Texas Instruments, and Intel are working on femtocells that support Wi-Fi, 3G, and 4G LTE.[10]

To meet consumer and enterprise demand for wireless data services, the density of wireless network infrastructure has expanded enormously during the last decade. It will continue to expand. In 2000, there were nearly 80,000 macro cellular base stations in the US, growing 210% to approximately 250,000 macro cellular base stations currently. If wireless carriers commit to seriously provisioning smaller cellular base stations such as femtocells, conservative estimates indicate that telecom operators might add 50,000 additional base stations per year to their networks between 2010 and 2015. Highly optimistic estimates indicate that wireless carriers' networks nodes might incorporate many more micro base stations, with nearly six femto cells being implemented for every macro cellular tower.

The efficient and economical deployment of thousands or even millions of small cells necessitates a drastic change in management and radio planning methodologies. Traditional macro networks are deployed with a semi-static configuration whereby cell planning to reduce inter-cell interference is done via simulation and spectral analysis. This approach is not feasible when the cell size is reduced and the number of cells increases drastically. The development of Self Organizing Network (SON) techniques, algorithms, and eventually standards is a critical step in LTE femtocell deployments.

Current femtocell techniques rely on algorithms that are self-contained within each base station, lacking coordination among cells and focusing primarily on power control. The next generation of SON will extend these algorithms to include coordination among cells as well as to take into account power control and parameters like cell loading and proximity of terminals to the radio. Organizations such as the Third Generation Partnership Project (3GPP) have already identified SON as a critical area for standardization, and work is already underway. SON will likely appear in later versions of LTE Advanced specifications.

Ultimately, femtocells are seen as not only boosting the capacity of the macro cell network by offloading indoor users, but as driving technical innovation in "self-organizing" base stations (i.e., cells that automatically manage and minimize inter-cell interference). Such innovation will likely drive further standardization of base station interfaces, so that someday MNOs can mix different base stations instead of purchasing from a single supplier, expanding competition, increasing economies of scale and scope, and pushing costs down further.

Ultimately if deployment of femtocells finally does take off, vehicles equipped with telematics systems will likely be connecting to them as they are parked, or move in and out of smaller coverage areas such as home driveways, parking garages, or even intersections and other confined spaces. Furthermore, femtocells could even support some non-safety critical spot applications in the distant future such as cordon or road tolling, road or parking access and control, and even parking spot usage recognition in areas where cellular coverage may be problematic. However, for these applications to work, MNOs will need to acquire highway right-of-way and other easements to allow buildout of small wireless nodes.

Quality as Job #1: 4G Future Performance

The wireline internet was never designed to support wireless services, where transport protocols function poorly because wireless links frequently fail, forcing retransmissions, and devices constantly change home addresses as they geographically "roam" across wireless subnets. 4G radio access and core networks are designed to manage these challenges and will likely improve wireless network performance—specifically, higher quality of service, reduced latency, improved reliability, and enhanced capacity.

Despite the ongoing expansion of wireless infrastructure capacity through upgrades to 4G, and continued cell splitting with the introduction of femtocells, latent demand will still likely leave MNO capacity constrained in the near future. MNOs will need to manage wireless traffic proactively to ensure that application performance does not suffer. There is one major approach to managing the deluge of data—traffic shaping. This strategy relies on the concept of identifying application needs and prioritizing transmissions, or even prohibiting or re-routing transmission to other networks, based on MNO operational and pricing policies. The key question is what the impact might be on future automotive connected vehicle applications and specifically how data transmitted from vehicles might be treated by future 4G "traffic shaped" networks.

MNOs recognize that revenue growth hinges upon their ability to deliver a wider range of mobile broadband applications and services, which require higher bandwidth and lower latency. The critical challenge for MNOs is to develop networks where high volume applications (such as video streaming) do not interfere with critical lower volume applications (such as 911 calls or automated crash notification messages). As cellular systems morph into a general purpose internet-based wireless network, they will be challenged to meet the needs of all applications with the single "best effort" quality of service class that is typical of the wired Internet. 4G describes and combines several Quality of Service (QoS) attributes (such as maximum acceptable delay, jitter, and bit error rate) into a minimum of four QoS categories such as conversational (e.g., Voice-over-IP), streaming (e.g., video), interactive (e.g., web browsing), and background (e.g., email and file transfer).

However, future quality of service implementations might need to be more sophisticated in prioritizing content for wireless networks than they would for a strictly wireline network because multiple real-time streams might look alike, but still need to be further differentiated. For example, a video stream can be delayed and cached on the terminal, but a 911 call or an automated collision notification cannot be.

Today's mobile networks are carrying many complex combinations of traffic that potentially defy simple classification of data packets into four broad quality-of-service buckets. Using Deep Packet Inspection (DPI), MNOs can automatically sort and classify packets according to a variety of criteria in real time. After classification, different traffic shaping policies may be applied to the packet and its associated stream such as prioritization, rate limits, or even blocking. Today's DPI systems can identify many protocols and traffic types while shaping traffic in real-time, at speeds of nearly 100 gigabytes per second. Internet applications are becoming more sophisticated, and application developers are burying data deep inside packets such as voice and other media mash-ups, significantly increasing traffic classification complexity.

Quality of service was not adopted in the early days of the wireline internet because there was little need for it, as most traffic was asynchronous file transfer, email, and web browsing. Wireline operators could over-provision link capacity, adding additional optic fiber, for example, as a way to combat congestion caused by peak internet traffic. Overabundance of wireline link capacity has allowed the wireline internet to move beyond bulk, asynchronous applications and has allowed introduction of real-time applications such as unicast video and audio streaming and voice-over-IP (VoIP).[11] Unfortunately, MNOs cannot add spectrum as wireline telecommunications providers do and, therefore, they must be careful of applications or users that can potentially act as "bandwidth hogs."

Traffic shaping technology will allow MNOs to introduce flexible tiered quality of service and pricing incentives that fit to particular applications. 4G core networks and even terminals will likely have embedded operational policies to differentiate and prioritize traffic, and then schedule it for transmission based on time sensitivity or quality of service requirements. Differentiating traffic can also be used to charge different tariffs for different services, to charge by bulk data limits or by on-peak or off-peak usage, or to exclude certain types of usage completely under more affordable data plans. This effort is dependent in part on long-term efforts by standards organizations such as 3GPP to create an IP Multimedia System as a part of later releases of LTE that will shape traffic on radio access networks for transfer to and from the fixed internet.

Designers of automotive applications must be cognizant of the effect that network congestion might have on their applications in the future. The design of the Internet precludes a state where latency is ever sufficient for all applications at all times, unless the multiplicity of applications is reduced.[12] The wireline Internet and its wireless extension are both dynamic shared bandwidth systems that rely on statistics to gauge quality of service for most applications. A great deal of analysis is needed to understand the impact of different applications to network capacity. For automotive applications where data is processed "in the cloud," such as off-board navigation, an understanding of the application's demands on the network may be important. Even though the LTE IP Multimedia System will be able to allow applications the same quality of service as voice calls, application programmers must have quite detailed knowledge of the network's traffic dynamics to avoid creating congestion or other pathological effects on application performance.

If MNOs still cannot cost effectively meet the growing traffic on their networks after network traffic shaping, then one alternative will be for MNOs to "shape" the terminal devices, requiring them to "offload" some traffic to less congested links such as Wi-Fi networks, or to roam to less-congested competitor networks. The emergence of multi-standard, multimode radios such as those found in app phones means that applications will have a choice of interfaces beyond cellular. These systems may incentivize development of vertical "roaming" across heterogeneous networks such as Wi-Fi, ZigBee, DSRC/WAVE, or other systems. Models where lower priority traffic (or lower profit margin traffic, depending on the pricing model) is "off-loaded" to these redundant links will take advantage of the LTE IP Multimedia System capability to implement traffic shaping in both the network and the terminal.[13] Future versions of LTE System Architecture Evolution (SAE) include an "anchor" for roaming between 3G/4G systems and wireless local area systems such as WiFi or DSRC/WAVE.[14] The concept would allow an MNO to establish interfaces to WiFi or DSRC/WAVE hotspots, maintaining a contact list of trusted nodes that LTE would use to off-load data traffic in the event the normal 4G network is congested.

LTE's future IP Multimedia System represents a different approach to the more traditional telecommunications architecture of a set of specific network elements implemented as a single telco-controlled infrastructure. Services will be created and delivered by a wide range of highly distributed systems (real-time and non-real-time, possibly owned by different parties) cooperating with each other as part of LTE IP Multimedia System. There will likely be some tension between opening up IP Multimedia System for third party development to make it more accommodating of new M2M and telematics applications, and keeping the architecture simple and easy to understand and efficient to operate.

Since 3G, MNOs have sought "flatter networks" with fewer jumps between the radio access network (cell tower to mobile terminal) and the core network (cell tower to the internet gateway), because generally the more hops on a network, the greater the chance for bottleneck failures and congestion. For example, LTE currently supports radio access network round-trip times of less than 10 milliseconds, but this does not measure latency as data hops to core wireline gateways or to other base stations. Flatter networks are also cheaper to build and maintain, and are critical in keeping MNO's operating cost-per-bit low, which is critical if tariffs are decoupled from usage (i.e. all-you-can-download, flat rate pricing) as is occurring with many wireless data plans. Again, there may be a conflict between keeping the wireless teleco infrastructure flat and simple to drive down MNO operating costs and adding complexity via more accommodating application-aware interfaces supported in the LTE Multimedia System.

Future 4G systems are being designed to meet the requirements of a wide variety of quality-of-service categories, but application developers will likely still need to experiment with future LTE IP Multimedia System features to ensure adequate performance and priority for applications. However, MNOs are uncertain what the impact of M2M and telematics will be on demand for their network services. In an environment where M2M applications are growing and "human" subscribers are a minority, network communications traffic will likely occur in bursts based on either timely automated routines or events that may be unpredictable.[15] Predicting the maximum capacity and provisioning networks to support M2M and telematics peak capacity may be a future challenge in maintaining QoS.

MNOs will need to work with the largest M2M application service providers to understand the implication of M2M on traffic patterns and capacity. MNOs and application service developers must be cognizant that traffic shaping and simple QoS classification and prioritization might not always work, or might work in ways not intended by application developers.[16] On future wireless networks, application developers must conduct extensive analysis to determine how an application's performance influences, or is influenced by, the variable conditions on the wireless network, and how it may interact with elements with a future LTE IP Multimedia System.

This is especially the case for mission-critical systems, but also for ones where there is a high expectation of reliability from the user. "End-to-end" performance management tools are used by network engineers to monitor the symptomatic impact of applications on end-user and service-level performance metrics, but most network management tools only understand the network in discrete, device-level pieces.[17] Consumers might have higher expectations of the quality and dependability of automotive OEM equipment, in contrast to shorter-lived, less durable consumer electronics devices. Therefore, automotive application developers might need to invest more time and effort to understand how they can build "end-to-end" quality and reliability into their services that rely on frequent off-board communications.

The Right Radio: What Will Gain Traction for the Vehicle?

Automotive electronic engineers designing future-connected vehicle systems will probably see LTE (and later LTE Advanced) as the most sophisticated available, affordable system, in particular because of its industry support, economies of scale and scope in terminal and networking equipment, and its geographic coverage in most regions of the world. In general, LTE is the choice for most MNOs, as most operators must integrate past generations of legacy infrastructure such as second and third generation (2G and 3G) radio access and core networks. Legacy 2G and 3G infrastructure is a deciding factor for choosing LTE over WiMAX, as capital and operating expenditures vary depending on whether a mobile network operator has an existing base of 3G or 2G infrastructure and users. LTE in general is the choice for many MNOs because it allows carriers to allocate spectrum across generations of technologies, allowing MNOs to re-farm spectrum for use from 2G and 3G to 4G as users dispose of or upgrade their terminals to the latest technology over time. Re-farming spectrum from legacy generations to 4G depends on a number of factors such as how many legacy terminals are still in use, and how consumers or enterprise decide to upgrade.

In the past, many M2M application service providers, and specifically telematics solution providers, have utilized SMS or 2G data services, mostly because data transfer needs were small and intermittent. Second generation wireless networks are also attractive for M2M use because of their nationwide and cross-border coverage footprints. Furthermore, the price of service and equipment for 2G has declined at a faster rate than 3G (or 4G) equipment and services. The lower cost of 2G was a critical feature for M2M given low financial break-even points for many enterprise M2M projects, or depressed willingness-to-pay on the part of consumers who were unacquainted with the benefits of new telematics services.

Furthermore, automakers faced another unique challenge: choosing a wireless communications technology that must remain secure and useful for the entire design/build/service lifecycle of an automobile. Telematics service providers assumed that 2G coverage and SMS would remain a part of the carriers' networks for years. It is estimated that nearly 90 percent of embedded modules currently deployed use 2G.[18]

However, traffic is expected to explode on cellular networks, and all MNOs are committing to aggressive deployment of 4G technologies, with new terminals (handsets, USB dongles, laptops, etc.) already becoming available as soon as infrastructure is deployed. Conversion to 4G would mean re-farming 2G spectrum by decommissioning 2G base stations that support current M2M and telematics services and converting them to either 3G or 4G wireless technologies.

The uncertainty of the MNOs' infrastructure and service provisioning strategies might influence, and be influenced by, the existence of a large number of legacy 2G M2M modules embedded in automobiles and other devices. For the most part, however, MNOs are unaccustomed to taking into account legacy consumer equipment when making decisions regarding new network technology rollouts, because handsets and other consumer devices have a relatively short product life. MNOs may try to entice MVNOs to purchase more expensive 3G or 4G M2M modules, with the assurance that even though upfront costs will be higher, use of 3G and 4G will result in lower on-going network tariffs that will lower total cost. According to Analysys Mason and the GSMA, the most likely scenario is that most operators will decommission their 2G networks in the next ten years, forcing an upgrade or replacement of thousands of modules to either 3G or 4G technology. Although decommissioning 2G may tarnish the reputation of the MNOs in the eyes of the telematics and the broader M2M application community, it is very necessary for MNOs given that the costs of acquiring new spectrum to support 4G are far greater than the expense to them if they were required to pay for a portion of obsolete M2M module replacement costs, or lose revenue from M2M application providers who may abandon them.[19]

Standard and Flexible: Business Models and Shared Resources

Theory and experience suggest that the success of any innovation hinges on two factors: the power of the core technology and its impact on existing business, plus its potential in new areas of growth. Basic technology constraints on widespread deployment of telematics and M2M have been largely overcome for many applications. The driving determinant is usually cost (specifically life cycle cost), which is driven by manufacturing and operational processes (provisioning, design, and certification based on application needs such as security, reliability, etc.) and, to a lesser extent, the cost of wireless telecommunications services.

In telematics, many opportunities exist to offset these costs, given that vehicle operators (the general driving public or fleet operators such as commercial freight, passenger, or transit carriers) are often "in the loop." Many applications can be tied to a "human subscription" and be subsumed under a single data plan that wireless carriers might provide to cover all of their subscribers' terminals (laptop, tablet, smartphone, or car) under a flat rate scheme. Telematics service providers have also contemplated innovative pricing schemes to adapt to consumers' depressed willingness-to-pay for niche vehicle-only services.

Wireless data service subscription models (such as charging by session, device type, application type, time, quantity of data, speed of transmission, or other flexible categorical pricing schemes) might help the telematics service provider's bottom line and counterbalance enterprise and consumers' low willingness-to-pay for telematics or M2M services.[20] Traffic shaping technology and strategies implemented by the MNOs might ultimately support this type of flexibility and spark demand for services by lowering introductory price points for some telematics applications.

Business process innovations are the key to reducing costs and making M2M services more accessible to a wide variety of applications. A "managed service infrastructure" provides a hardware and software platform that can support common needs across multiple sector applications (e.g., transportation, healthcare, energy) such as device activation, monitoring, and security, among others. A managed service infrastructure operated by a Mobile Virtual Network Enabler could, for example, establish a middleware platform that can provide service level agreement monitoring and device profile reporting. Innovation in this area will depend on standardization, which can create interoperability and compatibility and focus firm innovation and competition in areas of higher value-added equipment and services.

Standardization in M2M is occurring, but there are several disparate initiatives and no de-facto standard as of yet. Organizations include European Telecommunications Standards Institute (ETSI)/Third Generation Partnership Project (3GPP), International Telecommunications Union (ITU), Institute of Electrical and Electronics Engineers (IEEE), the International Standards Organization (ISO), and Global Standards Collaboration M2M Standardization Task Force. The Telecommunications Industry Association (TIA) Smart Device Communications Engineering Committee (TR-50) is currently developing an M2M framework that can work over any type of communications framework, and this group intends to publish a well-defined application programming interface for the industry shortly.

With standards, it is more likely that a managed service infrastructure could be developed and shared across many independent applications. A new, open infrastructure, object-oriented approach could ultimately lead to services and features common to many applications, thereby reducing complexity, development effort, and maintenance costs. There would also be significant operating cost savings because the resulting service infrastructure could be pooled across many independent applications. However, the biggest benefits would come from the ability to allow application data from telematics systems to be shared in a secure manner across any application and to allow any device to connect to any application. This means that new applications could be created and advertised on the basis of installed vehicle terminals, not on the basis of exclusive use by a particular telematics application service provider, MNO, or MVNO.[21]

Conclusion

The success of telematics is riding on the ease of use and practicality of integrating multiple generation technologies, from portable consumer devices to embedded equipment. This must, however, be done in a manner that reduces costs and complexity and meets basic needs of drivers for connectivity, beyond infotainment and mobility, extending to vehicle diagnostics, occupant crash protection, and crash-avoidance.

Connected vehicle applications are set to explode in the next few years because of the development of mobile application platforms such as Google Android and others, and enterprise machine-to-machine support services. Supporting these platforms beyond 2015 is 4G LTE wireless and its progeny. Automotive application developers will need to be cognizant of how application data is treated by 4G systems, and how innovations such as "traffic shaping" may improve quality of service for "off-board" (or "cloud–based") vehicular applications. Furthermore, "on-board" vehicular applications utilizing vehicle-to-vehicle, vehicle-to-infrastructure communications, such as femtocells, Wi-Fi and, for safety applications, Dedicate Short Range Communications/Wireless Access for Vehicular Environments (WAVE), might also likely be integrated into 4G. Later versions of LTE, such as LTE Advanced might establish and manage communication sessions that hop among many of the above mentioned wireless technologies, a concept known as heterogeneous or "vertical" roaming.

In the future, as vehicles add new technology such as electric powertrains, drive-by-wire chassis systems, autonomous advanced driver assistance features, and even future cooperative collision avoidance systems, these new advances might require new levels of maintenance, service, and diagnostics.[22] Diagnostics for a safety system are necessary to ensure reliability (low mean time between failures), availability (readiness for service), maintainability (low mean time to repair), safety (no risk of catastrophic failure), and security (authorization of trusted users to operate and maintain the system, plus system resistance to malicious attacks). Diagnostics for these new vehicle technologies might need to be monitored and analyzed "off board" to improve safety, vehicle performance, and future product quality. As far as choice of wireless technology, the ideal vision is to have vehicles communicating using any system that is available and secure or, if multiple systems are available, choosing the most direct, unencumbered, efficient or lowest cost path, based on the technical requirements of the application and the business needs of the application service provider.

It is likely that the auto industry will see computing platforms in vehicles in the next decade similar to ones found in mobile app-phones, although with device interfaces designed to ensure that they are accessible to the driver in a way that does not reduce attention to safety-critical driving tasks. With new computing platforms and interfaces, the fixed menu of applications that had been the hallmark of telematics packages in the past will change. Like app phones that allow users to choose à la carte from multiple applications, "app stores" create a platform that allows users to choose the mobility applications they want, making the value proposition for telematics more attractive in the eyes of consumers, especially when weighed against the additional cost of supporting the connectivity elements of these systems.

LTE will likely not have national coverage for some time beyond 2015, and deployment will likely start in urban areas first. Many in the wireless industry have talked of a targeted deployment strategy known as "inside-out" deployment. The "inside-out" strategy places LTE first in homes and offices with 3G/4G/WiFi femtocells, then 4G macro-base stations in major metropolitan areas, eventually expanding 4G outward to other cities and rural areas. Network infrastructure innovations, such as small, self- organizing cells and heterogeneous roaming across different wireless systems, might significantly influence and accelerate 4G deployment.

In addition to several choices of wide area communications such as cellular or mobile satellite services, vehicles will also likely have wireless systems devoted to safety-critical or highly mobile "spot" communications. Vehicle applications that are run entirely on-board and very localized, such as adjacent vehicle-to-vehicle cooperative collision avoidance or traffic signal preemption and intersection collision avoidance, will require very fast communication using peer-to-peer communications systems such as DSRC/WAVE.

Many of these peer-to-peer wireless systems found in future vehicles will likely interact closely with 4G, supporting each other to authenticate users for secure use of safety-critical "cooperative" applications (such as vehicle-to-vehicle collision avoidance) or to run diagnostics to ensure proper functioning of all vehicle- or infrastructure-based safety and mobility applications. 4G terminals in vehicles might even off-load data traffic opportunistically (possibly at reduced tariffs) to short range 4G femto cells, Wi-Fi, and DSRC/WAVE nodes in urban areas. DSRC/WAVE and WiFi might provide a number of inexpensive options to vehicle applications to offload data or communicate to the "cloud." Ultimately, automotive electronics engineers will need to recognize these possibilities and anticipate trends in wireless infrastructure to develop compelling, reliable, and cost effective vehicle-oriented applications.

Acknowledgements

The authors would like to thank the ITS America Staff for their contributions, in particular Tom Kern, Radha Neelakantan, Joerg "Nu" Rosenbohm, Adrian (Xu) Guan, and Rod McKenzie. The authors would also like to especially thank Shelley Row, John Augustine, Valerie Briggs, Walt Fehr, and Tim Schmidt at the U.S. Department of Transportation for providing support and comments for much of the research conducted. Also, the authors are indebted to a number of experts that provided advice and comments in discussions on the topic: Sheryl Wilkerson (Willow), Ralph Robinson (University of Michigan), Bakhtiar Litkouhi (General Motors), Craig Copland (ATX/Cross Country), Rebecca Hanson (Federal Communications Commission), Upkar Dhaliwal (Future Tech Wireless), and Dave McNamara (McNamara Technology Solutions). A portion of this material is based on work under Contract DTFH61-08-D-00011.

Acronyms

DSRC	Dedicated Short Range Communications
GPS	Global Positioning System
LTE	Long Term Evolution
M2M	Machine to Machine
MNO	Mobile Network Operator
MNVO	Mobile Virtual Network Operator
MNVE	Mobile Virtual Network Enabler
MSS	Mobile Satellite Service (or MSS Ancillary Terrestrial Component - MSS-ATC)
SAE	System Architecture Evolution (core network architecture for LTE)
WAVE	Wireless Access For Vehicular Environments (WAVE)
Wi-Fi	Wireless Fidelity (standard base on IEE802.11a,b,g,n)
Wi-MAX	Worldwide Interoperability for Microwave Access

References

1. Federal Communications Commission. "The National Broadband Plan: Connecting America." http://www.broadband.gov/.
2. Kim, Y. K., and R. Prasad. 2006. *4G Roadmap and Emerging Communications Technologies*, p. 55. Lavoisier S.A.S.
3. Lucero, S. 2010. *Maximizing Mobile Operator Opportunities in M2M: The Benefits of an M2M-Optimized Network*, p. 2. ABI Research (Report for Jasper Wireless).
4. Emmerson, B. "M2M: The Internet of 50 Billion Devices." *Win-Win*, January 2010: 5.
5. Emmerson, B. "M2M: The Internet of 50 Billion Devices." *Win-Win*, January 2010: 19.
6. Xiao, M. "Cellular Fundamentals." Xiamen University, Wireless Communications Engineering Lecture, November 18, 2004.
7. Web, W. 2011. "Wireless Communications: Is the Future Playing Out as Predicted?" In *Interdisciplinary and Multidimensional Perspectives in Telecommunications and Networking: Emerging Findings* edited by M. Bartolacci and S. Powell, pp. 9-18. Hershey, PA: IGI Global.
8. Ricknäs, M. "Femtocell Prices Have Dropped below 100 Says Vendor." *PC World*. March 30, 2010.
9. Mavrakis, D. "3G Femtocells Now Outnumber Conventional 3G Basestations Globally." *Informa Telecoms and Media Blog* press release, June 21, 2011.
10. Rasmussen, P. "Femtocells Still on the Sidelines, but New Opportunities Beckon," Europe: *Fierce Wireless:* June 22, 2011.
11. Bennett, R. *Going Mobile: Technology and Policy Issues in the Mobile Internet.* Information Technology and Innovation Foundation, March 2010.
12. Bennett, R. *Designed for Change: End-to-End Arguments, Internet Innovation, and the Net Neutrality Debate.* Information Technology and Innovation Foundation, September 2009: 24.
13. Paolini, M. 2010. "The Migration to Mobile 4G Networks: Leveraging the All-IP Network to Improve Efficiency, Subscriber Experience, and Profitability." Whitepaper prepared for Cisco by Senza Fila Consulting, p. 4.
14. Anristu Company. 2010. "Future Technologies and Testing for Fixed Mobile Convergence, SAE and LTE in Cellular Mobile Communications." Whitepaper.

15. Bastien, F. "Subscriber Data Management for Machine to Machine - M2M Applications Landscape." Presentation November 11, 2010.
16. Singh, M. "Lethal Cocktail: Traffic Offloading and Shaping Don't Mix Well." *Continuous Computing*, 2011, p. 3.
17. Eslambolchi, H. *Eslambolchi's Law of Telecom Complexity*. Dr. Hossein Eslambolchi's blog, April 8, 2009.
18. Sanders, L., I. Streule, and G. Monniaux. *Final Report for GMSA: The Embedded Cost of Mobile Owner Devices*. Analysys Mason: November 11, 2010, p. 20.
19. Sanders, L., I, Streule, and G. Monniaux. *Final Report for GMSA: The Embedded Cost of Mobile Owner Devices*. Analysys Mason: November 11, 2010, p. 5.
20. Bastien, F. "Subscriber Data Management for Machine to Machine - M2M Applications Landscape." Presentation November 11, 2010, p. 9.
21. Emmerson, B. "M2M: The Internet of 50 Billion Devices." *Win-Win*, January 2010: 22.
22. Najm, W. G., J. Koopmann, J. D. Smith, and J. Brewer. 2010. "Frequency of Target Crashes for IntelliDrive Safety Systems." Washington, DC: National Highway Traffic Safety Administration.

The Evolution of the Driving Experience and Associated Technologies

Tim D. Bolduc, Staff Engineer
Gerald J. Witt, Advanced Driver Interface Systems
Douglas L. Welk, Chief Engineer
Keenan A. Estese, Connected Vehicles and Applications
Delphi Electronics and Safety

The Evolution of the Driving Experience and Associated Technologies

Tim D. Bolduc, Staff Engineer
Gerald J. Witt, Advanced Driver Interface Systems
Douglas L. Welk, Chief Engineer
Keenan A. Estese, Connected Vehicles and Applications
Delphi Electronics and Safety

Abstract

The driving experience is in a state of perpetual evolution. Societal trends often lead to new usage scenarios for drivers. Throughout the history of the automobile, designers have incorporated technologies such as navigation systems, satellite radio, iPod and MP3 players, and cell phone connectivity to meet new driver demands. During the past few years, smart phones and wireless connectivity in general have changed the way people acquire information, communicate with others, and elect to be entertained. These changes have led to new driver demands for connectivity within the vehicle. The automotive industry is working aggressively to deliver systems that meet these needs while simultaneously delivering solutions designed to minimize and manage any additional driver workload associated with these new features. This paper examines these trends, discusses their ramifications on the driving experience, and provides an overview of technologies under development to manage these new features while helping to mitigate driver distraction.

Introduction

As with all successful enterprises, the automotive industry has worked to provide solutions that meet buyer demand. The vehicle continues to serve many needs in people's lives. It is first and foremost a mode of transportation. For many, it is also a navigational assistant as well as an optimal space to listen to news of interest, enjoy personal music collections, or to stay in touch with friends and colleagues via cellular networks.

As the vehicle evolved, the industry added features including navigation systems, iPod and MP3 controllers, satellite radio receivers, as well as many wireless features for Bluetooth enabled phones like streaming music, remote media playback control, phone book management, and hands-free calling. The automotive industry has a long history of working to design new features that do not compromise the primary task of operating the moving vehicle effectively.

The latest consumer trends highlight the need for the automotive industry to ensure that vehicles evolve further in these areas. New car buyers are seeking connectivity features that allow them to use personal nomadic devices such as smart phones as well as internet-based services and applications through their "connected" vehicles. Leading automotive manufacturers and their suppliers are designing systems to address this market need while working to ensure that the task of effectively operating the vehicle remains the highest priority.

Industry Trends

The driving experience is in a constant state of evolution and is influenced by broader societal trends. During the last century, consumers have continuously demonstrated a propensity to incorporate new technologies into the driving experience. The desire to remain informed and entertained while driving has led to an ever-growing assortment of technologies and services like broadcast radio receivers, 8-track tape players, cassette players, CD players, satellite radio receivers, navigation systems, and integrated telematics services like OnStar. In the more recent past, the driving experience has evolved to embrace personal nomadic devices like iPods, MP3 players, and cell phones. The market research firm, iSuppli, has forecast Bluetooth penetration rates in vehicles to climb from 40% in 2010 to 79% in 2015.[1] Bluetooth technology affords drivers the ability to manage phone calls in a hands-free manner, but also allows for more advanced features like phone address book management, text messaging management, and remote control of phone-based applications for music playback through the vehicle's speakers.

Each of these evolutionary advancements has led to an enriched driving experience. However, the desire to remain connected, informed, and entertained while driving has raised concern about drivers becoming distracted from the primary task of operating a vehicle. This concern is not new, but it has recently gained greater attention due to the popularity of smart phones across virtually all demographics.

During the past few years, smart phones and their accompanying ecosystems of developers have led to new paradigms in the way we acquire information, communicate with others, and elect to be entertained. Their impressive processing and memory capabilities, coupled with integrated sensors including cameras, accelerometers, and touch screens, make these devices compelling host platforms for thousands of innovative applications. The wireless communication capabilities of these devices, in conjunction with the ever-expanding footprint of reliable, cost-effective, high bandwidth network coverage, has resulted in smart phones becoming a critically important component of people's daily lives. For example, researchers from the International Center for Media and the Public Agenda (ICMPA), in partnership with the Salzburg Academy on Media and Global Change, recently published a global study of use of the internet by university students.[2] Two of the study's findings are that 1) students around the world repeatedly used the term "addiction" to speak about their dependence on media, and 2) a clear majority in every country admitted outright failure of their efforts to go unplugged.

The trend toward more constant connectivity is growing unabated. According to International Data Corporation (IDC), 303 million smart phones were shipped globally in 2010. IDC forecasts that this number will increase by 50%, to more than 450 million smart phones shipped in 2011.[3] The raw consumption of mobile data traffic is expected to reach 30 million terabytes in 2014—up from 2.3 million terabytes in 2010.[4] These trends heavily influence the driving experience. A 2011 J. D. Power and Associates study found that 86% of smart phone owners use their devices while driving.[5] A Pew Research Center report in 2010 found that 47% of adults who text have done so while driving a vehicle.[6]

Connectivity has become an important consideration in consumer decisions about buying new vehicles. A 2010 J. D. Power and Associates survey found that more than 77% of current smart phone users surveyed indicated interest in wireless connectivity systems for their vehicles. Approximately 56% of the same group expressed interest in mobile router functionality, and 47% were interested in having a vehicle with an in-dash computer.[7] According to Gartner, Inc., the majority of vehicle manufacturers will concentrate product development efforts for mature markets on enabling wireless data connectivity in more than half of their next-generation cars by 2012.[8] The penetration of embedded wireless network devices into automobiles is also expected to grow rapidly from approximately 400 million vehicles in 2011 to more than 900 million by 2015, according to Machina Research.[9] They predict that approximately 75% of vehicles on roadways around the globe will have embedded machine-to-machine functionality by 2020.

Balancing the Trade-Offs

When combined, these trends and innovations lead to a reality where innovation can occur extremely rapidly, often with much lower barriers to market introduction than are possible with today's automotive ecosystem. While there is great promise in this ability to evolve the driving experience in the field, the implementation must take into account concerns over the possibility of increasing driver distraction.

Some have argued that passing legislation prohibiting the use of devices like cell phones while driving is a viable way to address this concern. Although legislation can offer a level of deterrence, it can also have unintended consequences. The result of a recent study by the Highway Loss Data Institute (HLDI), an affiliate of the Insurance Institute for Highway Safety, questioned the efficacy of legislation. The report concluded that there were "no reductions in crashes after laws take effect that ban texting by all drivers. In fact, such bans are associated with a slight increase in the frequency of insurance claims filed under collision coverage for damage to vehicles in crashes." According to Adrian Lund, president of both HLDI and the Insurance Institute for Highway Safety, "Texting bans haven't reduced crashes at all. In a perverse twist, crashes increased in 3 of the 4 states we studied after bans were enacted. It's an indication that texting bans might even increase the risk of texting for drivers who continue to do so despite the laws."[10]

As an alternative to legislation, some automotive original equipment manufacturers (OEMs) and their electronics suppliers are developing and deploying technologies that can be used to help drivers manage the workload of the driving task while still enjoying new connectivity features. Some of these "active safety" technologies include features like adaptive cruise control, brake assist, lane keeping, lane departure warning, blind spot warning, forward collision warning, pedestrian and large animal detection, and infrared (IR) enhanced night vision systems.

Other technologies include new application programming interfaces that enable smart phone applications to use the controls and displays mounted in the vehicle. These technologies all strive to achieve a simplified, less-distracting user interface for the driver's latest "must have" applications.

- Ford's approach with its Sync system requires smart phone app developers to incorporate Ford's AppLink Application Programming Interface (API) into their applications. This API allows the smart phone app to be controlled by voice as well as by other head unit controls. This approach allows Ford to maintain tight control of the features that can be used, as well as how they are used by the driver.
- In a somewhat similar approach, Airbiquity offers its Mobile Integration Platform API to app developers and introduces cloud-based policy management to allow the vehicle manufacturer to modify the appearance and/or behavior of smart phone hosted applications when rendered and controlled by vehicle-based components.
- Apple has released its iPodOut protocol for use with modern iPhone and iPad products. This protocol allows the external device to render its modified screen content onto a vehicle display and to be controlled by iPod Accessory Protocol (iAP) commands. Apple currently retains control of the list of apps that are enabled for use with iPodOut.

Successful distraction mitigation depends on providing a safety-optimized cockpit and interaction intelligence via workload management systems, all integrated into a holistic driving experience. To safely manage the wealth of information and services provided by wireless connectivity, an optimized workload management system is imperative. Most importantly, the driver must "want" to dock his/her mobile device in the vehicle to fully experience connectivity in a safe and intuitive manner that is superior to distracting handheld or remote solutions.

The safety-optimized cockpit must also promote the concept of eyes-forward displays and controls enabling drivers to keep eyes on the road and hands on the steering wheel. Supporting cockpit technologies include high-mounted displays, head-up displays, voice recognition, synthesized speech, and steering wheel controls.

Real-time workload management will assist the driver in maintaining situational awareness and vigilance to the primary task of driving, as depicted in Fig. 1. Workload management will render visually and manually demanding features such as texting, email, and destination entry into voice recognition, text-to-speech, and speech-to-text interfaces, helping to reduce the workload to that of a conversation. Additionally, feature access will be granted based on the driving environment and situational assessment, and will drive appropriate interaction via the data fusion of active safety sensor data, driver state, and cockpit activity. Active safety will not only provide the necessary data for situational data fusion, but its warnings will be adapted to support the real-time needs of the driver in concert with workload management. Alert intensity will be proportionate to driver attention, gently warning the driver when vigilant, or intrusively alerting the driver when distracted. Warning sensitivity will also be increased to accommodate impeded reaction time. Media and navigation sources will be integrated much like AM and FM radio bands are today. Active distraction mitigation will draw the driver's attention back to the forward field of view should the eyes off road time be deemed excessive via the real-time monitoring of driver visual distraction. Comprehensive distraction mitigation and the associated thresholds should be controlled by software, which is defined via research-driven results such as those presented within the NHTSA SAfety VEhicle using adaptive Interface Technology program[11] (SAVE-IT).

Intelligent driver assistance systems as provided by workload management should yield high acceptability, reduced driver reaction time to safety-critical events, improved vigilance when compared to other alternatives, and a marked preference for intelligent integration of smart phone technology versus rogue handheld remote solutions. The holistic concepts of workload management, coupled with the intelligent management of smart phone connectivity features and functions, are expected not only to minimize the distraction potential of smart phone interaction, but also to improve the guideline-driven embedded nav-radio systems of today.

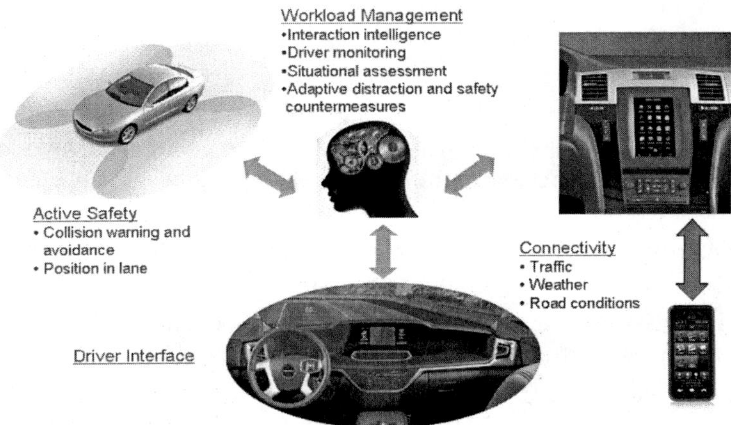

Fig. 1: Real time driver assistance via workload management

Summary/Conclusions

Throughout the history of the automobile, technical advances in adjacent markets have often become "must have" features while driving. The wave of innovation occurring in the smart phone market is influencing nearly every aspect of people's lives, including their drive-time experiences. Whether connected by a personal device like a smart phone or by an embedded network device in the vehicle, future drivers will expect and demand access to an ever-changing and ever more compelling list of personalized, timely, and high-value applications available via the internet.

These trends will continue to cause the driving experience to evolve through the use of new technologies designed to manage and mitigate the distractions that these new usage scenarios bring with them. Many of these technologies are deployed in vehicles today. Also, more are being advanced daily in the design labs of leading automotive manufacturers and their partners. For example, Delphi is developing its MyFi brand of connected infotainment products to address both the consumer's desire for connectivity features and the absolute requirement that these features be available in a manner that promotes eyes-on-the-road and hands-on-the-wheel driving.

Independent studies continuously indicate that consumers desire and are willing to pay for connectivity features in their vehicles. The voice of the customer is clear, as are the societal trends in connectivity. The automotive industry continues to work aggressively to keep pace with these new opportunities while simultaneously ensuring that the critical task of operating a vehicle remains the primary focus.

References

1. Trajectory Group. "The Connected Car Comes in Many Flavors." October 2010 (updated March 2011). http://www.slideshare.net/ralfhug/ralf-hug-connectd-car-comes-in-many-flavors-oct-2010.
2. International Center for Media and the Public Agenda in partnership with the Salzburg Academy on Media and Global Change. "the world UNPLUGGED." http://theworldunplugged.wordpress.com.
3. International Data Corporation. "IDC Forecasts Worldwide Smart Phone Market to Grow by Nearly 50% in 2011." March 29, 2011. http://www.idc.com/getdoc.jsp?containerId=prUS22762811.
4. Massachusetts Institute of Technology. "Technology Review." Vol. 114, no. 3: 73. http://www.technologyreview.com/communications/37382.
5. J. D. Power and Associates. "2011 U.S. Automotive Emerging Technologies Study." May 27, 2011. http://businesscenter.jdpower.com/JDPAContent/CorpComm/News/content/Releases/pdf/2011074-uset.pdf.
6. Madden, M., and L. Rainie. "Adults Text While Driving Too!" Pew Internet and American Life Project. June 18, 2010. http://pewresearch.org/pubs/1633/adults-texting-talking-on-cellphone-while-driving-like-teens.
7. J. D. Power and Associates. "2010 U.S. Automotive Emerging Technologies Study." June 23, 2010. http://businesscenter.jdpower.com/JDPAContent/CorpComm/News/content/Releases/pdf/2011074-uset.pdf.
8. Gartner, Inc. "Gartner Says Wireless Connectivity to be Main Focus for Vehicle Manufacturers by 2012." May 28, 2009. http://www.gartner.com/it/page.jsp?id=996912.
9. Machina Research. "Latest Research: 1.8 Billion Automotive M2M Connections in 2020." June 16, 2011. http://wirelessnoodle.blogspot.com/2011/06/latest-research-18-billion-automotive.html.
10. Insurance Institute for Highway Safety. "Texting Bans Don't Reduce Crashes; Effects Are Slight Crash Increases." September 28, 2010. http://www.iihs.org/news/2010/hldi_news_092810.pdf.
11. John A. Volpe National Transportation Systems Center. "Safety Vehicle Using Adaptive Interface Technology (SAVE-IT)." http://www.volpe.dot.gov/hf/roadway/saveit/index.html.

Wireless Charging of Electric Vehicle Converged with Communication Technology

Dr. In-Soo Suh
Professor
Cho Chun Shik Graduate School for Green Transportation

Wireless Charging of Electric Vehicle Converged with Communication Technology

Dr. In-Soo Suh
Professor
Cho Chun Shik Graduate School for Green Transportation

Abstract

As part of a global trend regarding new transportation, there has been an increase recently in research and development of electric vehicles (EVs) including commercialization, technical standards, and policy fields to overcome global environmental pollution and the exhaustion of oil resources. However, EVs have not yet been massively introduced to the market because of the technological limitations related to batteries as well as charging stations. The wireless EV fast charging system, either stationary or dynamic, is a technology break-through based on wireless power transfer technology, which secures the customers' safety and provides enough transmission efficiency for practicality. With the Shaped Magnetic Field in Resonance (SMFIR) technology, the EVs can receive electric power from a power line installed under the road surface, while the vehicle is stationary or in motion, with more than 80% of transmission system efficiency. In addition to the wireless charging system, communication technology can be incorporated to integrate the vehicle-infrastructure-customer interface. This paper introduces the concept of the intelligent wireless charging system and describes the technical architecture of the system combining billing, central operation, and management systems with communication technology.

Introduction

With more than one hundred years of internal combustion engine (ICE) age, many endeavors have focused on efficient energy conversions and applications due to the recent increased concern about petroleum depletion and global warming. The development and operation of greener transportation vehicles has gained more momentum than ever with the recognition of a global warming crisis and CO2-related regulations. The creation of EVs is a step toward reducing CO2. However, there are still critical issues regarding mass production of battery technology such as less-than-desired mileage per charge, heavy weight, long charging time, cost, and cycle life. Additionally, plugging in for high-voltage, fast charging is still a safety concern for public application.

Regarding conductive charging methods, Types I and II of SAE Standard J1772 properly address terminology, protocols, and equipment installation. DC fast charging is also under discussion in Type III of SAE J1772. In non-contact wireless charging, the committee for SAE J2954 began in 2010 to develop a standard, targeting application in 2015. While wireless technology improves transmission efficiency by optimizing magnetic shapes and more efficient power electronics devices, and the power consumption of consumer devices of application has recently been significantly reduced, we will soon observe that these two aspects create innovative applications in home electronics, in transportation vehicles, and in various industries.[1] Plug-in electric vehicles (PEVs) have shown limited market penetration due to the need for cable and plug charger, galvanic isolation of the on-board electronics, the bulk and cost of the charger, and the required larger energy storage system (ESS).[2] In comparison to conductive charging, wireless charging provides improved convenience and inherent electrical isolation. Therefore, it is safer in field applications. As already introduced by a group of researchers in KAIST (a college in Korea specializing in science and engineering), dynamic charging(i.e., charging EVs while in motion on the road) can reduce the required capacity of ESS on board, which can be another significant benefit of wireless charging technology.[3,4] This paper introduces and describes a practical, applicable wireless charging system developed by KAIST, including the billing and central operational management system.

Electric vehicles are charged wirelessly via the induced magnetic field from power lines that receive electric power from the power supply inverter. As shown schematically in Fig. 1, the system is composed of a power supply constructed on roads and on-board equipment in a vehicle. A pick-up device is mounted at the bottom of a vehicle, which collects the magnetic field and converts it into electricity. The power line is installed under the road to generate a magnetic field. The resonance between the power supply electric circuit and the receiving pick-up circuit is adjusted to increase the transmission efficiency. The wireless power transfer technology via electromagnetic field is named SMFIR, which was developed by KAIST. This technology enables EVs to charge the battery while at a stop or even when driving.

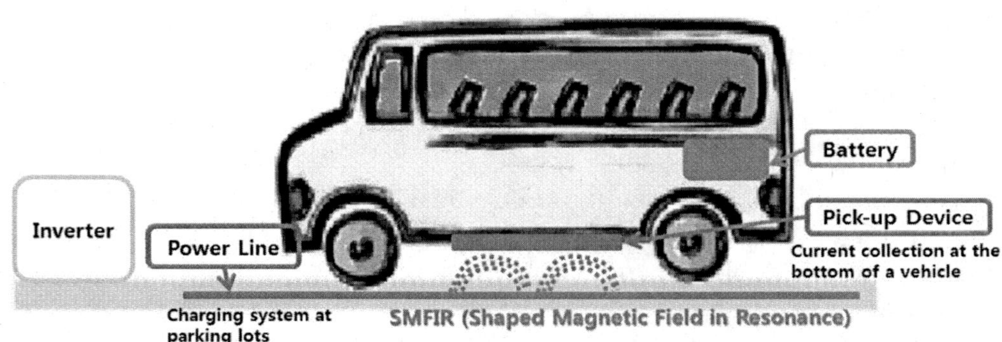

Fig. 1 SMFIR charging infrastructure and vehicle

The static wireless charging system can be applied in a parking lot with a similar installation layout of power inverter, power lines, and pick-up device. As shown in Fig. 2, power lines can be deployed as two loops, which are included in one long loop to optimize efficiency. A longer vehicle (such as a bus) can be charged at the red loop, or two different passenger cars can be charged at a bus parking lot simultaneously. For this charging system, an electricity billing system can also be included to create electric bills depending on the amount of electric power consumed.

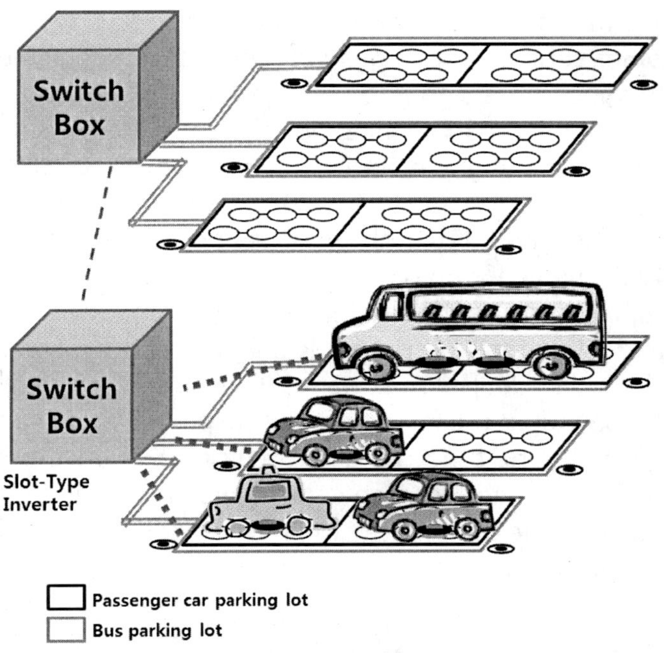

Fig. 2 An example of a concept layout of wireless charging technology

Table 1 summarizes several examples of wireless power transfer with published specifications in the field of surface vehicle, rail, and home appliances. In this paper, several distinguished features of KAIST's On-Line Electric Vehicle (OLEV) system will be discussed.

Table 1 Several examples of wireless charging technologies in published domain

Application	Organization	Charging Pattern	Specification
Vehicle	PATH, UC Berkeley	Charging during stop and driving	- Air gap: 7.5cm, Efficiency: 60% - Construction cost: 0.74-1.22M$/km
	Conductix-Wampfler	Charging at stop	- Air gap: 3cm - Inverter: 70kW
	Showa	Charging at stop	- Air gap: 10cm - Prototype: 30kW, 60kW, 150kW - Weight: 1kg/kW
	HINO	Charging at stop	- Air gap: 5cm - Efficiency: 83.2% at 51% SOC 84.1% at 42% SOC

Application	Organization	Charging Pattern	Specification
Rail	Bombardier	Charging during stop and driving	- Air gap: 6.5cm, Efficiency: 93% (750V DC 800A, maximum output 250kW) - Segment : 16.2m
	Sorrento Electronics	Charging at stop	- Air gap: 2.5cm, Efficiency: 53%, Output: 80kW - Linear motor car
Electronic appliances	Qualcomm	Portable equipment	- Efficiency: 60% - Frequency: 13.56MHz - 500~600mA electric current charging
	MIT WiTricity	TV	- Capacity: 60W - Efficiency: 40%@2m

SMFIR Technology and Wireless Charging System

SMFIR technology enables vehicles to charge wirelessly either when stationary or in motion. During the charging process, the power inverter supplies the electricity to the power cables installed under the road surface, which is converted to single phase 20 kHz, 200 amperes from the commercial grid electricity of 3-phase, 60 Hz, 380 or 440 V AC. The strong electricity in the power cable creates the 20 kHz alternate current of electromagnetic field. Then the pick-up device mounted at the vehicle's bottom collects the magnetic flux with the resonance phenomena of tuned coil sets, which is designed in the vicinity of 20 kHz natural frequency for maximum transfer efficiency. The pick-up device is a T-shaped iron core with turned coil sets in the middle. The schematic diagram of the magnetic field shape with SMFIR is depicted in Fig. 3 with dual elliptic shapes. The induced electromagnetic field shape can be controlled by the layout of the ferrite core in the power supply and the pick-up sides, air gap, core-to-core distance, and tuned coils in the pick-up. To ensure the maximum power transfer capacity and efficiency, those design parameters have been iteratively optimized using 2-D or 3-D modeling of the magnetic field in the power range capacity of 100 to 200 kW for the bus application.

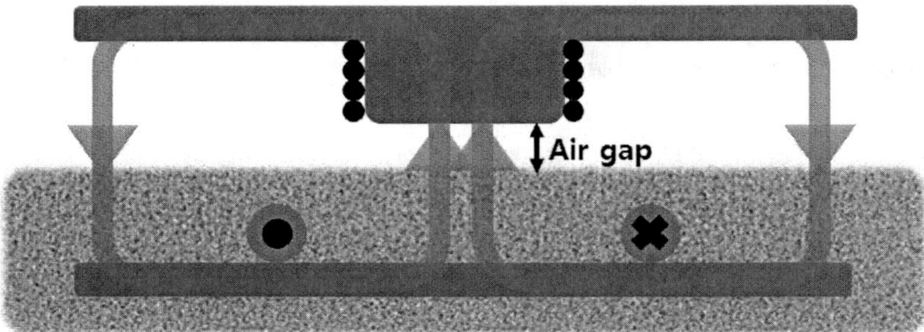

Fig. 3 Dual elliptic shape of magnetic field with SMFIR

As shown in Fig. 4, the magnetic flux amplifies by the E-shaped ferrite installed with power lines. When the amplified magnetic flux reaches the pick-up device on the vehicle, the induced current is generated in the tuned coils sets. The pick-up device collects the power, and the collected current is used to charge the battery or to operate the driving electric motor directly.

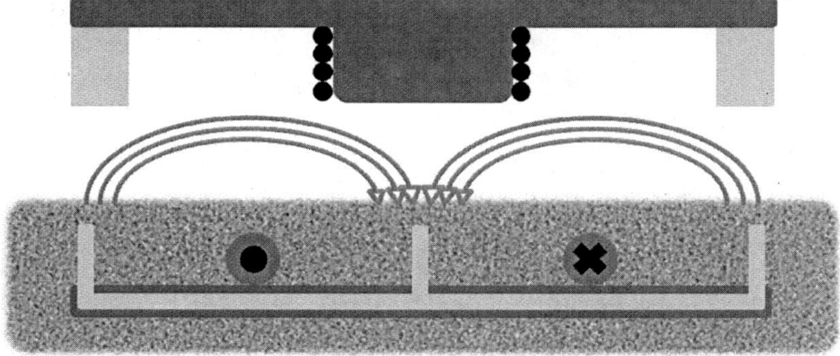

Fig. 4 Schematic layout of optimized magnetic field with SMFIR charging process

The technology demonstrates that the wireless power transfer efficiency is higher than 80% while an air gap is 200mm. The air gap is defined as the distance between the road surface and the bottom of the pick-up device. It is necessary for the vehicle to be equipped with pick-up device and battery for OLEV wireless charging. OLEV is the new concept of the EV system adopted by SMFIR technology, which can be charged wirelessly.

Compared with conductive charging, wireless charging with SMFIR has many advantages as summarized in Table 2.[4,5] The OLEV system can be superior in cost, safety, limited mileage, and charging time when compared with PEV or the battery swapping concept. With dynamic (in-motion) charging, the required battery capacity can be only one-fifth of the typical PEV battery capacity, while the status of charging swing (SOC) can also be minimized. The inherent nature of electricity isolation between the power supply system and vehicle eliminates safety concerns during the plug-in process as well as on-board isolation difficulties. The wireless charging system also demonstrates significantly improved weather-resistive charging conveniences.

Table 2 Comparison between conductive and wireless charging

Contents		Pure Electric Vehicle	Battery Swapping EV	OLEV
Power train structure		Charging Station, Battery, Motor	Charging Station, Battery, Motor	Battery, Motor, Pick-up
Operation down-time of vehicle		Run → charging → run	Run → swapping → run	No down-time
Battery	Weight and price	Heavy and expensive	Heavy and expensive	Reduced to about 20% of PEV
	Safety	Safety concerns	Safety concerns	Safe
	Driving distance	Limited	Limited	Conditionally unlimited
Charging	Safety	Possible electric shock (contact)	Safe (Battery changing)	Safe (non-contact)
	Number of stations	At least 5 times the number of existing gas stations	At least 2 times the number of existing gas stations	None (with on-line charging)
	Time	Long	Time to change	No extra

The wireless power transfer technology can have a wide range of applications as shown in Fig. 5. It can be applied to automobiles, home appliances, robots, seaports, railways, portable equipment, and other areas.

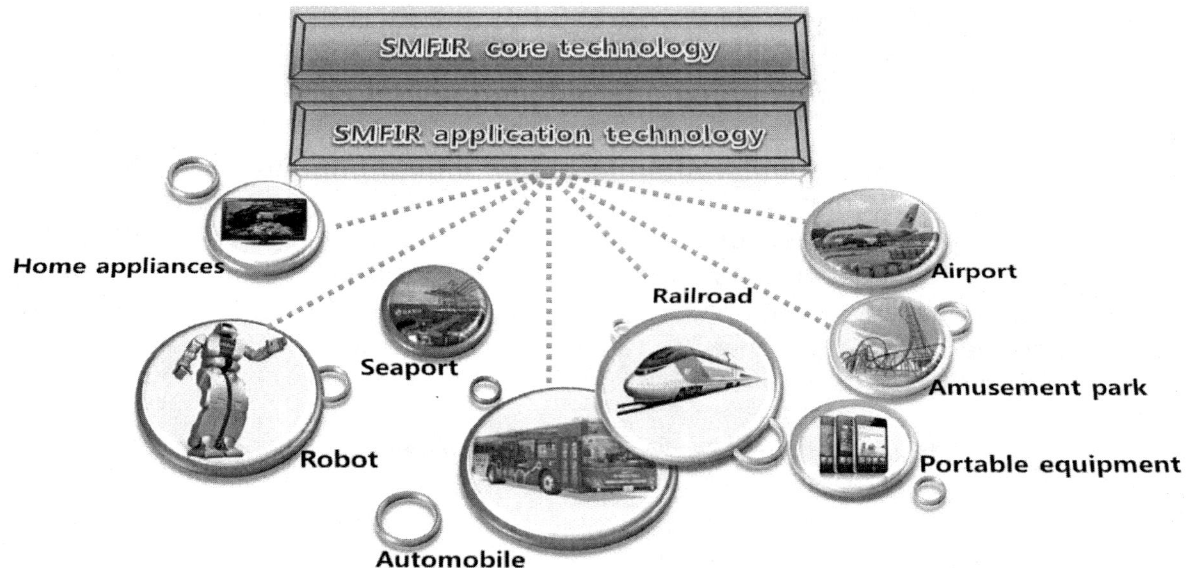

Fig. 5 Application areas for wireless power transfer technology

Billing System with Communication Technology

This section describes the electricity billing system combined with wireless static charging. Although it relates to static charging in a parking lot, the basic concept of the billing system can be applied to dynamic charging as well. The recent IT convergence concept plays a significant role in developing a user-friendly billing system for EV charging. One example of the billing system with wireless power transfer is shown in Fig. 6, which demonstrates applying WiFi, RS 232C/484, and transmission control protocol/internet protocol (TCP/IP). When the vehicle is charged at a parking lot, charging information can be transferred to servers through communication networks for the purpose of operational management.

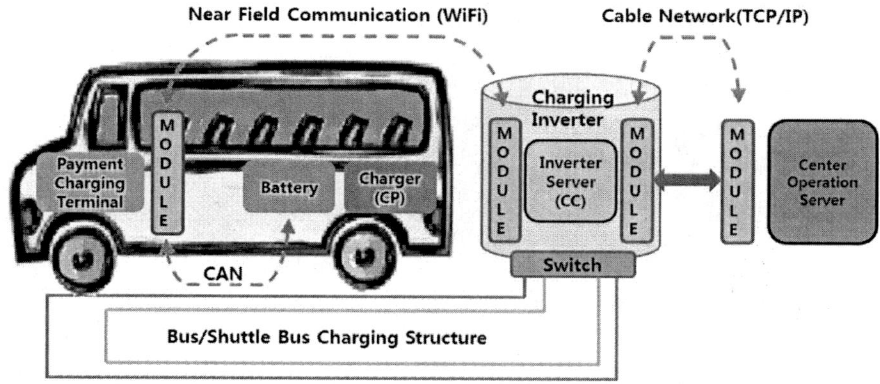

Fig. 6 Wireless charging system with billing and payment system combined

When the driver pulls onto a parking surface equipped for charging, the vehicle identification sensor taps into the vehicle controller area network (CAN) in order to communicate about the battery management system (BMS) information so that the power inverter can identify that the vehicle is ready for charging. The payment charging terminal installed inside the vehicle provides the man-machine interface. The driver can select a desired payment option with pre-charged card or credit card (or other applicable local automatic payment system or smart card) with the option of choosing a pre-determined energy quantity or dollar amount. The inverter server determines the complete protocol communication between the power inverter and BMS. The inverter then turns on the power supply, and the payment option by driver and vehicle status check are completed. The charging station server monitors and manages necessary information from the grid and the power inverter. The power supply cable can have a different set of layouts to accommodate various types of vehicles, and a specific layout can be selected upon the vehicle type identification.

Fig. 7 portrays how communication is established with the charging server, the inverter server, and the payment charger in the interface of the billing system, which is the intelligent wireless charging system with the electricity billing system. Adding the payment charger and the card reader, the charging information can be exchanged by the serial port. For example, the information includes a unit price, the total charged power, or the periodic charged power. It can be shared in the wireless charging system composed of four functional sections. In the first part, the inverter charge information is processed to compute the amount of the supplied electricity. The next part is for the battery information, which is the charge of the battery SOC (State Of Charge) level equipped in the customer's EV. Next is the user interface, showing the customer the information processed in the system. The last functional part, the fee charging component, connects the billing system with the other parts of the wireless charging infrastructure. This system can be incorporated to Dedicated Short Range Communications (DSRC) protocol as set by the U.S. Department of Transportation in the IntelliDrive program.[6]

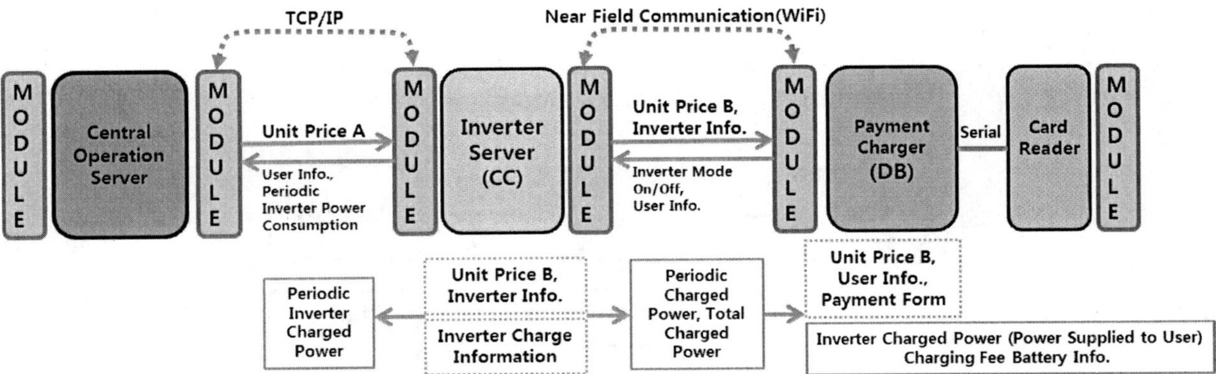

Fig. 7 Communication network of the billing system

The communication system must make use of vehicle-to-vehicle (V2V) technology for alerting emergencies and vehicle-to-infrastructure (V2I) technology for automatic vehicle identification to determine if the vehicle enters the power lines. DSRC is one of the communication technologies using the radio frequency that can link vehicles or a vehicle and the road. Generally, DSRC has been developing applications for the Intelligent Transportation System (ITS) based on the 5.9 GHz Wireless Access in Vehicular Environments (WAVE), which can adapt well to the wireless charging network for EVs. Because of the wireless standard based technology, the transfer speed of DSRC is competitive. However, the power consumption is about 250mA, which is slightly higher than other technologies. Furthermore, the transfer range of DSRC is from 1 to 1000m (3.28 to 3280 feet), which is efficient considering the safe distance between vehicles, and which makes it possible to transfer much more information.

Through V2I communication technology, information related to vehicles, driving environments, or traffic conditions is exchanged. It can transmit and receive the real-time vehicle information while the magnetic sensor has recognized only that the vehicle enters the power lines. Also, the traffic conditions within a specific range of the vehicle's location are collected by the DSRC, and the main server can make a decision about how much electricity is needed based on those conditions or other factors. V2V controls the communication among vehicles in motion to improve safety. By selecting multi-hop networking, vehicles share the warning information and each vehicle carries out the messages as the base station.

Central Operation System

The central operation and management system is required to complete the wireless charging infrastructure. As shown in Fig. 8, the central management system is the highest level governing the charger server, power inverter, pick-up device, and billing system. Information on the quantity of charged electricity is gathered and the charge is calculated on the varying unit price of electricity based on the advanced metering infrastructure (AMI) function of smart grid activity. The final charge is provided to the customer through the billing system. After the charging and payment processes are finished, the system maintains the database for operational purposes, acting as the central connection between the charging facility and the billing system.

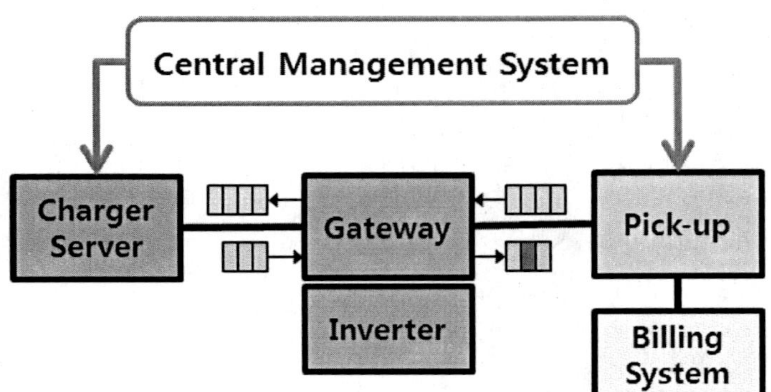

Fig. 8 Wireless charging system and communication network concept

The algorithm of management is comprised of the three phases from the vehicle entrance to the payment process, which are visualized in the user interface screen. For the first phase, the vehicle enters the charging lots and a driver selects the total quantity of power charging. The next phase is the data processing, by which the driver's choice can be sent to the distributor.

At the same time, the billing data uploads to the billing system to create a bill. All data are then transmitted into the central management system and the customer decides payment methods (cash, credit card, etc.). Therefore, the three sequential steps create one cycle from the vehicle's entrance to its departure. Smart Metering is one of the important concepts in the central management system. Briefly, the electricity rate is defined by the power per dollar (kW/$), which can vary depending on the charging duration, time of day, or other factors.

Wireless Charging Demonstration Project

To verify the system's functionality, a demonstration project was conducted with the SMFIR technology. A charging facility was constructed with a vehicle, a power inverter, and power lines under the road surface. The billing and central operation systems were also included. The layout of the physical demo site is shown in Fig. 9.[7] From the distribution box, the power is supplied into the inverter as 3-phase 60Hz 380V AC. The inverter converted the 3-phase power to single-phase 200 Arms 20 kHz and its capacity is 100kW. The converted power from the inverter is fed to the power lines to create a magnetic field. In this test bed, two different lengths of power supply line were installed to accommodate two different types of vehicles on one parking space. When a vehicle is pulled onto the space for charging, the vehicle identification sensor detects the vehicle type and sends a signal to the power inverter to select proper power line layout for charging.

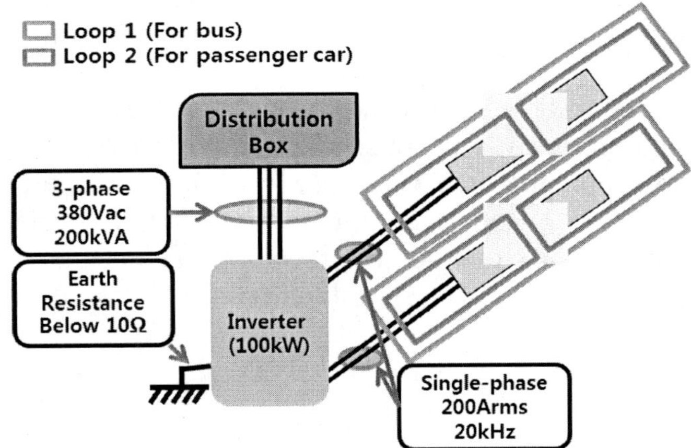

Fig. 9 Schematic layout of charging facilities

The vehicle was an OLEV microbus, 25 passenger capacity, retrofitted from the conventional diesel engine operated vehicle (see Fig. 10). The OLEV bus had a set of pick-up devices, Li-polymer battery, and other EV powertrain. The vehicle in the demo project had two sets of pick-up devices with a maximum pick-up capacity of 40 kW and a battery capacity of 28 kWh.[7]

Fig. 10 A photo from the test bed with wireless charging facility

Conclusion

Innovative wireless charging with SMFIR enables the introduction of EVs to the market in massive scale due to improved convenience and safety measures, while achieving practically applicable transmission efficiency. Combined with recent developments in communication technology, the user interface can also be significantly enhanced, as introduced in this paper. The fast charging system with 40 kW of power capacity and wireless power transfer has been successfully designed and demonstrated with the flexibility to apply to different types of vehicles such as a buses, micro buses, and passenger vehicles. The potential opportunity for more enhanced communication interface with DSRC has also been discussed.

Although this paper emphasizes the static charging system, the technology in wireless power transfer and integrated communication convergence can be applied to dynamic charging, which will lead to waiving concerns such as battery limitation and charging distance on electric vehicles. With wireless power transfer technology further developed in phase with standard activities and regulation clearance, the technology can be a core approach to future transportation systems that is especially applicable to mega urban city design.

References

1. Fuji-Keizai USA, Inc. *Wireless Power: Current Status and Future Directions: R&D Competition, Commercial Markets and Future Applications*. Dublin: Research and Markets, October 2010.
2. Miller, M. J. *Wireless Plug-in Electric Vehicle (PEV) Charging*. Oak Ridge National Laboratory, Project ID VSS061, May 2011.
3. Suh, N. P., et al. "Design of On-Line Electric Vehicle (OLEV)." Plenary Presentation, CIRP Design Conference, April 2010, Nantes, France.
4. Suh, I. S. *KAIST On-Line Electric Vehicle as a Future Transportation Technology with Low Carbon and Green Growth*. International Forum on Public Transportation Policy hosted by Korean Transportation Institute, March 30, 2010, Seoul, Korea
5. Suh, I. S. "On-Line Electric Vehicle Development." Presentation at the 2010 International Forum on Electric Vehicle, June 2010, Daejeon, Korea
6. U.S. Department of Transportation, Research and Innovative Technology Administration. 2011. *Connected Vehicle Research*. http://www.its.dot.gov/connected_vehicle/connected_vehicle.htm.
7. Suh, I. S. et al. 2011. *Intelligent Wireless EV Fast Charging with SMFIR Technology*. Society for Design and Process Science.

Challenges and Benefits

Autonomous Driving - A Practical Roadmap

2010-01-2335
Published
10/19/2010

Jeffrey D. Rupp and Anthony G. King
Ford Motor Company

ABSTRACT

Successful demonstrations of fully autonomous vehicle operation in controlled situations are leading to increased research investment and activity. This has already resulted in significant advancements in the underlying technologies necessary to make it a practical reality someday. Not only are these idealized events sparking imaginations with the potential benefits for safety, convenience, fuel economy and emissions, they also embolden some to make somewhat surprising and sometimes astonishing projections for their appearance on public roads in the near future.

Are we now ready for a giant leap forward to the self-driving car with all its complexity and inter-dependencies? Humans will need to grow with and adapt to the technological advancements of the machine and we'll deeply challenge our social and political paradigms before we're done. Even if we as engineers are ready, is the driving public ready?

Putting a man on the moon was achieved through a series of logical extensions of what mankind knew, with necessity driving a search for technical solutions in the usual as well as unusual places, much as the Defense Advanced Research Projects Agency did with their Grand Challenges. This paper addresses the autonomous vehicle vision in terms of the current state and some of the practical obstacles to be overcome, and proposes a possible roadmap for the major technology developments, new collaborative relationships, and feature implementation progression for achieving those ambitions.

1.0. INTRODUCTION

The desire for the ultimate personalized, on-demand, door-to-door transportation may be motivated by improved personal convenience, emissions and fuel economy; yet there are also potential safety benefits from the pursuit of autonomous vehicles. This paper describes some of the practical obstacles in achieving those goals, and explores the use of near term applications of technologies that will be by-products of pursuing them. This includes a partial history of autonomous vehicle development (Section 2), potential consumer acceptability issues (Section 3), followed by a development roadmap and discussion of some variables to be addressed before autonomous vehicles become viable (Sections 4 and 5), and ends with a consideration of collaborative relationships that could assist in acceleration of development and issue resolution (Section 6).

2.0. THE CURRENT STATE - PUTTING THE HYPE INTO PERSPECTIVE

There has been escalating excitement about fully autonomous vehicles in the robotics community for some time and the excitement has now spilled over to the automotive industry. The idea of a self-driving, road-ready vehicle sparks the imagination, and is a familiar concept due to repeated exposures in popular culture; be it movies, cartoons, television, magazines, books or games.

An exhibit at the 1939 World's Fair in New York[1] presented a vision where cars would use "automatic radio control" to maintain safe distances, a depiction of transportation as it would be in 1960, then only 21 years into the future. One of the earliest attempts at developing an actual vehicle was led by Dr. Robert E. Fenton who joined the faculty at Ohio State University in 1960 and was elected to the National Academy of Engineering in 2003[2]. It is believed that his pioneering research and experimentation in automatic steering, lane changing, and car following resulted in the first demonstration of a vehicle that could drive itself. Since then,

OEMs, universities, and governmental agencies worldwide have engineered or sponsored autonomous vehicle projects with different operating concepts and varying degrees of success.

Most recently, the Defense Advanced Research Projects Agency (DARPA), an agency of the United States Department of Defense, sponsored three autonomous vehicle challenges. While a number of media friendly successes resulted in good 'photo ops', those in technical fields and many others readily appreciate the magnitude of work required to mature these vehicles into a viable, real world, design.

2.1. Contemporary Error Rates -- We're Way Off

In the months preceding the inaugural DARPA Grand Challenge in 2004, William "Red" Whittaker of Carnegie Mellon's Robotics Institute, with over 65 robots to his credit, stated "We don't have the Henry Ford, or the Model T, of robotics", "Robotics is not yet mainstream; it's not yet a national conversation."[3] His contributions and those of his students over the next few years would move the needle significantly, but his comments suggest the true nature of the challenge.

The error rates of robotically piloted vehicles today are still very high compared to human-piloted vehicles. At the 2005 DARPA Grand Challenge (DGC2) 5 of the 23 finalists successfully finished the 132 mile course, while two years later, at the 2007 DARPA Urban Challenge Event (UCE), 6 of the 11 finalists finished a 60 mile course. The mean mileage between significant errors (failure) at these events was 120 miles for DGC2 and 100 miles for UCE[4]. The errors cannot be attributed to a single primary cause, rather, multiple simultaneous causes and interactions including sensing, interpretation of the scene and simplification of its full complexity, simplifying assumptions and non-representative tradeoffs built into the algorithms, as well as unintended software bugs and hardware durability. Compare robotically piloted vehicle errors to that of human drivers, who averaged 500,000 miles driven between crashes in 2008[5].

Despite humans being 3-4 orders of magnitude better at driving than robots, crashes of varying severity occur regularly. In 2008 in the United States alone, there were 34,000 fatal crashes and 1.6 million injury crashes. Autonomous vehicles may need to be better drivers than humans, exhibiting fewer errors, to gain acceptance. The error rates inherent in today's autonomous vehicles are unacceptable for real world deployment in the present and will be for some time to come.

2.2. Progress Has Been Slow

Recalling the many predictions of a self-driving car over the last four decades, it is obvious that autonomous vehicles have taken and will take far longer than expected, especially when it comes to operational safety. Fully autonomous vehicles today are the product of laboratories, test tracks, and prize winning competitions, mainly conducted under favorable conditions with minimal and controlled uncertainties and no penalty for error. With limited success even in ideal situations, industry has little choice but to methodically split the problem into attainable steps, learning and developing the necessary enabling technologies along the way.

The combination of radio detection and ranging (RADAR) functionalities was patented by Christian Hülsmeyer in 1904[6], building on work from the mid-1800s by physicists James Maxwell and Heinrich Hertz. The majority of the development since then has been driven by maritime collision avoidance and military defense applications, including important signal processing extensions such as target velocity estimation based on frequency shift as proposed by physicist Christian Doppler. Despite this early start, it wasn't until 1999, with seven years of focused target tracking and controls development as well as electronics miniaturization, that Ford Motor Company launched the world's first-to-market radar-based ACC system with braking for an automotive application, on a Jaguar XKR.[7]

More than a decade later, advances in sensing technology critical for autonomous vehicle applications are just now accelerating significantly. Functionality of automotive forward-looking radars is increasing, even while prices are decreasing, with a drop of 75% over two generations expected in one case.[8] The progression to today's state of the art dual mode electronically scanned systems has allowed industry to use the resulting increased accuracy and availability to expand to new customer functions.

Digital camera systems have similarly been in existence for quite some time, with a patent application for "All Solid State Radiation Imagers" filed in 1968[9], and are now progressing more rapidly too. CMOS imagers have demonstrated increasing sensitivity, dynamic range, and pixel count, while costs have decreased due to the large volumes of consumer electronics applications. More recently, advancements in machine vision algorithms have enabled the evolution from lane tracking to significantly more complex vehicle and pedestrian detection and tracking functions.

Fusion sensing systems are also starting to see more automotive applications as well. Combining multiple sensing modalities, fusion leverages the orthogonality that can be established where the strength of one complements the weakness of another. This can create a sensing system with robustness and reliability greater than the sum of its parts. Ford developed and launched a radar-camera fusion system for Collision Avoidance Driver Support (CADS) functionality on the Volvo S80 in 2007. This was further expanded on the 2011 S60, overlaying a fused camera / forward looking multi-mode radar, with a multi-beam infrared and ultrasonic sensors, enabling collision warning and full auto braking for vehicles and pedestrians for collision avoidance, a world first, in addition to ACC, Lane Departure Warning, and Driver Alert (driver impairment monitoring) functionality.[10]

Other sensing technologies are also under development to better describe and interpret the external environment. Although automotive lidars, especially for ACC, have fallen out of favor, the development of 360° scanning and flash designs may bring about their resurgence. Detailed on-board maps are now available to help predict the road attributes ahead. Even as the number of radars and cameras in the vehicle proliferate, the industry also recognizes that on-board sensing could be significantly augmented through direct communication with other vehicles and the infrastructure. Research in the area of vehicle-to-vehicle and vehicle-to-infrastructure communications will be critical to any future cooperative transportation network. Despite these advancements, the verdict is still out as to the form of the ultimate sensing solution.

The majority of today's situation assessment algorithms enable only advisory and warning systems, as these systems are more easily implemented than fully autonomous control; using sensor data, the algorithms interpret the environment, predict the future, and provide some related driver support. With this limited approach, most performance errors merely result in annoyance. The environmental sensing system and control algorithm requirements are not as stringent as needed for autonomous operation, where the machine makes a decision and takes control of the vehicle. In the latter case, an incorrect decision may possibly result in a wrong action, possibly causing a collision when one may not have occurred otherwise. While designing a system that reacts positively (e.g. automatically applies the brakes prior to a collision) is readily achievable, the more difficult part of the task is to design the system to seldom make a mistake, and have the reliability and robustness necessary to appropriately respond to real world noise factors. The autonomous systems that exist today in controlled laboratories and test tracks are just not ready for the uncontrolled uncertainties of real world conditions. Automotive engineers are proceeding slowly to help ensure that appropriate level of performance exists before introduction.

2.3. Reluctant Consumer Acceptance of Autonomous Control

One need read only a few blogs in order to appreciate that consumers are uncomfortable with a machine making decisions for them and you can easily conclude that some drivers do not trust their vehicle taking even limited autonomous control. An independent analysis is available that describes the phenomenon of decision trust and the attributes affecting safety feature purchase.[11] Furthermore, the lack of third party endorsements for more than the most basic CADS functions (i.e. Forward Collision Warning; further enumerated in Section 4.2, Use Cases) has created little feedback for these technologies and therefore little customer enthusiasm and 'pull', and the lack of government mandates has created no 'push'.

Governmental and public domain agency action may help accelerate acceptance and adoption, or at least access and usage, of autonomous technologies, and several organizations around the world are considering regulation. Anti-lock braking systems were introduced in 1971, and reached 86% market penetration only after 37 years, in 2008. Compare that to Electronic Stability Control (ESC), introduced in 1995. Although the industry already had an implementation plan, the U.S. National Highway Traffic Safety Administration (NHTSA) accelerated penetration by mandating standard ESC in all new vehicles by 2012, less than 20 years later. NHTSA has included Forward Collision Warning and Lane Departure Warning in the ratings for the Active Safety New Car Assessment Program. The European Commission is considering mandates for Collision Mitigation Systems on light commercial vehicles. Non-governmental organizations such as the Insurance Institute for Highway Safety and the Consumers Union (publishers of Consumers Report magazine) have started to address CADS technologies, raising consumer awareness. Insurance companies are considering lower rates for vehicles with CADS features.

It is interesting to note that market adoption rates may have some cultural influence. Take the ACC system for example, a fairly straightforward extension of traditional cruise control that provides longitudinal control of the vehicle using brake and throttle to maintain distance to a vehicle in front. Ten years after initial introduction, it is finally getting significant mass market recognition, but the penetration rate in North America is only a fraction of that in Japan where the market seems to have a greater percentage of early adopters, allowing for rapid technology evolution. An independent study detailing these differences is also available.[12,13]

2.4. Today's Feature Implementation Progression

Although the adoption of CADS functions in private vehicles has been slow to date, the world is on the cusp of more widespread implementation of limited autonomous control. Technology will continue its rapid advance and as consumer acceptance expands, the industry will see systems that warn the driver of hazardous conditions, support driver actions, provide limited autonomous control with driver command, and even take some fully autonomous action to avoid a potential collision. The nature, direction, and pace of CADS feature introduction and progression can be inferred from the following list:

- Longitudinal support:

1958	Cruise Control (non-adaptive)
1971	Anti-lock Braking System (ABS)
1991	Ultrasonic Park Assist
1999	Adaptive Cruise Control (ACC)
2003	Forward Collision Warning (FCW)
2003	Collision Mitigation by Braking (CMbB),
2006	Stop & Go ACC (S&G)
2006	Full speed range ACC
2008	Low Speed CMbB (collision avoidance, City Safety™)
2010	Full Autobraking CMbB
2013 (*est.*)	Curve Overspeed Warning (electronic horizon-based)
2015 (*est.*)	Curve Overspeed Control (electronic horizon-based)

- Lateral support:

1971	ABS
1990	Variable steering assist, cross wind compensation, etc. (electrical)
1995	Electronic Stability Control
2001 (*Japan*)	Lane Departure Warning (LDW)
2001 (*Japan*)	Lane Keep Assist (LKA)
2002	Roll Stability Control (RSC)
2003 (*Japan*)	Lane Centering Aid (LCA)
2004 (*Japan*)	Intelligent Parking Assist System (IPAS)
2005	Blind Spot Information System (BLIS)
2006	Active Parking Assist
2007	Driver Alert, Driver Impairment Monitoring
2012 (*est.*)	Lane Change Merge Aid (LCMA)
2013 (*est.*)	Emergency Lane Assist (ELA)

- Integrated lateral and longitudinal support:

2010	Curvature Control (stability control-based)
2014 (*est.*)	Traffic Jam Assist (TJA) - S&G ACC + LCA

With the continuous evolution and improvement suggested by this feature progression, it is clear that many benefits from warnings and limited autonomous control are being realized, and more soon will be. Beyond this, incremental benefits can be reasonably attained only by advancing to a more complex and potentially intrusive level of functionality, one more closely associated with fully autonomous driver-support features. As suggested previously, consumer paradigms may need to shift again, and the governmental and social infrastructure may need to adapt. The key factor in establishing consumer comfort with these technologies may be empowerment of the driver in making the final control decision, say, overriding the function of the CADS feature.

3.0. A LOOK TOWARD THE FUTURE

3.1. Uncertainty, Unpredictability and Human Error

According to a World Heath Organization study from 2004, traffic accidents result in approximately 3,300 deaths every day, equaling over 1.2 million fatalities each year worldwide. By 2020, annual fatalities due to vehicular accidents are projected to increase to 2.34 million, assuming continuation of current trends. Already the leading cause of injury mortality, road crash injury is likely to become the third leading cause of disability-adjusted life years (DALYs) in the same time frame, trailing only heart disease and unipolar depression.[14]

The pursuit of autonomous vehicles, where drivers are supported in the driving decision making process, has a positive correlation with the pursuit of fatality-free, and even collision-free, transportation. Humans are fallible; driver error is the primary cause of about 90% of reported crashes involving passenger vehicles, trucks, and buses.[15] A misconception links these human errors solely as "... evidence of lack of skill, vigilance, or conscientiousness"[16] or insufficient training, since highly trained and skilled experts, such as doctors and pilots, are also susceptible to making errors, some with serious consequences. Frequently, errors result from poor reactions to unpredictable events and incomplete information as factors in the decision making processes. These probabilistic external factors typically form complex interactions creating random non-repeatable events. One study of airline pilots found that"... small random variations in the presence and timing of these factors substantially affect the probability of pilots making errors leading to an accident."[17]

Given these uncertainties, it seems unrealistic to assume that a decision making process, be it human or machine, will make the appropriate decision 100% of the time. Moreover, we must be cognizant of the fact that drivers are not machines and contemporary machines were shown previously to have not attained any where near the levels of holistic human cognition. Further, human reaction to the same exact external input will vary from individual to individual, and will therefore continue to be subject to unpredictable outcomes.

These external and internal uncertainties characterize the system inadequacies in which errors occurred, where the driver and the vehicle are only a portion of the overall transportation system. Rothe describes how the concept of a living system, one that adapts to change and achieves a new balance, can be applied to a driving scenario.[18] He suggests that an interactive relationship exists among the various system factors - biological (health/illness), psychological (doubt), social (seclusion), societal (norms), economic (lost wages), legal/political (arrest), other drivers and vehicles, the road infrastructure, and information regarding their status (weather and road conditions). Each of these factors set the stage for the other with recursive feedback between them. Focusing on a single factor merely distorts the situation without resolving it.

The implication from this is that a better understood and more tightly coordinated overall system will result in reduced levels of unexpected future events, and thereby a reduced likelihood of collisions. Nearly error free decision making is a very hard problem but it needs to be solved before an autonomous vehicle system that provides 'Full Driver Assist' is ready. Predicting when it will be feasible is merely guess work, but a roadmap would still be useful in approaching it in a comprehensive and systematic fashion.

3.2. Autonomy in Other Transportation Modes

The Shinkansen railway system in Japan provides an example of a positive attempt and outcome. Running on separate track from conventional rail, the lines are built without crossings, use long rails that are continuously welded or joined with expansion joints that minimize gaps due to thermal conditions, employ Automatic Train Control for on-board signaling, have early warning earthquake detection so trains can safely stop, and enforce strict regulation with stiff fines to prevent trespassing on the tracks. From the train sets, to the tracks, the operators, the information availability, and the governmental regulations, this tightly controlled system is designed to reduce the amount of uncertainty and enable a high reliability of safe decision making. The result: no injuries or fatalities due to derailment or collision in 46 years of operation, and only one derailment (with no injury) caused by an earthquake in 2004, while carrying over 150 million passengers a year (in 2008).[19]

The Shinkansen system demonstrates that fatalities may not be an inevitable consequence of transportation after all. A major difference lies in the train operators themselves - besides being highly trained, their number is but a mere fraction of the billions of personal-vehicle drivers in the world today. Thus, tight control over the system includes control over this uncertainty: the variance of individual driver (operator) reactions to external inputs. In the quest for further reductions in collisions in private vehicles it is inevitable to eventually seek to replace human unpredictability with something a bit more predictable. The result may not be purely an electronic substitution, but rather a driver 'subsystem' that involves both the human and the electronic system. The electronic system informs and aids the human in the ways it is better suited, by leveraging its strength (e.g. estimating range and closing velocity), and leaving higher level tasks for the human 'driver' to perform. It's an orthogonal decision making mode, similar to fusion of multiple modalities of sensing (e.g. radar and vision). Each has its strengths and weaknesses, but when properly combined results in a more reliable and robust solution.

Consider another self-driving (autonomous) vehicle, one that has existed for centuries. A ship's captain is on board, but may never touch the wheel; he is in command but not necessarily in direct control. He has a surrogate system, in this case human, that is 'programmed' to carry out 'lower level' control functions, whether that human be a helmsman, quartermaster, or engine room operator, relieving the captain of the burden of continuous interaction. Similarly, you hire and 'command' a taxi as a system (car + driver) by requesting a destination, but there is no direct control.

Beyond those analogies, there are many 'self driving vehicle' applications in existence today. These are autonomous vehicles in a very real sense, some having greater autonomy than others. Commercial airplane pilots engage the autopilot and monitor the systems until direct intervention is needed, whether induced by tower commands or an emergency. Automated train systems, such as those within an airport terminal network, ferry people without an onboard pilot, but are still controlled by humans at a central station. The military has significant autonomous vehicle assets in operation today including Unmanned Air and Ground Vehicles (UAV and UGV respectively), which can be directly controlled by a remote operator, or programmed for autonomous operation to patrol a certain area for a set number of hours, for example. In the past few decades, modern sailing vessels have replaced mundane operator tasks with computer control; an autopilot can now navigate from waypoint to waypoint and seamlessly control throttle, rudder, and roll stabilizers.

All these are examples of vehicles with autonomous control, but still not completely without human oversight. If a pilot is not directly on board, then there is an operator monitoring remotely. There is no vehicle or transportation or mobility system that doesn't have human oversight of some sort. And we should expect the human operator to be 'in-the-loop' for a very long time.

Moreover, these semi-autonomous systems rely on operators trained specifically for driving. To become a commercial airline pilot, for example, one must first obtain a commercial pilot license after 250 hours of flight time, with allocations dedicated to specific conditions and maneuvers. Additionally, a commercial pilot needs an up-to-date first- or second-class medical certificate, an instrument rating and a multi-engine rating. Thousands of additional flight hours are needed to even be considered for hire at a commercial airline. Once hired, additional training begins. Typically a 10 week course ensues, followed by a few weeks in the simulator, where the trainee experiences just about every emergency and anomaly imaginable. Once this training is done, initial operating experience is gained by flying some 25 hours with a special instructor pilot, followed by another flight test. Now the pilot can become a crew member. In order to become the captain of a major commercial airliner, a pilot must then obtain an airline transport pilot certificate which requires passing a written test, and logging 1,500 flight hours including 250 hours as the pilot in command. Similar levels of training are required to pilot a ship, control military UAVs, or control NASA's unmanned vehicles. Current driver training for operating an automobile is not nearly so stringent.

3.3. Do We Want a Driverless Car?

When people talk about fully autonomous vehicles, a common image is that of a driverless car, like the autonomous trains in an airport or DARPA challenge robots. Do consumers want a car without a driver, a car that can go somewhere without you like a military mission, whether delivering a package or picking up the kids after school with no one in control on board? There may be a few cases where a consumer wants someone or something else to do these tasks, but we already have services in place for that - package delivery services, buses, carpools, taxis, etc.

Instead, consumers of private autonomous vehicles may not want a driverless car, but rather a car that drives itself. You are in the car, and the car transports you, your belongings and your family, but you don't necessarily want to directly pilot it. You don't want to be locked into the detailed, sometimes tedious, moment to moment tasks of driving, but instead merely want and need to direct where it goes and how it gets there. You want command, but are willing to relinquish the detailed control to automation so that you can do something else; listening to music, placing a phone call, watching a movie, or just enjoying the scenery. What the consumer really wants is not a self-driving car, but an autonomous vehicle system that provides Full Driver Assist.

Recall the first mainframe computer, first PCs, the first PDA, and then the first cell phones. These devices were going to make our jobs and lives easier. And they have - not by doing work for us as originally thought, but by helping us work more efficiently. At first they were all clumsy devices, difficult to use, and not well accepted. Eventually they are integrated into the connected world in which we live and evolved into productivity tools that enable us to work in more places, more of the time. Similarly, the advent of autonomous technologies in vehicles will result in drivers that are more engaged in some aspects of the driving process rather than further removed, providing them with greater capability in managing the overall process. The driver would now be much more akin to the captain of the ship, biased toward the tactical, strategic, interactive, and predictive roles while leaving the role of the helmsman, lookout, navigator, and even quartermaster to the vehicle systems. Handling this type of automation in everyday life, however, requires that the consumer paradigm change.

3.3.1. Driving to a Seamless Experience

Smartphone owners can buy a special application (app) for just about anything, from checking the weather to checking your bank accounts and paying bills, from playing games to updating your social network and checking sports scores, and so on. There are dozens of apps just for social networking - one for each online site - plus apps for email, contacts, text messaging, and instant messaging. In today's smartphone implementation, the entire task of staying in touch with a social network is an exercise in opening and closing apps, which is a clumsy and overly complicated interface at best. Soon there will be a single app where you can see all your friend's updates on the social networking sites, while tying it seamlessly together with the contacts, photos, email, and text messages on your smartphone.

Like consumer electronics, the automotive industry is now tackling these issues; focusing on improving the in-vehicle experience by combining these apps into seamless experiences. MyFord Touch™, Ford's new driver-connectivity technology, complementing SYNC®, Ford's device and off-board service connectivity technology, is an example of integrating and simplifying the experience of entertainment and connectivity in the vehicle. Through the digital cluster displays, large touch-screen interface and voice interaction, the system allows the driver to naturally command the vehicle to play new music, seek traffic, direction and journey-related information, answer calls, make calls, and even listen to text messages through multi-modal interfaces. Software application programming interfaces (APIs) will soon be available to allow apps like Pandora and Stitcher to be controlled through the voice-controlled SYNC® system to stream audio to build a consistent, user-friendly interface within the vehicle itself.

This development progression repeats a trend that has occurred time and time again. Compare these steps for starting a Ford Model T[20] with today's 'turn the key' or 'push the button' ignitions:

1. Pull the choke adjacent to the right fender while engaging the crank lever under the radiator at the front of the car, slowly turning it a quarter-turn clockwise to prime the carburetor with fuel.

2. Get into the car. Insert the ignition key, turning the setting to either magneto or battery. Adjust the timing stalk upward to retard the timing, move the throttle stalk downward slightly for an idle setting, and pull back on the hand brake, which also places the car in neutral.

3. Return to the front of the car. Use your left hand to crank the lever (if the engine backfires and the lever swings counterclockwise, the left arm is less likely to be broken). Give it a vigorous half-crank, and the engine should start.

Development focuses on the task the consumer is trying to perform, and works to improve the overall user experience associated with that task. Through integration, the functional evolution simplifies the operation and significantly enhances the efficiency in performing that task. Historically, the movement towards a simplified, seamless experience to improve operating efficiency has been a key to widespread adoption of new technology, stimulating a series of consumer paradigm shifts. Similar to a smart phone, the technologies discussed in Section 2.4, Today's Feature Implementation Progression, may be considered standalone apps as well, but in a vehicle environment. Many of the highest technology features have had limited take rates possibly due to perception of cost, complexity and uncertainty of performance, but we expect this will benefit from development into a more seamless experience. Traffic Jam Assist is a technology that operates the distance control of ACC S&G in conjunction with the lateral control of LCA at low speeds. A later step will be to integrate all CADS functions into a comprehensive Full Driver Assist functionality, simplifying the web of complex CADS functions into a coordinated holistic system - user-friendly, easy to understand, and available to all consumers.

When done well, this advanced development can result in recommendations by opinion leaders at many levels, improving the familiarity and comfort level with the technology, further speeding adoption and penetration into everyday life. But what does Full Driver Assist really mean to consumers? What tasks do automotive consumers wish were more efficient?

3.3.2. Of Desires, Expectations, and Values

America has always been a country where motoring nostalgia is heavily intertwined with the freedom of exploration. This explains American's love affair with the car; with hands on the steering wheel, foot on the accelerator, and hair blowing in the breeze while cruising down Route 66. Americans are in their cars a lot - an average of 87 minutes per day according to an ABC News survey.[21] Some automakers have recently focused on remaking car interiors like a comfortable and luxurious living room, but driving is not all for fun.[22] Commuting to and from work comprises over 27% of vehicle miles traveled, more than any other category. The next highest category was social/recreational travel, including going to the gym, vacations, movies or theater, parks and museums, and visiting friends or relatives; i.e. using the vehicle as a means to get to a destination. These two categories alone comprise over 50% of all vehicle miles traveled. A recent study by Northeastern University indicated that, given past history, one can predict anyone's travel route and location with 93% accuracy.[23] These studies imply that people are repeatedly visiting, or commuting to, the same locales with significant regularity.

So do people enjoy the daily driving routine? The study by ABC News indicates that nearly 60% of people like their commute, but only if the trip is relatively easy. Nearly 4 out of 10 state the primary reason they like their commute is that it gives them quiet or alone time, and nearly a quarter identified that their commute is easy and has little congestion or traffic. For city dwellers with more than a 30-minute commute or experience traffic congestion, the percentage who likes their commute drops into the 40's. To further understand consumer behavior, it's necessary to understand the human emotion and values. A great majority of drivers, according to this study, at least occasionally feel very negative emotions while driving, with 62% feeling frustrated, 56% feeling nervous about safety, and 43% even feeling angry. But the same survey also says that 74% often feel independent, while 48% often feel relaxed while driving. Interestingly, independent and relaxed are not really emotions, but relate to core human values. The Rokeach Value Survey (RVS) identifies 18 terminal values, which are values every human strives to experience at least once in their life (and more often if possible), and 18 instrumental values, which are the preferred means of achieving those terminal values.[24] Independence is an instrumental value, and relaxed can correlate to inner harmony, a world at peace, or comfortable life terminal values.

These values seem to at least partially explain, if not directly motivate, people's desire to drive. They explain the high consumer demand for infotainment in the car -- drivers want to enhance relaxation through music or conversation. Infotainment systems, as a relaxing agent, will become even more important as traffic congestion worsens. Hours spent in traffic delays have increased 50% from the last decade and continue to increase[25], so it is expected that the number of people feeling relaxed while driving might actually decrease, even with infotainment systems in the vehicle. On the other hand, Ford and MIT's AgeLab, in conjunction with the U.S. Department of Transportation's New England University Transportation Center, have been working since 2004 to develop vehicle systems that detect the stress level of the driver at key points in time.[26] A recent extension of that project intended to identify specific stress-inducing driving situations, apply biometrics to monitor driver reactions and evaluate methods to incorporate new stress-reducing or even stress-optimizing features.[27] These features include the Blind Spot Information System with Cross Traffic Alert, Adaptive Cruise Control and Collision Warning with Brake Support, MyKey, Voice-Activated Navigation, and SYNC®.

Additionally, the RVS values discussed previously explain why only 5% of trips are on public transportation. Although one can just as easily feel relaxed on a commuter train as in a vehicle, 93% find traveling by car more convenient. It is this convenience that keeps drawing drivers back to the road; the freedom to leave whenever you want; the convenience of getting you from exactly point A to point B without changing modes of transportation. Having your own personal vehicle translates to independence, eliminating the need to rely on someone else to accomplish your own tasks or pursue your goals.

What do drivers want? They want a utilitarian appliance that moves them from door-to-door on their terms; they want to be more effective in the driving process, and they want luxury comforts. They use descriptors such as 'productive', 'efficient', 'relaxing' and 'personalized'. An autonomous transportation device with independent supervisory control would fit the bill, but they also want the ability to drive the enjoyable drives which may add excitement and enhance a sense of freedom. A successful vehicle will likely need to seamlessly blend full assist and fully manual modes of operation and probably everything in between to satisfy consumer needs, expectations, desires, and values.

3.3.3. Consumer Paradigms

In order to build the future of personal transportation that people want, the associated consumer paradigms must change. There is precedence for the shift necessary for adoption of new technological innovations. When Nicolas Joseph Cugnot introduced one of the first self-powered vehicles in 1769 (which was commissioned by the French army), not many imagined that this curiosity would spawn a technological gold rush for the next century and a half in a race to provide 'auto-mobile' vehicles to the masses. Instead there were concerns about their safety and usefulness, as this early vehicle could only travel at 2.5 mph for 10 minutes at a time, and crashed in its first demonstration. Technology progressed, and by the first half of the 1800s there existed a small market for steam-powered auto-mobile vehicles. However, in 1861, the British Parliament was sufficiently concerned about public safety to enact The Locomotive Act that severely limited operation of motorized vehicles on-road. Although this stopped most motorized vehicle development in Britain, innovation continued elsewhere, especially in Germany, France and the United States. As the automobile moved into the mainstream and garnered ever more press coverage, consumers became more comfortable with and confident in the technology. This Act was partially repealed in 1896, and automobile development accelerated at the turn of the century with the advent of electric and internal combustion propulsion. By 1913, Henry Ford was building Model T's that every working man could afford, the result of standardized manufacturing and internal combustion engine technology.

The evolution from the driver-guided to the autonomous personal vehicle will parallel the evolution from the horse-drawn to the auto-mobile carriage: a period of initial caution and low acceptance, initial innovation and invention, use by early adopters, followed finally by rapid innovation and expansion, mass market penetration, and standardization. New technology will deeply challenge the social and political paradigms of the day, but now, as always, humans will adapt. As before, full consumer acceptance will not occur until consumers observe early adopters for a sufficient amount of time to trust that the system can operate safely and has a mature level of robustness and functional tuning. The wall of resistance to limited autonomous control is just starting to fall. With consumers showing signs of increasing comfort with automation, expect acceleration in the implementation and penetration of vehicle CADS technologies. Each generation of CADS implementation builds consumer confidence in the technology, and eventually consumers will accept autonomous control as naturally as they accept a self-powered (auto-mobile) vehicle.

4.0. DEFINITION AND ROADMAP FOR A FULLY AUTONOMOUS VEHICLE

Successful development of something as complex as a fully autonomous vehicle will be most readily achieved by those taking careful evolutionary steps, rather than one revolutionary leap. The DARPA Challenges served to jump start work on autonomous vehicles in the commercial sector, and fed new learning back to the military-industrial complex that has been working on the same problem for decades. These competitions and demonstrations provide glamour and some important lessons, but the technologies developed will not be directly applicable to the consumer market for quite some time, if ever. They just are not the practical next steps to putting something into production for public sale; these solutions leap right past more fundamental problems.

However, there's a place for the revolutionary vision, partly to show the world the march towards autonomous control, but mostly to motivate the effort and the long-term investment required. Industry and society both need high visibility demonstrations to sustain enthusiasm through the arduous hours of detailed engineering and analysis necessary to turn a dream into reality. We need to take time to understand true consumer values, and then engineer the technology and infrastructure for the reliability and robustness necessary to enact a safe and secure driving experience, one that inspires consumer confidence.

An on-demand, door-to-door, personalized automated transportation system may very well be achieved some day, but there are many lesser autonomous functionalities that customers will value that can be implemented much more quickly. As the industry researches and engineers towards Full Driver Assist it needs to follow a spiral development model, spinning off technologies and capabilities as they mature, bringing the consumer along step-by-step, little by little. These spin-offs cannot be limited to only the latest and greatest technology implementations. They must also include low cost solutions that can be implemented on lower cost vehicles for global implementation.

What follows is one promising roadmap for realizing a fully autonomous vehicle, or more precisely a Full Driver Assist-capable vehicle. It begins with an overarching design philosophy followed by customer-valued Use Cases that build upon existing collision avoidance and driver support features, which should be sequentially achieved, with appropriate operational reliability and robustness before proceeding to successive levels.

4.1. Design Philosophy

Until we have proven sufficiently reliable machine automation in a highly complex, continuously varying, unpredictable environment, one filled with both human and autonomous agents, the approach should be to keep the driver in the loop, as well as in the driver's seat. The driver should have the responsibility to engage the Full Driver Assist feature in a manner similar to how Adaptive Cruise Control (ACC) is currently engaged; by selecting certain operating parameters such as headway and vehicle speed.

During hand-off transitions, the driver will be expected to maintain vigilance and readiness to take control of the vehicle and will need to be supported in doing so. To accomplish this, the Human Machine Interface (HMI) must evolve from the current set of least/latest credible/imminent hazard warnings intended to minimize nuisance alarms, to providing more immersive situational awareness throughout the driving experience. Experience with automated aircraft cockpits[28] reveals that operators are often uncertain about its 'behavior'. What is it doing now? What will it do next? How did I get into this mode? I know there is a way to get it to do what I want, but how? The potential for automation success increases when several situations are created:

• Timely, specific feedback is given about the activities and future behavior of the agent relative to the state of the world,

• The user has a thorough mental model of how their machine partner works in different situations,

• Automated systems take action and act consistently with prior direction from the human operator.

The driver has legal responsibility for control of the vehicle and must have the ability to override the system by adding or subtracting steering input, applying the brake or adding throttle. He will have the ability to request or make certain maneuvers (e.g. initiate a lane change), and may be requested to confirm appropriateness and acceptance of a system recommended maneuver.

4.2. Use Cases

Although potentially interpreted as a simple roadmap or a checklist of sequential developments, each step may very well require extraordinary advancement in order to attain the necessary operational reliability and robustness in increasingly complex operating scenarios. As discussed in Section 2.1, Contemporary Error Rates - We're Way Off, autonomous vehicles will likely need to be better drivers than humans, exhibiting even fewer errors and more favorable error modes before they gain initial acceptance, let alone widespread implementation.

Use Case 0.0 - Status Quo

This case exists in the majority of vehicles on the road today. There are no on-board radars or cameras to measure the external environment, and no algorithms to provide information, advice, warning, or control.

In this case, the vehicle operator is left to his own preferred behaviors, behaviors that can change from day to day or moment to moment based on many and various external and internal factors, varying from relaxed to assertive and even unaware driving. Opportunities exist to provide timely advice or assistance to the driver in making the most appropriate decision in the given situation. Such decision making would require vehicle systems that are equipped with algorithms that can learn from the past driver's experience, identify hazard situations, and accordingly implement the corresponding emergency maneuvers.[29] We can expect more on-board algorithms for driver and situation learning, anomaly detection, probabilistic decision making, and more intensive interaction between the driver and the electronic vehicle control systems in the future, resulting in an increased level of intelligence of the electronic vehicle control systems.[30,31]

The addition of external environment-sensing capabilities to vehicles enables the following use cases:

Use Case 1.0 - Information, Advisory and Warning

This set of use cases comprises advisory and warning CADS functions that help the driver make better decisions. The CADS function provides information and advisories to the driver about the road environment as well as warnings about potentially hazardous conditions, such as the possibility of an impending collision, without any autonomous vehicle control actions being taken.

Use Case 1.1

In this use case, the CADS functions address the road environment. The information is not critical to the driving task, but will help the driver make informed decisions in the near future. These advisory functions could include speed limits, sharp curve ahead, blind spot information, ultrasonic park aid, etc.

Use Case 1.2

In this use case, the CADS functions address potentially hazardous conditions, such as the possibility of an impending collision or low mu conditions ahead. These warning functions include Forward Collision Warning, Lane Departure Warning, Lane Change Merge Aid, etc.

Use Case 2.0 - Emergency Control

This set of use cases comprises autonomous emergency countermeasures that help the driver mitigate or avoid a potential collision. It is useful to separate autonomous emergency action from normal steady-state vehicle control because the control logic tends to be considerably different. Whereas emergency action is taken with the focus on collision avoidance, normal driving focuses more on passenger comfort and smoothness. This emergency action is only taken when there is an error in the normal driving state, whether internally or externally imposed; an autonomous emergency action could be taken, regardless of whether the car is under driver control or fully-automated control. Many functions that are a part of this use case have been deployed in vehicles around the world, albeit at fairly low take rates.

Use Case 2.1

In this use case, the CADS functions support driver actions to avoid a potential collision. These functions include brake assist, brake pre-charge, and limited autonomous braking to reduce the collision speed.

Use Case 2.2

In this use case, the CADS functions autonomously take corrective action to avoid an otherwise unavoidable collision, only acting at the last possible moment. These autonomous collision avoidance functions include ESC, RSC, LKA, and autonomous braking such as that introduced on Volvo vehicles as City Safety™ (launched in CY2008) and Collision Warning with Full Auto-Brake (with up to 25kph speed reduction, launched in CY2010).

Use Case 3.0 - Steady State Control

This set of use cases comprises the first stage of Full Driver Assist in normal steady state driving. CADS functions in this family comprise limited autonomous control for a short interval at the driver's command, allowing the driver to focus on other aspects of driving. These functions are designed typically for a specific driving scenario, and the driver will need to take over once the expected scenario is compromised.

Use Case 3.1

In this use case, the CADS functions take limited autonomous control in a single axis when activated by the driver. Functions in this use case, many of which are in production today, include ACC (longitudinal control, freeway driving), LCA (lateral control, freeway driving), S&G (longitudinal control, traffic queue), etc.

Use Case 3.2

In this use case, the CADS functions take limited autonomous control in multiple control axes when activated by the driver. Functions in this use case include Traffic Jam Assist (a pre-emptive assistance during traffic jams, i.e. S&G ACC plus low-speed LCA), combined with autonomous driving from expressway entrance ramp to exit ramp, where the driver gets onto the freeway and enables the system to drive to, but not exit at, the desired ramp.

Even this use case can have phased introduction, starting with short intervals, i.e. 'take the wheel' until circumstances change appreciably. This would be 'on demand' by the driver, but with system concurrence that would take into account traffic density and road geometry, with the vehicle driving in automatic mode at posted speeds without lane changes.

The short interval can be extended further to full entrance-to-exit ramp driving, lane changes and even passing, but which might be limited to roadways that the vehicle has already successfully driven passively and analyzed as 'self-drivable' to verify road markings, GPS availability, number of lanes, etc. The system may still ask the driver for confirmation, possibly having started a conversation with the driver via SYNC®, "Of the standard options (provide list) which would you like?", and extend to "I recommend changing lanes, shall I go ahead and do that for you?" or "Do you concur that it's ok to change lanes now?"

Additional extensions of this use case can include auto-park, latch, and platooning functionality.

Autopark is where the driver and passenger depart the vehicle and engage an autonomous valet parking routine in a known infrastructure space with administratively restricted access for pedestrians, etc. Latch is where a vehicle strictly follows a selected forward vehicle at a standard following distance, initially at a low speed (e.g. TJA), then gradually at higher speeds. Platooning, the automatic following of a 'certified' lead vehicle, such as a commercial bus or truck, is further enabled by V2V communication with and between the lead and following vehicles, characterized by latch functionality and close quarters/shortened following distance for fuel economy benefits.

Use Case 4.0 - Transitional Control

This use case is highlighted by new functionality that helps the driver negotiate challenging traffic. This includes scenarios where vehicles come together in potentially conflicting intent and space. Support is provided either through information, advice, warning, or automatic control, both as late evasive actions as well as early smooth coordination and cooperation.

Use Cases 4.1 and 4.2 - Freeway and Intersection Blending

The first case aides the vehicle activity at a freeway on ramp and off ramp, extending the steady state control from freeway ramp-to-ramp to include merging and exiting. This includes anticipation of the exit and the pre-positioning of the vehicle in the appropriate lane, i.e. actively pursuing a lane change, as opposed to passively recognizing a lane change opportunity. This also includes a second case for turning and merging into similarly flowing traffic at an intersection.

Use Case 4.3

This use case is characterized by aiding the driver when traversing intersections with opposing flow traffic. The functions will inform, guide, or even control by assessing whether crossing traffic will collide, pass in front, or pass behind; thus determining the safe margin for a left turn across oncoming (head-on) traffic as well as the safe margin for entering into traffic from a branch intersection, such as turning left across oncoming traffic from the left or simultaneously merging with oncoming traffic from the right.

Use Case 4.4

This use case addresses convenience support at an intersection. More specifically, this includes the automated slowing and stopping for a stop sign, yield sign, traffic light, prioritized junction (e.g. driveway connection with roadway), or other traffic management system or protocol in a preplanned comfortable fashion when there is no preceding traffic that would otherwise govern free flow. This is in contrast to emergency-based intersection transition functionality.

Use Case 4.5

In simple terms, this use case involves the 'safe stop', appropriate as a bootstrap function in the event the driver becomes totally disengaged, unresponsive, or incapacitated with respect to performing further driving tasks. This function communicates an emergency situation to surrounding traffic followed by the slowing, stopping, and parking of the vehicle on the side of the road. This is a marginally preferred alternative to continuing non-stop without driver intervention or stopping in-lane.

Use Case 5.0 - Revisiting Known Destinations and Routes

This use case is highlighted by the extension to all roads, no longer biased to limited-access expressways. However it is still restricted to roadways that the vehicle has already visited and passively assessed; where the vehicle is familiar with these surroundings and only has to confirm, rather than recognize and analyze, the proper way to interact with this new environment.

Use Case 5.1

This use case is limited to areas frequently traveled, for example from home garage to work parking lot, and therefore has high confidence in familiarity and low likelihood of change in the nature and condition of the infrastructure, accompanying traffic flow, etc.

Use Case 5.2

The next increment could be related to a vacation or holiday destination, say a weekend or summer cottage or condominium; a place it has already been but with longer distances and less frequently visited, introducing the greater possibility of changes since the last time it drove there. The ability to recognize changes in infrastructure and nature of traffic flow is correspondingly increased.

Use case 5.3

A special use case would be the local shuttle scenario. The uniquely tailored character of this scenario would provide the first opportunity for full drive-for-me functionality. This use case would be a limited pre-implementation feasibility demonstration and learning opportunity only, where the new HMI and situational awareness and autonomous controls can be further developed for reliability and robustness. Besides the driver being on board, there would also be a specially trained test co-pilot who is there only to intervene on the driver's behalf if warranted. The driver would be observed for tendency toward non-driving activities given this level of driving support and HMI. If the vehicle runs into a scenario it hasn't encountered before, or has not been designed to handle, or when sensing becomes blocked and the vehicle goes into 'limp home' mode, the driver can take over and continue the shuttle delivery manually, etc.

A shuttle such as this could be administratively managed by and wholly contained on a private road network, such as at the Ford Research & Engineering Center in Dearborn, Michigan. In this case it could build on the current Smart Intersection,[32] which would allow for greater adaptation of the vehicle and infrastructure for experimentation in terms of infrastructure communication, dedicated localization targets at road edges and intersections, etc.

<u>Use Case X.0 - Traversing Unknown Routes and the General Case</u>

Here is where we put it all together, pursuing the idealistic fully autonomous functionality. Autonomous, Full Driver Assist functionality is extended to situations that have not been sensed, analyzed, or hardcoded previously. The vehicle is capable of traveling anywhere; to places it has never been before, handling scenarios never encountered before -- it's ready for the all new experience.

In order to proceed to this level, the engineering staff will have learned through all preceding technology development cycles and use cases. The sensing hardware/software, as well as assessment software, will have been shown to be reliable and robust in the prior use cases, and are now stretched to modes where safe, real time learning is permitted, enabled, and successfully achieved using advanced machine learning algorithms. Fully autonomous functionality should achieve at least the same outcome as the human driver when encountering new situations, but with the greater diligence and situational awareness, as well as rapid recognition of subtle novelty that a machine can have.

Learning safely will depend on continuing development of HMI concepts through successive use cases. Cases that now merely communicate unlearned situations to the driver will be continuously succeeded by more complex, autonomous designs that further offload the driving task as a design ideal. The focus will be on the development of models and algorithms that are not only able to learn but also to summarize identified relationships and facts to a higher level of abstraction. The goal is to integrate this part of the multi-attribute decision-making mechanism under different conditions and situations which is a necessary condition for autonomous driving.

As previously discussed and shown in the market, CADS warning and emergency functions have been introduced in phases of gradually increasing effectiveness:

• CADS 1 - capability sufficient to warn only for moving cars/trucks/motorcycles,

• CADS 2 - capability to warn and provide relatively small autonomous braking action for stationary, as well as moving cars/trucks/motorcycles,

• CADS 2.1 - capability for large autonomous braking in reaction to vehicles ahead (special low speed case),

• CADS 2.2 - capability to both warn and initiate a large autonomous braking action when an alternative steering path is not available,

• CADS 2.3 - warning capability for unintended lane departure or potential impairment based on the driver's lateral control performance, and

• CADS3 - capability to both warn and initiate a large action in reaction to both moving and stationary cars/trucks/ motorcycles and pedestrians.

In this use case, we build upon the level of effectiveness of the already available CADS functions and incremental use cases listed previously, and now extend them to the general case. The general case includes warnings and large autonomous actions (longitudinal and lateral) for hazards of all types including trees, poles, and other undefined or unexpected (e.g. debris in the driving lane) hazards, not just a smaller set of pre-classified types. The goal is to do this with early recognition and small actions for a smooth, seamless experience, vs. a panicked, last moment, large evasive emergency maneuver.

Intersection traversibility and cooperation, initially limited to conventional 3 or 4-way orthogonal configurations, is now extended to the n-way configuration. Scenarios may develop in such a way that the vehicle cannot brake to avoid a stopped car or large animal entering the lane, requiring an assessment whether it is safe to change lanes, e.g. whether there is parallel or oncoming traffic. Assessment of a 'safe alternative path' that may not be the designated driving surface, but which is suitable in emergency situations, such as the road shoulder, is also added. Implied in earlier use cases is the notion that late warnings of impending undesirable situations (a 'stop, don't do that' warning), will gradually be replaced with earlier advice, followed by increasingly stronger recommendations and requests for a positive desirable alternative action ('do this instead'), providing specifics the driver should focus on.

The CADS functions are also extended to the general case, including the full variety of weather and road conditions. Extreme weather conditions include snow where boundaries between driving and adjacent oncoming and non-driving surfaces are completely obscured. Road conditions include rural roads with painted lane markings only on the centerline, markings that may be faded, sporadic, or nonexistent, and gravel roads where the lane and road edge has no geometrically defined transitions whatsoever. Other extremes include off-road trails, stream fording, and open-spaces such as countryside, dunes, desert, tundra, etc.

5.0. SOME CONSIDERATIONS FOR BUILDING THE SYSTEM

Creating a system for autonomous personalized transportation involves more than just replacing one sub-system with another, replacing a driver function with an automated one, or completely replacing the human driver with a computer, let alone a robot. It will involve creating new subsystems, as well as new ways of integrating them; sub-systems that deal with interpretation of complex and cluttered driving environments, prediction of uncertain actions of other agents, and human-machine interaction ensuring sufficient situation awareness and engagement of the driver. The list of elements discussed here is by no means comprehensive, but highlights important areas of early development focus. As mentioned previously, the journey along the development roadmap will likely provide greater insights and uncover more proposals to be added to the list.

5.1. The Role of the Operator

Humans typically express the need for retaining control (beyond their fundamental legal responsibility), feeling that is safer and more secure than giving an unknown black box full authority over a highly complex task that, with an error, could seriously jeopardize their life or health. Since automation is classically described as better suited for dull, dirty, and dangerous activities, a driver in the autonomous personalized transportation mode will most benefit from Full Driver Assist functions. These functions offload moment-to-moment driving tasks, such as moving the driver from direct control of the throttle, brakes, transmission gear selector, and steering wheel, to predominantly a command mode. The driver then becomes an operator, who is still in charge, but in supervisory mode, like the orchestra conductor who commands all the instrumentalists (stop/start, faster/slower, louder/softer), but does not play the instruments himself. Even though the operator may be less involved in the moment-to-moment, direct control of actuators, the operator will need greater awareness of the situation, system status, and behavioral intent than is currently available to properly supervise the vehicle's actions. Through Full Driver Assist, the driver is provided additional time and can thereby have more confidence in performing a more appropriate role in the overall system, one that is partially tactical but becomes mostly strategic in nature.

Today, the automotive industry is providing driver support systems in private vehicles to help the driver in critical situations. Warnings, followed by preparation of actuators for operation, are used in sequence in an effort to guide the driver towards a collision avoidance response. Even with the best driver support systems, not all human responses will be ideal; some will inevitably be sub-optimal, not taking full advantage of the support system. The industry is therefore beginning to provide limited autonomous emergency actions in an effort to avoid or reduce the likelihood of an imminent collision. Many, if not all systems allow the driver some override capability versus the autonomous actuation, such as steering away to preempt, cancel or counteract an auto-braking function, if that is preferred. In a similar vein, limited autonomous driving support such as ACC has been introduced, with strict limits on control authority (longitudinal control only, limited deceleration levels, warns driver when control limits have been reached). On the other hand, allowing the driver to override the autonomous system would allow the driver to mistakenly override it as well; yet employing this method allows the earlier introduction and benefit of these autonomous systems.

When will we be ready to override human action with machine action? Flight control logic in modern aircraft already limits pilot input authority to a level which the plane's computers determine is within a safe operating regime. However, transportation modes that currently employ higher levels of autonomy vis-à-vis private road vehicles have one thing in common: very limited interaction with other operators. Airplanes are typically spaced a mile apart or more. The tightest train schedules place trains at least a few minutes apart, and the separation experienced on the ocean, without a harbor pilot aboard, can be even larger. This limited interaction significantly reduces the exposure to the unpredictability of the human reaction / interaction. On the other hand, consumers have an intuitive understanding of the complexity of interaction among vehicles sharing a road. This will likely slow their acceptance and adoption of fully autonomous vehicles.

Given that autonomous vehicles will change the very nature of driving, it is conceivable that the licensing of vehicle operators will need to change along with it. Today we have graduated driver's licenses with legal limitations, and as a driver fulfills certain requirements, more capability gets 'turned on'. Driver training today is mostly limited to several hours of on-road instruction, followed by real-world driving practice to build experience.

More specialized training may become the future norm. This training could include education on advanced CADS systems so that drivers will be better equipped to use the more advanced autonomous driving systems, similar to the pilot training required to fly a significantly autonomous commercial airliner. At some point, we may transition the first autonomous systems to only those in the driving public who have undergone specialized training, earning a certification and a special license to operate an autonomous vehicle. Ultimately, as autonomous vehicle technology matures and becomes more common, an even higher level of training and certification may be required to drive a vehicle in the totally manual, autonomous-off mode.

5.2. Communicating with the Operator

The Human Machine Interface is critical to continued operator engagement, and human-centered design will be essential for ensuring the HMI is properly designed for two-way interaction. The system must communicate everything the human operator wants to know in order for them to be comfortable with the autopilot driving the vehicle. Its effectiveness would be enhanced by knowing something about the operator's state as well.

The ultimate HMI for the autonomous vehicle may be the Brain-Machine Interface (BMI), first demonstrated experimentally in 1999.[33] The Full Driver Assist BMI application would benefit from operational feedback, proprioceptive-like cues, but on a vehicle basis. Similar to the notion that an autonomous vehicle will be available in just a few years, recent public demonstrations have combined with the magnitude of BMI's potential resulting in an enthusiasm that outreaches its readiness. Then again, there are many valuable and arguably necessary intermediate steps before that is realized in common practice.

Today's HMI systems focus mainly on general warnings that only give limited directionality and context. Continued research will be required to understand the best warning methods given the technology of the day, typically audible and visual. A recent study showed that haptic indications work well too, acting almost as a subconscious indication to induce mode changing. When warned at a point that a mode change was not expected, i.e. when a warning was given well before a problem arose that would be difficult to respond to, the operator reacted well to the inducement. When warned at the point that a mode change was proper and expected, the operator continued appropriately without distraction.

To enhance the human response, the HMI must evolve from generating warnings to providing a more immersive, situation-aware, experience. Improved situational awareness is important even in today's limited automatic control features such as ACC, where automatic control in benign situations reverts back to human control when the situational requirements exceed the control authority of the system. Emergency handoff, especially without proper context, is ill-suited to human behaviors. Human attention could waver during autonomous control and the operator may not be prepared to take decisive corrective action.

To improve awareness, the HMI could provide continuous feedback. Steering responsiveness or resistance could be altered as the vehicle gets closer to the lane boundary in order to provide feedback on lane position. Sound could be piped in to the operator correlating to the traffic conditions. With more traffic, there could be greater subliminal presence of sound. If a threat is increasing, then perhaps a localized and directional high frequency sound could be provided, getting louder as the threat grows.

Augmented reality displays (e.g. full-windshield Head-Up Display or wearable display) might be employed to provide directionality and improved awareness by highlighting objects of interest or displaying other scenario information. To achieve the even grander levels of autonomy sought by some, insight into HMI designs that allow the driver to take on more tasks, yet still be engaged, would be required. For the dull driving task, the augmented reality display could be supplemented with driver gaze monitoring to provide pertinent information as the driving scenario becomes critical, when the operator needs to be focused back onto the road. Warnings would still have their role as the last resort, but given an immersive situational awareness the driver would be more involved, informed and active in his role, so when it is time to hand over from autonomous to human control it's not a surprise, the context is understood and it will be a mutual decision. The autonomous system could request confirmation of readiness or willingness for handover of control. This request could be orchestrated so as to preserve a fall-back option of transitioning the vehicle to a non-moving and safely-positioned state suitable for an indefinite period of time (e.g. park it at the side of the road) if the driver doesn't respond or chooses not to accept handover from the autonomous control.

Another goal for a more advanced HMI would be to ensure greater awareness of *evolving* threats such that multiple simultaneous threats can be understood and prioritized, minimizing the need to respond to more than one at the same time, by dealing with the most critical earlier than necessary. In the meantime, other threats could mature or diminish, but all would be strung out sequentially and dealt with before any become critical for response, much the way an air traffic controller would handle it.

As mentioned previously, the autopilot may also need to determine whether or not to rely upon the interruption and guidance of the on-board human. For example, if the driver is in a sub-optimal awareness state (e.g. intoxicated), the computer may need to pursue a completely different task, such as preventing the operator from starting the car. The machine should also protect for the situation where the driver is in perfect operating condition, but misjudges the situation, such as when estimating the closing velocity of a vehicle (something that humans have difficulty doing), not seeing the 2nd car in the line of traffic, missing the car approaching from the right when looking to the left, etc. As the capability is developed, the HMI should include both direct and indirect driver monitoring and interpretation of operator state to ensure properly coordinated driver assist.

The transition from 'driver' to 'operator' will likely take decades, but it has already begun as previously discussed. Tomorrow's HMI designs should help guide and nurture this transition, but large step changes in HMI design may slow consumer acceptance. Therefore designs should evolve smoothly and gradually. Before the autonomous personalized transportation system is realized, the semi-autonomous systems (e.g. CADS) must gradually raise driver familiarity and comfort level for the warning, control, support and interventions of partial automation.

5.3. Deriving Situational Awareness

Real-time, up-to-date information is another critical element of the system. This includes information about the dynamic states and intended action of other vehicles; road hazards, environmental information (including weather, road conditions, natural disasters, etc), or road infrastructure information (e.g. traffic lights are not functioning ahead). The types and amount of information available to road vehicles today lack the reliability and comprehensiveness required to meet the demands of an autonomous personalized transportation system. It is improbable to think that these systems alone could predict other non-autonomous vehicle intentions or their likely future state, and little help is currently available from infrastructure-based information flows.

The radars, cameras, GPS/INS, and map data implemented in today's vehicles are key building blocks for the future; and many more advances are in the foreseeable future. Monocular vision systems may lead to stereo. Lidars may reappear in earnest with scanning multi-beam designs. Flash lidars or 3D cameras may mature enough to enable low cost long-range sensing providing dense range and intensity maps with integrated night vision capability. The numbers and coverage of these sensors will expand to encompass 360 degrees around the vehicle, with longer range and improved positioning and classification.

Additionally, sensors are needed to determine vehicle position relative to proper path. Current localization methods, however, are not precise at all times. For example, GPS positioning accuracy may fall below necessary levels due to atmospheric inconsistencies, drop out zones (due to a tunnel, tree canopy, etc.) or multi-path (urban canyons) failure modes. Alternatively, localization through a comparison of geographic and infrastructure artifacts detected by an on-board sensor to self-generated or publicly available 3D maps may also become important. This technology was demonstrated during the DARPA Grand Challenge 2 and improved in the Urban Challenge Event; subsequent study suggests capability with a single beam scanning lidar within centimeter levels of accuracy. Moreover, 3D maps are on their way, with a number of companies recently discussing their development publicly.

Vehicle-to-vehicle (V2V) and vehicle-to-infrastructure (V2I) network communications can be considered a sensing element that will significantly improve the accuracy and timeliness of information when fused with other on-board environmental sensing. V2V and/or V2I communication (V2X) will enable visibility of other vehicles when direct line of sight is blocked. It will also enable new information to be passed to vehicles, including traffic, weather, and road conditions, and information about the states of other vehicles. Infrastructure information may include environmental sensing of the road network through sensors on the roads, such as placing lidar localization targets in areas with GPS blackouts, or through compilation of the on-board sensing data available from other vehicles connected to a V2V network. If the detection or prediction of low mu conditions prior to encountering them is not yet possible, communicating the experience of a preceding vehicle to others approaching the hazardous area by V2X is a good alternative. The information update and flow would need to be seamless, not only from vehicle-to-vehicle, but also to/from the government, industry, and private sources. New invention and coordination is necessary to make sure the data is the most recent and relevant to autonomous personalized transportation vehicles.

Ultimately, sensing will need to evolve to 'general case' detection, tracking, and classification. Sensors today interpret the world by looking for patterns that match known objects, some of which use a training set and classifiers. Automotive radars are designed to look for a vehicle, which is why they initially worked only on faster moving objects in the driving scene. On the other hand, when humans see the world, they

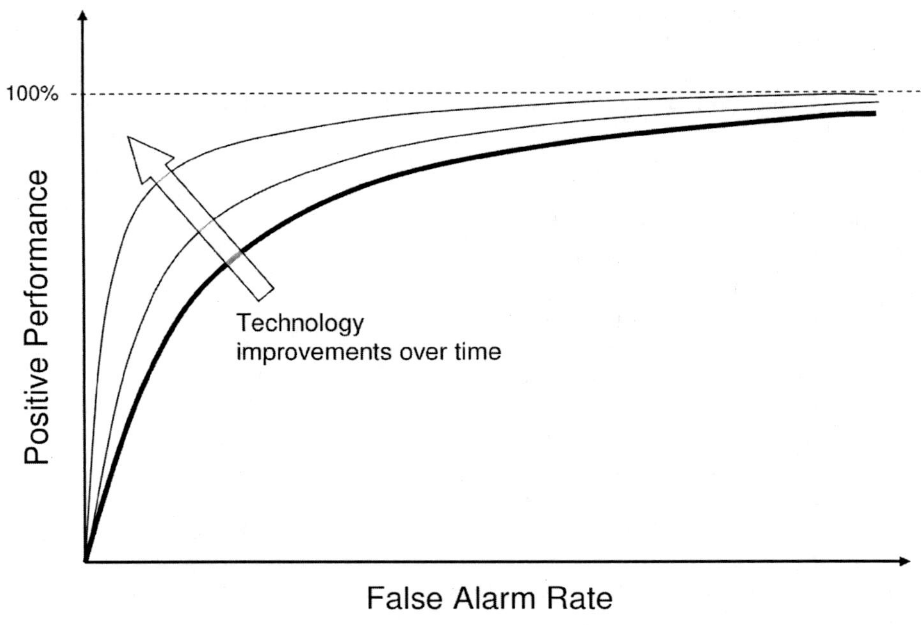

Figure 1.

also look for other cues that help determine whether or not the object ahead is of interest, or if the road is safe to traverse. Beyond just a measurement, there is a level of interpretation and judgment that must be implemented with the sensing system. This would allow estimation of lane and road boundaries when they are not really visible, due to faded, snow covered, glare-obscured conditions or judgment that an object in front, be it a vehicle, bicycle, pedestrian, tree, or moose, may be of interest; or even the gut feeling humans get that the scenario ahead may become a threat and the system should be wary. Knowing that sensors can physically measure much more accurately than humans, we should strive not only to replicate the human sensory perception capabilities, but also to exceed them. An important aspect of this is the use of multiple modalities of sensing in order to address the important problems of sensor reliability and validation of the sensor readings. The common sense verification mechanism that naturally accompanies human perception should be replicated in autonomous vehicles as algorithmic preprocessing validation of the measured data and capability for inferring and predicting new events through associative and case-base reasoning.

5.4. Limits of Situational Awareness

Sensors for situational assessment or awareness (SA) are statistical in nature, merely returning a digital representation of the external environment that must be interpreted for accuracy. Not only do the accuracies of the target characteristics have to be interpreted (e.g. relative range, range rate, and azimuth as well as classification, etc.), but whether the detection itself is valid also needs verification. Both radar and vision systems provide ample targets for interpretation. So it becomes a matter of trading off the true vs. false detection rate (i.e. positive performance vs. false alarms for a collision warning system) for a given modality and specific hardware capability, and then tuning along the curve for an appropriate level of reliability and robustness as shown in Figure 1. As SA technology improves, the tradeoff relationship improves, thereby shifting the curve. This is not much different from when the human acts as a SA system, with cognitive systems that include inductive reasoning, which by their nature, occasionally reach erroneous conclusions even when the basis for it is true.

Humans will never attain perfection, yet we allow them to perform challenging activities, tacitly accepting the consequences. How much better does a machine have to be than the human it would replace, before society allows that replacement to happen? Without knowing the answer, we can still utilize the machine as a situational awareness tool, not feeding an autonomous decision and control system, but in a limited capacity as a driver's aid. Machines are less susceptible to distraction so can provide a benefit given their greater diligence alone. Perhaps it is not a matter of how good an SA or decision-making machine is, but more a matter of how well it learns. Maybe it will be sufficient to allow replacement when it performs *and* learns at least as well as a human, i.e. without making the same mistake twice. Perhaps to break through into a truly autonomous decision making machine, it must be required to, even designed to, learn from and not repeat the mistake of other machines that previously made such an error? The industry has much development ahead before making that determination, but future SA systems should be conceived with consideration of these limitations in mind.

Perhaps the single greatest challenge to effective situational awareness is the speed at which the vehicle must travel to be considered a valued mode of transportation. Initial robotic successes were characterized by the very slow, seemingly deliberate, pace at which the sensing platform traversed the environment. With increasing velocity comes a need for increased sensing range, speed of situational interpretation, hazard detection, classification, and path planning, as well as reliable dynamic control.

5.5. The Vehicle and Artificial Intelligence

The artificial intelligence (AI) that commands the autonomous control system must also evolve, but the evolutionary path is still unclear. Should it be nondeterministic, implementing stochastic type algorithms of learning, optimization, decision making, planning, and goal formation under different situations and states that are not generally known in advance? We don't really know how useful that will be in the long run, but that may be a function of how strong the match must be between the pre-programmed and actual event. Does it need to be more human-like to be self-sufficient, being intuitive, adaptable, and strategic in its functionality? On the other hand, it is important to remember human fallibility; we're not even sure yet how much involvement the operator should have in the system.

We can say that whatever the AI, it needs to handle some level of unexpected environmental perturbations, because chaos exists even in a tightly controlled system. The AI needs to handle any intentional system compromise, for example, dealing with external hacker attacks and false signals. It needs to handle unknown objects in the external environment, like a new type of vehicle on the road that doesn't communicate. It needs to handle unexpected internal failures such as electronics and software faults. The AI really needs to make use of information whenever and wherever it's available, making judgment as to which information to use and when.

Moreover, the AI needs to be able to make decisions spanning both physical safety and societal norms, accounting for the social, political, and cultural complexities inherent in human decision making. Even in a task as simple as a lane change, the decision making logic is complex. When is it safe to make a lane change? When is it appropriate to make a lane change? When is it socially acceptable for an autonomobile to make a lane change? Is it ever acceptable for one autonomobile to cut in front of another, say in an emergency? And in mixed mode operation, one driver may feel comfortable handing control over to his autonomobile, but are other drivers in the adjacent lane ready? All this presumes learning specific driver's actions and preferences in the operation of the vehicle. The models are later used by the intelligent control system to invert the mapped relationships and advise the driver for the most appropriate actions under specific circumstances. All these questions impose requirements on the AI system that are well beyond the capability boundaries of the existing decision making systems and suggest a wide range of challenging research problems.

5.6. The Road Infrastructure

Infrastructure may also require modification to support future autonomous operational modes. As we transition towards full autonomy, we must accept that mixed mode operation may be the norm for a long time, with both human and computer pilots interacting on the road. Some thought needs to be given to this transition - given the uncertainty of human reaction and the interactions that result in random events, we may look to minimize this uncertainty by some day providing special autonomous-only traffic lanes, much like the High Occupancy Vehicle carpool lanes demarked <HOV> today. These lanes could have very limited access, with known access locations, allowing only autonomous pilot-enabled vehicles to enter.

When enough vehicles on the road have autopilot capabilities, we may progress to having some roads, such as limited access highways, be autonomous only; while human drivers could still operate on secondary roads. Eventually, we may transition to virtually all roadways being autonomous only, with only a few exceptions, such as scenic Route 66, preserved for nostalgia's sake.

5.7. The Regulatory Environment and Beyond

While government and regulatory environments will need to adapt to enable the autonomous future, and will likely play a key role in their success, non-regulatory ratings can drive OEM strategies with the same rigor. These latter ratings include government ratings such as NHTSA's New Car Assessment Program (NCAP), as well as third party ratings such as the Insurance Institute for Highway Safety's Top Safety Pick. Many vehicle manufacturers emphasize their performance on these ratings as a communication strategy for vehicle safety; hence these ratings have considerable clout and could even be considered defacto regulations.

Collision avoidance technologies are the fundamental building blocks for autonomous vehicle operation and have been subject to 3rd party influence since NHTSA's NCAP action in 2002 (which applied the fish-hook performance test criteria to ESC systems) which was followed by EuroNCAP braking requirements in 2006. These actions have reverberated around the globe, with Korean, Japan, and China NCAPs all enacting dynamic rollover requirements.

Based on recent history, some NCAPs evolve into regulations. In the preceding example, the US began mandatory phase-in of requirements for ESC by the 2009 model year, a 14 year lag from introduction to regulation. In contrast, regulatory phase-in of passive restraints, a combination of automatic seatbelts and airbags, began in 1986, while a full phase in of airbags began in 1996. A shorter delay is not necessarily preferred even though it can create an earlier 'pull'. A longer delay provides more time to evaluate different technologies and let them mature.

This path is not universal with respect to steps or timing either. In 2010, the US launched a new NCAP Assessment for collision avoidance, with the addition of a FCW and LDW protocol and test methodology. Just prior to that, Japan elected to proceed directly down a regulatory path for collision avoidance, kicking off "if fitted" requirements for CMbB systems, as well as convenience based technology like ACC and Reverse Parking Aid systems. EuroNCAP also just announced the "Advanced Award" (formerly referred to as Beyond NCAP) to supplement the overall safety star rating of the vehicle if the vehicle has Blind Spot, Driver Distraction, or Lane Departure Warning capabilities or Advanced Emergency Braking Systems (AEBS). This can result in near-instantaneous rating assessment of the newest technologies.

These are likely just the first stages of many more requirements to come. Industry is closely watching the US and the EU for regulatory movement in collision avoidance beyond stability control. The US Crash Avoidance Metrics Partnership is a collaboration between several OEMs and NHTSA, researching crash imminent braking system test methods and requirements, among other things, which may result in new NCAP or regulatory requirements. The European Union has already begun to shape commercial vehicle regulations for AEBS and LDW systems, with the United Nations Economic Commission for Europe planning to develop technology requirements in the near future.

Many in the automotive industry are looking for harmonization of these new requirements, with the hope that ISO standards, which exist in either a released or draft form for many of these new features, become the foundation. If harmonization attempts are unsuccessful, the OEM base will face a substantial challenge as it drives toward global technology platforms. Regionally unique requirements could result in key enabling technologies that are unique at a fundamental level. Considering the preceding SA tradeoff discussion (Section 5.4, Limits of Situational Awareness), this could result in one market having a stringent false positive reliability requirement, while another elects to have a high degree of positive function capability, and a third market implements a more simplistic feature presence-based rating or regulation.

Make no mistake, governmental action can stimulate and encourage development of technologies, especially in infrastructure intensive areas, but it should also be careful to not regulate in ways that are restrictive to innovations with societal benefit. All things considered, however, CADS and autonomous vehicle research and development could greatly benefit from the inclusion of governmental agency and legislative partnerships.

6.0. NEW COLLABORATIVE RELATIONSHIPS

Several key factors affecting the pace and extent of innovation are the generation of new concepts, available investment levels, and available time to mature them to a meaningful implementable level.

The solution to complex problems such as Full Driver Assist can only come from the synthesis of many diverse inputs, from diverse sources, and through cooperative relationships. The large investment that will be required presents its own challenge, and that burden is well suited to collaboration as well. Achieving new goals typically requires new skills, developed on the job or gained through additional education, yet both require significant time. Alternatively, skills can be immediately brought into the team by partnering outside your own enterprise.

The traditional supply base is focused primarily on solving today's problems; that is where the majority of demand is, where their expertise is, and where they can be profitable. Yet suppliers also earmark a portion of their budget for R&D to solve future problems. How to spend that investment is a challenging question, with some suppliers extending today's knowledge and others branching out in new directions. Maintaining a regular dialogue with suppliers on trends and new directions ensures alignment and efficiency, but gaps can arise when there is a discontinuity, such as that presented by Full Driver Assist. Sometimes disruptive (i.e. beyond evolutionary) technologies, whether they're from traditional or non-traditional sources, are required.

Disruptive technologies may come from traditional suppliers, but also from other industries, percolating from advanced engineering, fundamental university research, or wherever inspiration may arise, even nature. This opens the door to new entrants in the technology supply base and all should be considered. Looking in non-traditional areas can be like early gold prospecting; you eventually find what you were looking for, but you would probably dig a number of empty holes first.

The following is a partial outline of collaborative relationships that have been or are being explored, but they are presented in a generic and partially fictionalized way. For the purposes of this paper, it is less important to discuss a specific set of corporate relationships, and more relevant to illustrate the breadth and variety of partnerships and technologies, both traditional and non-traditional.

6.1. Traditional partnerships

6.1.1. Tier 1 and 2 suppliers

Long standing chassis and body electronics suppliers are essential contributors to the rapid development and proliferation of new collision avoidance and driver support system technologies. They have proven their capability through the years, but now their out-of-the box creativity is being tested. An opportunistically timed new feature or functional capability breakthrough has the potential to extend their market share overnight in a highly competitive and otherwise mature market.

6.1.2. Pre-Competitive OEM Partnerships

Most notable in this category is the Crash Avoidance Metrics Partnership (CAMP), a research consortium of automobile manufacturers and suppliers engaged with the United States Department of Transportation for the advancement of promising new active safety technologies. This has been a highly effective and productive relationship, having generated numerous concepts, requirements, specifications, and field operational test results on track for eventual implementation.

CAMP's role in the development of V2V and V2I safety communications could serve as a model for Full Driver Assist. Since 2002, CAMP has organized multiple OEMs to work cooperatively on this technology with NHTSA and other parts of the US DOT. The work has ranged from basic testing and analyses to building applications to developing necessary standards and then working together to get these standards adopted. The OEMs currently working together at CAMP (Ford, GM, Honda, Hyundai/Kia, Mercedes, Nissan, Toyota and VW/Audi) are completing the standards necessary for a NHTSA deployment decision in 2013. To support this NHTSA decision, the OEMs working together at CAMP are also building vehicles with this technology for Driver Acceptance Clinics and for model deployment.

To support full commercial deployment of V2V and V2I safety communications, OEMs and the government needed to come together to define the enabling pre-competitive elements, such as infrastructure requirements, as well as message protocols, content, and security, etc. OEMs will need to be able to trust the wireless messages that their vehicle receives from vehicles manufactured by their competitors to provide warnings to the drivers of their vehicles. The level of cooperation and trust for Full Driver Assist applications will need to be examined and, if appropriate, mechanisms such as CAMP should be utilized.

6.1.3. Academia

Also common are relationships with colleges and universities ranging from a one-time grant to formal multi-year alliances. These can in turn leverage research funding from governmental science and military sources, industrial military sources, health care providers, etc. as well as collaborative relationships with other universities.

One quickly finds that university faculty, students, research staff, and affiliated technical institutes working in areas directly relevant to Full Driver Assist form a rather small community, yet draw upon knowledge, skills, and experience from non-automotive ground (construction, agricultural, industrial) and marine vehicles, general/commercial/military aviation, planetary exploration applications, medicine, and brain & cognitive science.

6.2. Non-Traditional Partnerships

Non-traditional partnerships are especially important in tough economic times. You can readily find a partner on a pay-to-play basis, but you easily exceed tight budgets with aggressive long term research when there is a priority on near term results. Non-traditional partnerships often arise when both partners have budget challenges and are motivated to find an equal equity partner, one that brings intellectual capital to move new concepts forward. These can be very strong relationships when they are born from mutual dependence, toward a shared ultimate goal/vision and well aligned with individual goals. The title for each of the following examples serves to capture the essence of these unique relationships.

6.2.1. The Mental Athlete

Formal contests, or any competitive context, can provide motivation and a means for a technical staff to perform at very high levels of creativity on a very short time scale. These contests are common in academic circles and range from toothpick bridges, baking soda cars, and science fairs for the younger set, to high performance and fuel-efficient ground vehicles, concrete canoes, and energy and space efficient homes for those more learned.

This approach to innovation is especially powerful when the team constituents are multi-disciplinary and blended from academics, OEM, suppliers, etc. This has likely driven the recent expansion to include competitions aimed at motivating professional participants as well. These competitions investigate topics ranging from human powered flight, to commercial space flight and space exploration, to ultra-high fuel efficiency, education, health care, and beyond.

Those well suited for this high energy, high stress, instant feedback, creative environment can find themselves supporting professional competition or time sensitive high-stakes consulting teams (e.g. Formula 1 racing, or oil rig fire control, mine collapse rescue, etc.). The downside is that this high level of energy is difficult to sustain for indefinite time periods, and can result in burn-out if continued for too long.

In the Full Driver Assist context, the most notable examples have been contests sponsored by the Defense Advanced Research Projects Administration (DARPA), namely their two Grand Challenges and their Urban Challenge for autonomous vehicle operation. These have drawn hundreds of teams from around the world and brought the notion of 'driverless cars' into mainstream media with widely publicized demonstration events, all while technical advancements (primarily software) are finding their way into further research activities behind the scenes.

6.2.2. The Start-up

Every once in a while a group of engineers has an idea that is ahead of its time, at least within their current context, which warrants a parting of the ways. This has happened several times in the robotics community, and in one case, the engineers decided to spin themselves off from their military contractor parent and start their own company, rather than bookshelf their ideas. Specializing in situation awareness, path planning, threat assessment, vision/image processing, proprioception, search/processing prioritization, and real-time computing, these individuals are highly regarded in the robotics community, regardless of their venue, and they have made good on their vision.

An OEM seeking to push the envelope can learn from such an organization, working together to explore different theories and rapidly prototype complex sensing and control systems with great utility. Their story ends with their former parent organization re-recognizing the value of their abilities, accomplishments, and vision, and ultimately reacquiring them.

Another form of the startup, graduating university students, is also common and possibly more predictable. Typically graduate and undergraduate work is extended into a focused product or services business model by those funding their research. This presents a ground floor opportunity and can be especially powerful if they're also building upon a Mental Athlete collaboration model - first hand knowledge and proven under fire.

6.2.3. The Hobbyist

How often does it happen that someone turns their hobby into a new business and becomes a new entrant in a highly competitive field? It only has to happen once, in the right technology, and you have the makings of a potent collaboration - if you are in on the ground floor.

In one case, a hobbyist applied curiosity, a little inspiration, and a lot of perspiration to develop a new sensing device. This device wasn't entirely novel, but it was uniquely capable nonetheless. It solved a much larger portion of the general case SA problem than had previously been accomplished, addressing road departure and safe path detection, planning, advice, and control.

This sensor is currently being used as an instrument grade research tool and is being produced at low volume for architectural applications, among other things. It has put incumbent sensor suppliers on notice, illustrating that there is a disruptive technology opportunity. Perhaps with additional packaging, manufacturing, and robustness development, this technology will become suitable for automotive applications.

6.2.4. The Gamer

They may 'only' write software for video games, but a serious skill set may be overlooked without a little more investigation. The gamers are really solving an image-processing problem, in their own unique way in some cases, and it is that diversity of knowledge, concept, and approach that can be leveraged. If you find a connection and can draw out their best efforts focused on your problem, the progress could be quite amazing.

6.2.5. The Coach

If you want to teach someone (or an intelligent vehicle) to drive, you might start with someone who is a professional driver, or even better, a professional driving instructor or coach. You, or the intelligent vehicle, need to get that seat-of-the-pants/'been there done that' experience, but without repeating their entire driving history. You need someone to distill and convey it to you efficiently and effectively. Furthermore, advanced driving skills are perishable for humans, so coaching isn't necessarily a one time event.

You (the intelligent vehicle) need to learn the vehicle's nominal character, its limitations, and how it behaves beyond its limits. If this could be done online or in a virtual environment, it could be done in a repeatable way, without the peril of hazardous situations, and in a concentrated fashion. This leaves out the nominal driving mileage and focuses the time on key events and experiences. This might ultimately enable novice drivers to start out with the wisdom of a mature driver, and an intelligent vehicle might embody the natural understanding, presence, and anticipation of a professional.

6.2.6. The Improviser

You need a test method to characterize a collision scenario in a repeatable way, without harm to the test drivers or test vehicles, and you need to ultimately validate such a system. Enter the Improviser. You tell him/her your story and before you know it, something has been discovered in the barn, the hangar, or the tool crib that with a bit of blacksmithing, a few extra wires, and a handful of plastic wrap, perfectly fills the bill. You don't teach someone to do this; this type of person just happens.

6.2.7. The Biologist

The application of chaos and complexity theories in the field of biology is not new, but their application to the human driving condition is. There are inhabitants of planet earth that are wired differently than humans: insects can perform collision avoidance on a time scale, within physical proximities, and with innumerable distractions and clutter, that a professional athlete or intelligent vehicle would be envious of. To understand how to mimic and embed the instinctive as well as cognitive processes observed in nature in future intelligent vehicles, you would do well to diversify your automotive team with this atypical skill set.

7.0. SUMMARY/CONCLUSIONS

It is fanciful to consider practical Full Driver Assist capability achievable in the near or even midterm. Amazing capabilities have been achieved and demonstrated in the carefully controlled environment of the test track, even in the glare of the TV lights. But are we ready to turn this loose on the mainstream consumer? Ultimately the argument of when, or even if, we will ever be ready is moot, as the benefits from the journey itself is worth it regardless the answer.

Having provided a summary of the current challenges and a roadmap for future work, it is fitting to revert to history for some perspective. It has been said that we put mankind on the moon in one giant leap. President Kennedy set forward a visionary challenge and in less than a decade we were there. Why? "We set sail on this new sea because there is new knowledge to be gained … and used for the progress of all people."

Necessity drove a search for solutions in all conceivable places, the usual and the unusual, but the first moon walk was achieved through a set of logical extensions of what mankind knew. Many challenges remain - more than forty-five years later we still don't have regular commercial service to the moon, earth orbit, or even the upper atmosphere. While our undertaking may not be as grand as putting a man on the moon, perhaps our task is more difficult - there is no road rage in space.

REFERENCES

1. Federal Highway Administration, "The Dream of an Automated Highway," ttp://www.fhwa.dot.gov/publications/publicroads/07july/07.cfm, July/Aug 2007 Vol. 71 No. 1

2. Annual Report, Department of Electrical and Computer Engineering, Ohio State University, 2003.

3. The Great Robot Race, McGray, Douglas, Wired, December 2003.

4. A Perspective on Emerging Automotive Safety Applications, Derived from Lessons Learned through Participation in the DARPA Grand Challenges, McBride, J., Ivan, J., Rhode, D., Rupp, J., Rupp, M., October 2008. Journal of Field Robotics, Volume 25, Issue 10, pp. 808-840. ISSN 1556-4959.

5. NHTSA, "Traffic Safety Facts; 2008 Traffic Safety Annual Assessment----Highlights," http://www-nrd.nhtsa.dot.gov/pubs/811172.pdf, June 2009

6. Reichspatent Nr. 169154, Verfahren zur Bestimmung der Entfernung von metallischen Gegenständen (Schiffen o. dgl.), deren Gegenwart durch das Verfahren nach Patent 16556 festgestellt wird, November 11, 1904. Translation: *Methods for determining the distance from metal objects (ships, or the like), whose presence from the procedure under 16556 patent is found.*

7. Delphi, "Manufacturer Products: Safety Electronics, Active Safety, Delphi Adaptive Cruise Control," http://delphi.com/manufacturers/auto/safety/active/adaptive-cruise-control/, 2009.

8. Smyth, Louise, "Adapt to Survive," Vision Zero International: 27-30, June 2010

9. Stupp, E. H., Cath, P. G., and Szilagyi, Z., "All Solid State Radiation Imagers," U.S. Patent 3 540 011, Nov. 10, 1970.

10. Volvo Car Corporation, "Pedestrian Detection with full auto brake - unique technology in the all-new Volvo S60," http://www.volvocars.com/en-CA/top/about/news-events/Pages/press-releases.aspx?itemid=17, March 2010

11. Frost & Sullivan, "Active and Passive Safety Systems in the US: Customer Desirability and Willingness to Pay (Passenger Vehicles)," Sept. 2005.

12. Frost & Sullivan, Analysis of North American Market for Advanced Driver Assistance Systems, December, 2009.

13. Frost & Sullivan, Strategic Analysis of Japanese Passenger Vehicle Safety Systems, June, 2010.

14. World Health Organization, "The Global Burden of Disease, 2004 update", WHO Press, Geneva, Switzerland, ISBN 978 92 4 156371 0, 2004.

15. Research and Innovative Technology Administration, John A. Volpe National Transportation Systems Center, "Focus," http://www.volpe.dot.gov/infosrc/highlts/02/mayjune/d_focus.html, May 2002.

16. Dekker, Sidney, The Bad Apple Theory pp. 1-14, The Field Guide to Human Error Investigations, 2002. TJ International, Padstow, Cornwall, Great Britain. ISBN 0-7546-4825-7.

17. Dismukes, R.K., Berman, B. A., Loukopoulos, L. D. "The Limits of Expertise", Ashgate, United Kingdom, ISBN: 978-0-7546-4964-9, 2007.

18. Rothe, P. J., James, L., Nash, J., Freund, P., Martin, G., McGregor, D., Frascara, J., Redelmeir, D., "Driving Lessons, Exploring Systems that Make Traffic Safer" University of Alberta Press, Canada, ISBN-10:0888643705, 2002.

19. Central Japan Railway Company, "About the Shinkansen", http://english.jr-central.co.jp/about/index.htrnl, retrieved: Aug 2010.

20. Car and Driver, "How to Drive a Ford Model T," http://www.caranddriver.com/features/09q3/how_to_drive_a_ford_model_t-feature, July 2009

21. Langer, G., "Poll: Traffic in the United States; A Look Under the Hood of a Nation on Wheels", ABC News, Feb 2005.

22. NHTS, "Summary of Travel Trends: 2001 National Household Travel Survey," http://nhts.ornl.gov/2001/pub/STT.pdf, Dec. 2004.

23. Song, C., Qu, Z., Blumm, N., Barabási, A., "Limits of Predictability in Human Mobility", Science 327(5968): 1018-1021, Feb. 2010, DOI: 10.1126/science.1177170.

24. Rokeach, M., "The Nature of Human Values", Free Press, New York, ISBN 0029267501, 1973.

25. Shrank, D. and Lomax, T., "2009 Urban Mobility Report", Texas Transportation Institute: The Texas A&M University System, Jun 2009.

26. Massachusetts Institute of Technology, Center for Transportation & Logistics, "Driver Wellness, Safety & the Development of an AwareCar," http://agelab.mit.edu/system/files/file/Driver_Wellness.pdf, Dec. 2009.

27. Ford Motor Company, "Ford and MIT Team Up To Improve Safety by Reducing Driver Stress", http://media.ford.com/article_display.cfm?article_id=31682, Dec. 30 2009.

28. Joint Cognitive Systems: Patterns in Cognitive Systems Engineering, Woods, David D., Hollnagel, Erik. Published by CRC Press, Boca Raton, Florida, 2006. ISBN 978-0-8493-3933-2.

29. Cacciabue, P., Modeling Driver Behavior in Autonomous Environments, Springer-Verlag, London, 2007.

30. Lu, J., Filev, D., "Multi-loop Interactive Control Motivated by Driver-in-the-loop Vehicle Dynamics Controls: The Framework," Proc. of the IEEE Conference on Decision and Control, December 16-18, 2009, Shanghai, China.

31. Pentland, A., Liu, A. (1999) "Modeling and Prediction of Human Behavior", Neural Computation, Vol. 11, pp. 229-242.

32. Ford Motor Company, "Ford's new smart intersection 'talks' to cars to help reduce collision, fuel-wasting congestion", http://www.ford.com/about-ford/news-announcements/press-releases/press-releases-detail/pr-ford26rsquos-new-smart-28610, July 10, 2008.

33. Chapin, J.K. et al. (1999) Real-time control of a robot arm using simultaneously recorded neurons in the motor cortex. Nature Neuroscience, 2, 664-670.

CONTACT INFORMATION

Jeffrey D. Rupp
Ford Motor Company
Product Development Center - Safety Engineering
Room GB-A83, MD 129
20901 Oakwood Blvd
Dearborn, MI 48124-4077
jrupp@ford.com

ACKNOWLEDGMENTS

The authors would like to thank those who have provided invaluable input and critique for, as well as the investigation and advancement of the technologies which are the subject of this paper: Tom Pilutti, James McBride, Dimitar Filev, Louis Tijerina, Mike Shulman, Alex Miller, Dev Kochhar, Andy Sarkisian, Venkatesh Prasad, Stephen Rouhana, Doug Rhode and Roger Trombley. The authors also gratefully acknowledge those who have facilitated the process of creating this paper in innumerable ways: Nancy Burke, Haleh Ghalambor, Stephen Kozak, Randal Visintainer, and Loralee Rupp.

The Line Within: Redrawing the Boundary of Connected Vehicle Systems Engineering

2010-01-2322
Published
10/19/2010

Robert Gee
Continental Automotive Systems, Inc.

ABSTRACT

The interdisciplinary and structured integration of subsystems into a functioning whole is at the root of Systems Engineering. Until recently in the automotive market, much of this has been specific to an automotive sub-domain such as Telematics, Infotainment, Chassis Control, or Engine Management Systems. In the realm of Telematics and Connected Vehicles, the recent trend has been outward from the vehicle, focusing on expanding connectivity and data sources. Systems Engineering for Telematics now includes multiple transports spanning PAN, WLAN, and WAN communications, and beyond that has grown to include entities on the far side of the network link, including data servers, aggregation portals, and network security.

Although it was not trivial for Continental to develop the embedded Telematics connectivity subsystems for products such as GM/OnStar®, Ford SYNC®, BMW Assist™, and Mercedes Tele Aid®, consumer and regulatory expectations are rendering inadequate the artificial boundary of an embedded connectivity domain for new automotive systems.

For example, reducing vehicle weight is a common approach in the effort to improve fuel efficiency, and weight targets have been cascaded down to each subsystem and module. However, for each 100 pound weight reduction for passenger vehicles (and without corresponding changes to other vehicles or additional safety technologies), NHTSA and other studies have indicated the effect of hundreds of additional fatalities per year [1, 2, 3] in the United States. With both safety goals and US CAFE fuel efficiency goals to meet, the design and interaction of many previously unrelated subsystems in the vehicle become key factors, and in particular, the new interaction between vehicle connectivity subsystems with vehicle safety and performance subsystems.

For this discussion, we take a systems view of the evolving field of vehicle connectivity, review the historical trends, introduce a framework to analyze several human constraints, and use the framework to identify ideal characteristics in a modern vehicle system.

INTRODUCTION

Basic engineering theory begins with the concept of drawing a system boundary. Everything within this boundary is the subject of calculation, and the external interfaces provide sources and sinks, such as for mechanical forces or data.

FIRST GENERATION

For Telematics, a subset of the overall vehicular field of Connectivity, the line has traditionally been drawn around the realm of an embedded cellular device (Network Access Device, NAD), satellite positioning (for example, GPS), and a Telematics Service Provider (TSP). Such First Generation Telematics systems were feature-limited, offering non-driving-related features such as Information Call (I-Call), Breakdown/Diagnostics Call (B-Call), and manual Emergency Call (E-Call) during normal vehicle operation. The Telematics features did not directly affect vehicle safety and performance, and only in the event of a crash did the system exhibit an increased level of integration, whereupon the airbag deployment would trigger an automatic emergency call on behalf of the vehicle occupants. Although this first generation system was end-to-end with both a specific TSP and embedded vehicle system, the over-the-air protocol was fixed and well known before product shipment, and the in-vehicle functions generally did not change over time.

SECOND GENERATION

Second Generation systems introduced what began to be called Connectivity, adding infotainment features such as Bluetooth® interfaces for hands free calling and media streaming, radio data services like traffic, video and Rear Seat Entertainment (RSE), and other information. Thus for the Second Generation, the system boundary encircling the consumer-oriented use cases and some Connectivity- and Infotainment-related hardware systems grew larger, but the system still focused on features oriented toward the vehicle occupants and less so the act of driving.

THIRD GENERATION

Today, Third Generation systems are being introduced. They are growing the boundary surrounding what is connected to the vehicle, whether wirelessly or by wire. Mobile devices are now a lasting part of the mix, sharing applications with the vehicle and in some instances becoming part of the vehicle Human-Machine Interface (HMI). Content providers, data aggregators, application stores, network carriers, and service providers are adjusting their market positions to solidify their roles in the value stream, adding more back-end hardware and cloud-based services. As a result, more information choices are available to vehicle occupants than ever before.

Third Generation systems have driven an interesting expansion of automotive Systems Engineering. Systems Engineering grew out of the traditional need for cross-disciplinary engineering to design the hardware and software, in order to ensure the success of a system. Initially, OEMs defined the feature set of the system, but as the key product features become driven by updateable software and services, and in particular open services accessible across different market segments, it has become increasingly important that the system be designed with human interfaces in mind, specifically accounting for human needs, desires, and capabilities. For example, fundamental human needs include transportation, communication, and safety. Human desires, on the other hand, often work as a detrimental factor - compelling the use of mobile devices with non-driving-related services in the vehicle, in some cases resulting in unsafe practices. Finally, human capabilities restrict what can be performed by one person, reducing the potential benefits of the many available features.

EXPANDING THE ROLE

Systems engineers have often had a supporting role in technical marketing and product conception. The shift toward services, though, increasingly adds the cross-disciplinary aspects of business and marketing to the systems skill set, whereupon the systems engineer will need to have strong capabilities in marrying the areas of market trend analysis, business case development, partnership coordination, and customer roadmaps, in addition to his traditional knowledge of hardware and software engineering. Those who believe that the architecture of a successful, distributed system can rely solely on technical engineering skills run the risk of inappropriate consideration of consumer needs and market trends, thus ill-designing the current system and limiting desired growth in the unanticipated directions. Therefore, Systems Engineering must necessarily consider both business and technical aspects when designing a modern, connected system.

TRANSPORTATION

There is a singular reason that consumers need vehicles: transportation. We define transportation as the act of carrying passengers and cargo from Point A to Point B. Infotainment subsystems have improved the comfort of the experience by providing features like navigation, traffic advisories, a mobile phone book, and points of interest (POIs). Improved engine and chassis subsystems have increased vehicle reliability and agility, increasing the likelihood of reaching the destination without a breakdown and allowing the vehicle to be guided deftly away from potential hazards.

However, the separate development of these trends is not without its consequences. Expanding and ever more compelling options for infotainment carry the risk of driver distraction, while increased engine power and smoother suspensions create nimbler cars that require higher levels of driver attention than before.

The time has come for a new generation of Connectivity systems. The need exists to redraw the system boundary past the vehicle data bus that links a vehicle Connectivity system with the rest of the vehicle subsystems, in order to find a way to balance consumer interests with safety and performance.

HUMAN CONSTRAINTS

As a means for understanding human behaviour in relation to vehicular products, we have developed and used a "Four Corners" conceptual framework that can aid in the discussion of human constraints. The reason behind such a framework is that it is not sufficient to merely create an item of interest to a consumer, but such an item must satisfy what a consumer can and will choose to use, as well as account for limitations of which the consumer is not aware. Comprising this Four Corners framework, the first three corners are the limitations of time, money, and attention. The fourth corner is the added factor of aging and its effects on the first three.

"Green field" Systems Engineering, in which there is no single, omniscient customer defining the requirements, may benefit from use of this framework to help optimize key characteristics of the system, in order to enhance the likelihood of product acceptance and success.

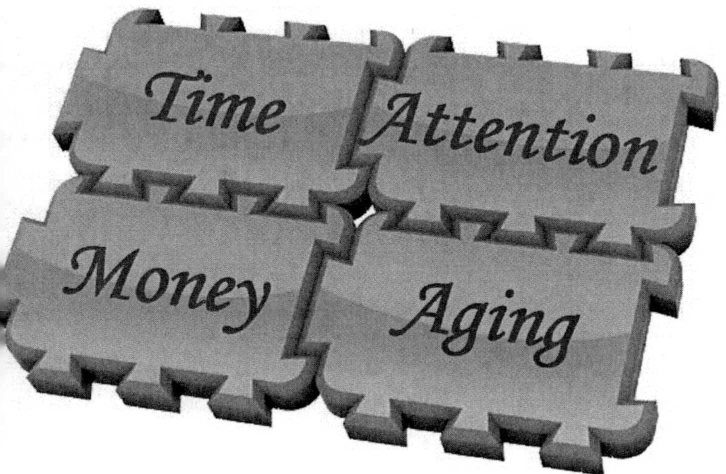

Figure 1. The Four Corners Framework

THE TIME ELEMENT

First, the Four Corners framework considers the effects of time on product perception and value. There are only 24 hours in a day, and of these, only a limited number are available for discretionary activities once one accounts for sleeping, eating, and other necessary activities. Some time will be spent at home, some watching television, a large portion may be spent at work, some exercising, and about an hour will be spent in the car each day[4] or on other transportation. As some services are now mobile and easily accessible via smart portable devices, it is also interesting to consider those times spent shortly traveling from one point to another, such as walking from the office to the car.

The value of a product, as affected by time, is related to the benefit attained over a period of time. Under this paradigm, the ideal product would be one of high benefit that is available at all times and under all circumstances, although few use cases would support such an ideal configuration. As noted earlier, communications is one such core need, and thus of high benefit. Small, affordable mobile phones have enabled that benefit to be used at any time of the day, resulting in an industry where 1 in 2 persons worldwide has a mobile phone[5,6], with annual unit volumes around 15 times that of the automotive market[6,7].

Some other features, such as navigation which may also incorporate traffic information and points of interest, have a high enough benefit and so are considered valuable even if they are used for only limited amounts of time. This allows for successful products like Portable Navigation Devices (PNDs) that might be used for just one or two trips per week[8]. The Four Corners framework time element may also help to explain the prevailing interest in PNDs relative to vehicle-embedded navigation systems: PNDs are perceived to be accessible nearly any time of the day and in any mode of transportation, while embedded systems are only operational during the typical daily hour the consumer is in a specific vehicle. Furthermore, a person can perform searches and route planning anywhere, including away from a vehicle and without requiring access to a personal computer. Therefore, the problem for many features that are embedded in vehicles is that they are inherently available for less time than their equivalent counterparts on non-vehicular devices.

Nevertheless, the average consumer chooses to spend one hour each day in their car due to the need for transportation. Transportation, being the primary function of a vehicle, is also the one thing a well-designed vehicular system can affect in ways that cannot be performed when the consumer is outside of their car. Therefore, it could be expected that the value of embedded transportation-related functions should not suffer the same detrimental business effects as embedded navigation due to the short time that those functions are available within the vehicle.

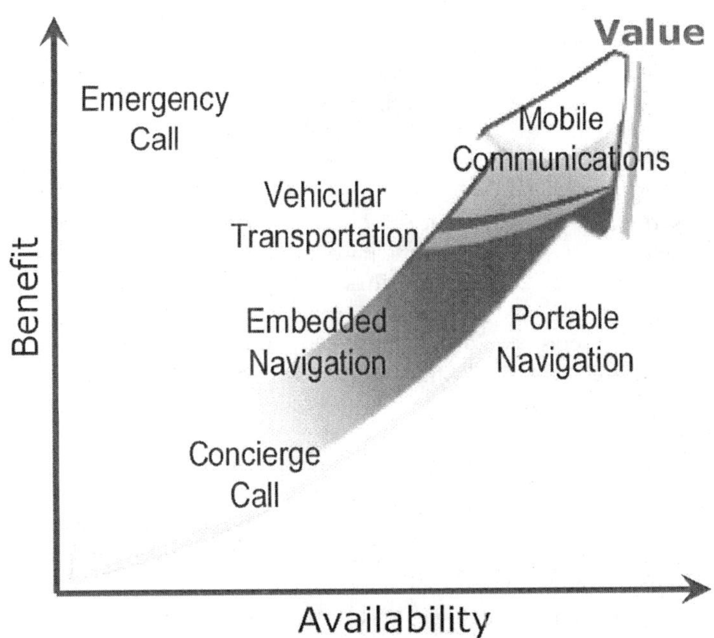

Figure 2. Value Relative to Benefit and Availability

THE MONEY ELEMENT

The second element in the Four Corners framework is money. As another personal resource, we may consider a person's available money in the same way as the hours in a day: for most people, there is a limited amount, reduced by both necessary and optional expenditures. If we postulate that many people spend as much on discretionary items as they can afford[9], then in order to achieve an additional, ongoing revenue flow from consumers for a new product, the perceived value of that new product must be enough to displace a different option that the consumer has previously selected.

Using this consideration, the value for a duplicated feature (both embedded in the vehicle and available from other devices) is lower than if the feature were not duplicated, as consumers would likely want to optimize their free cash for as many distinct features as possible. A product that offers similar features to another, but is more readily available for a longer period of time, may be expected to have a higher perceived value based on the Four Corners time consideration.

Further, we can postulate from this framework that while there may be additional value to tailoring functions for safer or more convenient operation in the vehicle, in the end the consumer will analyze his limited time and limited financial resources to determine which he would select: either the function available for more time during the day, or tailored for more safety in the vehicle. It would certainly work to the detriment of the in-vehicle system if the in-vehicle function offers only moderately enhanced functions or HMI, particularly if offered at a higher price for shorter periods of time than a comparable portable option.

There is, however, a more efficient solution by combining the business models for both in-vehicle and extravehicular services. This would ideally keep the total cost similar to a single instantiation of the feature, but allow the consumer to benefit from both instantiations. Doing this in a way that is tailored for use in both in-vehicle and extra-vehicular environments could be the ideal situation for enabling replicated features within the vehicle. Being tailored for in-vehicle use may also mean using those services to improve transportation, particularly where the service performs differently when driving in the vehicle.

Therefore, the primary lesson learned from the money element of the Four Corners framework is that with an increasing number of choices for discretionary spending, there is an increased need to combine business models from the automotive domain with extra-vehicular features, thus reducing the competition for scarce income. Furthermore, this trend need not be down the slippery slope of ever-decreasing product prices, but the value may be increased by finding a means to incorporate this service in a transportation-enhancing manner while in the vehicle.

THE ATTENTION ELEMENT

For the automotive industry, the third and most critical element of the Four Corners framework is attention. Time and money are important for the selection of services and features as described above, but after a driver starts the engine and during the time his car is moving, the driver will primarily avail himself of the services and features previously selected. Attention, then, is a limited human resource that has the greatest impact on the transportation activity as it directly affects safety.

Driver workload is a commonly-used term, but for the purposes of this discussion, we will consider driver workload to be a subset of the attention element. While driver workload focuses on how many things a driver may perform concurrently, and the effects of these ongoing actions on any additional activity or changing situation, the concept of attention also takes into consideration elements that may be mutually exclusive. An example of the difference can be illustrated by the multiple audio choices available in vehicles today.

In a moderately-equipped vehicle, the choices may include all of the following: AM radio, FM radio, satellite radio or digital radio, iPod® or other audio player, USB memory stick with audio files, embedded storage with music, and CD player. With only a single vehicle occupant in 80% of the cases[4], it is clear that only one of these seven options would command the driver's attention at any given moment, and we have not even considered that persons sometimes prefer to drive without any such audio. This illustrates the concept of attention, whereby a person can only pay attention to, and thereby benefit from, a limited number of things at a particular time. Referring back to the Four Corners framework, the value of each of these audio options to the end consumer would increase if the benefit could be applied for longer periods of time. A non-automotive example would be the hundreds of television channels available from cable and satellite providers. With so many choices available, the average incremental value for each of these infotainment channel options is relatively low as the consumer surfs between so many selections.

But there are even more available features: in addition to audio entertainment, the driver may be on a telephone call, navigating to a destination, receiving news and information alerts, searching for points of interest, reading or listening to their e-mail or text messages, listening to an audio book, or talking with another vehicle occupant. Because some of these tasks may be concurrent with other activities, cognitive workload becomes an issue as driver attention is further divided and shifted away from the primary task of driving.

THE FOURTH ELEMENT

Aging, the fourth element of the Four Corners framework, tends to influence the other three elements in nonlinear ways, and we have found that this element suggests several characteristics for successful automotive products. As a person ages, the element of time can be either relaxed or constricted, depending on a person's changing position and responsibilities in their job, the myriad tasks when raising a family, and the course workload when getting an education - each of these will directly affect the amount of time spent in the car and the number of demands on a person's attention. Money may be short for the youngest drivers, more plentiful near middle age as salaries expand the discretionary options, and may again be restricted in retirement when on a fixed income.

For the youngest drivers, aging can help to improve mental maturity and experience[10,11]. However, aging after physical maturity will result in an ongoing and increasing detrimental effect on the capability to handle a cognitive work load. First, as new features continue to be added because of advancing technologies, the aging driver faces more difficulty in learning those features. In the prime years of their job, the consumer may find themselves with spare money to buy additional infotainment features, thereby adding to the driver's cognitive work load. As the years go by, a person's reaction times will decrease, and thus distraction may become more apparent when multitasking across non-driving-related tasks.

Attention and aging may thereby suggest an increasing need for intelligently interconnected services that provide their benefits with a minimum amount of driver attention and manual intervention. Examples of such interconnected services include zero or one click applications, appropriate sharing of data between applications without additional manual steps, and "intuitive" services requiring little or no training or memorization.

PREFERRED CHARACTERISTICS

Overall, the Four Corners framework suggests that the ideal automotive product would be one with high benefit over most of the time a user is in the vehicle, would relate to benefits outside of the vehicle, would avoid monetary budget contention by leveraging existing consumer spending decisions, and would minimize its presence in the ongoing competition for driver attention. Adding the aging constraint to broaden the audience to multiple age levels, it would be further suggested that such a system should minimize or eliminate the learning curve (such as being usable without instructions, memorization, or training), minimize discretionary costs, and offer a graduated range of support for drivers who are already overburdened with other available activities.

However, merely listing the ideal characteristics of a product, which could be determined through other means, is not the primary reason for using the framework. Instead, we use Four Corners to provoke the necessary conversations when analyzing the value and viability of a proposed new product.

CONSEQUENCES OF AN INTERCONNECTED WHOLE

Understanding the historical evolution of connected vehicle systems[12] and with a framework to assist in analyzing human behaviour in relation to such systems, we can now apply this information to the modern in-vehicle experience.

By 2013 in the US market, the typical vehicle is expected not only to have a bevy of broadcast and local infotainment features, but forecasts indicate that over half of the new vehicles will include wireless connectivity, whether via connection through mobile devices or embedded NADs, WiFi, or other transceivers[13,14]. This high connectivity adoption rate provides expanded options for linking the vehicle with external data sources.

INFOTAINMENT DIRECTIONS

Consumers are anticipated to pay for enhanced vehicle HMI elements, via speech recognition, large center-stack displays, reconfigurable cluster modules, screen-based tactile interfaces (including resistive, capacitive, and force feedback mechanisms), and other HMI devices. While all of these HMI elements are primarily used to control applications, they can also be used as sensors, giving vehicle a better sense of where a person's hands are, and where their eyes are likely to be looking, by detecting the activation of the various interfaces.

SAFETY TRENDS

The number and capability of vehicle sensors and actuators is increasing year over year, as an effect of the improvement of safety and driving systems. Some are by mandate as with stability control, which brings information about wheel rotation and gyroscopic data while enabling autonomous brake control over individual wheels. Others are by choice as with full speed range adaptive cruise control (ACC), which adds forward-facing sensors and can control both the vehicle's accelerator and the braking systems in response to dynamic traffic conditions. Lane-keeping assist (LKA) adds side sensors and steering actuators, which can be used to help the driver keep the vehicle within the marked lane. Aiding the actuators is a range of supporting sensors, including RADAR, cameras, LIDAR, et al, resulting in a large amount of available data from these vehicle sensors.

ENGINE SYSTEMS

Vehicle engine power and nimble high-speed driving performance seem to be on the rise in US consumer preferences again; even so, there is also a burgeoning undercurrent of eco-awareness tempering consumer decisions, supported by ever more restrictive emissions and fuel economy standards. It is this combination of trends - eco-awareness and increased engine performance - that raises the level of instrumentation and electronic management over the engine systems.

MINIMAL CONNECTIONS

While each of these subsystems has shown ongoing improvements in its specific area, most deployed systems have not been optimized for the vehicle system characteristics suggested by the Four Corners framework. The reason is that each subsystem has evolved in a typically safe engineering way: by limiting the system boundary to only closely-related functions (for example, the embedded infotainment subsystem, which includes HMI, links external mobile devices, and data from servers), and by minimizing the interactions with elements beyond that boundary (limited CAN, LIN, and MOST® bus functions). However, by model year 2013, the year for which many vehicle electronics modules are now being designed, the basis of each of these connectivity, safety, and performance subsystems will be well-proven, resulting in the possibility for expansion of the system boundaries in order to create a more coordinated vehicle.

If we temporarily do not consider external connectivity, the vehicle is almost an archetype for the concept of an interconnected whole. Each element in this closed system affects the characteristics of the whole, whether by changing weight or balance, consuming energy, affecting EMC, or providing functionality. The question is whether the tightly-woven subsystems are intentionally designed to draw the maximum benefits out of such interconnectivity, or whether the design is merely an allocation of weight, space, and power consumption across modules.

We can already observe some results of the unintended influence of these design approaches on the driving experience:

- Consumers buy faster, more-responsive vehicles, but they also buy and use new and additional distracting features while driving, whether on mobile devices or embedded with their infotainment subsystems. However, when all such devices are in the same vehicle and are concurrently active, there is typically no coordination across those different subsystems to offset the dangerous combination of speed/power and driver distraction.
- To be most effective, active safety subsystems need to have information about two things: driving situation and driver intent. However, those active safety subsystems with the power to intervene cannot do so in many cases, because driver intent and activities are not well known.
- Safety subsystems, such as for lane departure and forward proximity, often provide an HMI separate from the infotainment HMI, thus increasing the number of visual, audible, and haptic elements presented to the driver. This also applies to many diagnostics subsystems (the "Check Engine" light or other dedicated warning symbols).
- Despite a bevy of vehicle sensors and available data, infotainment subsystems do not understand and account for the dynamic status of the vehicle, environment, and driver.
- While navigation and infotainment subsystems may use map and server-based data to inform the driver of the road ahead, the rest of the vehicle is typically blind to this knowledge.

Excluding external connectivity, we can see that even rudimentary coordination between safety, engine, and user-oriented interior electronics subsystems can result in benefits for all such intelligent subsystems.

Revisiting the forecast that over half of all vehicles will have wireless connections to external data sources, and with transportation being the linchpin that enables the entire automotive industry, the question to be asked is what can and should vehicle systems do to improve transportation in accordance with the Four Corners framework.

DRIVER ADVOCATE®

To tailor and improve the driving experience, the vehicle must first gather relevant data and then act upon it on behalf of the driver. We call this concept Driver Advocate. This approach is one in which the vehicle actively works to improve safety and drivability, acting as a partner to the driver, and neither only as a substitute (autonomous driving) nor relegated to the role of an assistant (either warnings only or action when commanded).

In the first through third generations of Telematics and Connectivity, such relevant data was in most use cases limited to information sourced within the boundary of each vehicle subsystem and was not available from server-based sources.

Illustrating Driver Advocate in these early generational systems: in the infotainment realm, radio volume might be reduced or muted when a turn-by-turn instruction is announced, or when the user begins a hands-free phone call. For collision avoidance systems, the brake controller might only apply partial braking until a user signals his intent by touching the brake pedal, whereupon the system might autonomously perform maximum braking if a collision is imminent. The same applies to stability control, which may use the wheel angle as an indicator of desired direction, then control each wheel to achieve that path. Powertrain systems shift gears based on sensor data from the engine and transmission, as affected by the driver's use of the accelerator pedal.

While each of these functions are useful, they are but a small set of what is possible with today's available information and electronics systems.

ENOUGH TIME

Using the Four Corners framework to help the analysis, we take each of the four elements in turn and look for system-wide improvements. We start with time. The framework postulates that the user will perceive maximum value when a beneficial function is available for the longest possible period of time. For driving-related functions, this means at least the typical 1 hour per day in the vehicle.

Traditional Telematics includes E-Call, B-Call, and I-Call. While some of these features can have a very high benefit, statistics suggest that the usage rate is very low - for instance, an airbag-triggered E-Call occurs, on average, less than once for each vehicle[15]. For the maximum value, the improved functions need to be ones that are beneficial and active during the periods that the passengers are in the vehicle. The features that would be nearly always active are ones that affect how the vehicle moves over the road, how efficiently the engine runs, and what actions the vehicle can perform to actively take a defensive driving posture relative to all surrounding objects and road regulations.

SAVING MONEY

Next, we look for synergies in business models to account for the money element. Designing each subsystem individually results in features that are only paid through the cost of the subsystem itself; i.e., the capability of a chassis control subsystem to maintain vehicle stability or avoid a collision is not affected by the money spent on the infotainment subsystem. So in order to gain a new feature, a consumer must traditionally pay for more hardware.

However, because an increasing number of users are voluntarily choosing to spend money for data connectivity, that particular funding battle need not be fought by the automotive industry. For example, many smart phone users today have access to up to 5 gigabytes of data per month, which includes traffic and incident information, recent roadway segment speeds, map data, traffic cameras, and more. Thus, the question becomes how to leverage that data connectivity to additionally improve the performance of the powertrain and safety subsystems.

STAYING FOCUSED

In the ongoing contest for driver attention, it is imperative to recognize that vehicular improvements, particularly ones that are utilized throughout the driving period, must not add to the driver's workload. Such additional or enhanced functions should add no new HMI, and in the ideal case, would result in a simplification or reduction of the HMI cognitive load as perceived by the driver.

A driver need not necessarily know when a Driver Advocate function intervenes on his behalf. A zero HMI example is when the powertrain system uses Electronic Horizon (eHorizon) map data in a way completely invisible to the user, but potentially resulting in fuel efficiency improvements of 3 to 5%. Another example is deferring the notification to a driver of an incoming call or message when the system detects a potentially unsafe driving situation - driver notification of the message would occur only when the critical situation, which may include blind spot status or unsafe distance to the vehicle ahead, has abated.

Examples of near-zero HMI include slight, corrective resistance in the steering wheel when a driver begins to drift out of a marked lane, or using an accelerator force feedback pedal (AFFP) to provide slight upward pressure on the driver's foot when the driver should slow down. A well-known example of near-zero HMI is the Antilock Braking System (ABS): the driver has signaled his intent to stop, the vehicle detects wheel slippage, and the system's only responsibility is to maximize friction by reducing the slippage.

To reduce driver workload, one must first consider what is essential to the HMI, and then work to eliminate non-essential elements. A screen is needed for navigation, and on the latest vehicles, for control of many interior functions. A cluster module is needed for key driving data such as speed and diagnostics alerts, and in some cases as a quick-glance aid for other functions (odometer, navigation turn indicator). There are, of course, many other necessary HMI elements such as accelerator and brake pedals, steering wheel, discrete knobs, buttons, and levers. Non-essential elements should be eliminated, much as the current generation of smart phones has fewer buttons than ever before.

For any new driving feature, the question is whether the HMI can be almost entirely hidden, as is the case with ABS activation. If a feature must involve the driver, it should make use of an existing, essential HMI element, such as warning notifications using existing displays or spoken warnings rather than adding new lights, unintelligible sounds, or vibrations where no other HMI exists.

BETTER FOR AGE

To accommodate the aging element, we look for zero or near-zero cognitive workload when using any of these advanced functions. This means near-zero training requirements, no memorization, and seamlessly intuitive transitions across multiple active functions without user annoyance or confusion.

An example of this would be integrating an accelerator force feedback pedal with other vehicle systems. An AFFP was added to the prototype ContiGuard® C2X connected safety vehicles, which have been driven by over two thousand drivers without special training. Most of the drivers were not told that an AFFP system was installed. Instead, they were typically given the instructions, "Just drive and follow the rules of the road." In every case, the drivers responded appropriately by automatically reducing the pressure on the accelerator pedal when prompted by the vehicle. Most interesting and surprising, though, was the user feedback. Even with all of the safety and communications mechanisms built into those C2X vehicles and demonstrated using everyday driving scenarios, users were most delighted because the vehicle simply "helped me to slow down at the right time" via the AFFP. The complex technology was distilled down to a simple HMI, so in the end, it was the users' perception that the vehicle was helping them to drive better that was most memorable.

TWO AREAS FOR IMPROVEMENT

However, we have noted that for the vehicle to become a full-time partner and advocate for the human driver, two areas are of essential importance: situational awareness and interpretation of driver intent. The reason is that in order to take action, the vehicle must understand as much as possible about its precise environment, including changes therein, and also about what the driver wants to do. This is different than autonomous driving, wherein the vehicle is solely in control of the driving task, and where there may not be an active human driver. This is also different than traditional, non-autonomous vehicular systems, in which nearly every action is directly initiated by the driver so the vehicle would need less knowledge of its surroundings. Acting as the driver's partner, the vehicle must be strong in both areas, thereby knowing better when to assist the driver and when to assume control.

The approach for a system designed to address these two problem areas becomes the definition for the next generation of Connected Vehicle Systems (CVS).

4TH GENERATION CONNECTED VEHICLE SYSTEMS

Where 3rd Generation Connectivity expanded the systems boundary to actors outside of the vehicle, we define 4th Generation Connected Vehicle Systems by the removal of the artificial boundaries between internal vehicle subsystems. Therefore, automotive products can benefit from coordinated feature development using all of the vehicle's capabilities.

AVAILABLE VEHICLE KNOWLEDGE

We begin with the assumption that information known by one vehicle subsystem can be shared with other vehicle subsystems. Such information could be:

Infotainment

• Activation of the infotainment HMI, including use of buttons, dials, and touch screens, to detect occupant activity and as indicators of driver attention

• Higher-workload moments in infotainment applications (dialing the phone as compared to just hands free conversation, verbally responding to a text message as compared to having a message read to the driver, selecting a particular audio playlist instead of just listening to the music, entering a navigation destination as compared to navigation guidance, etc.)

• Satellite positioning (such as GPS) and Dead Reckoning using vehicle sensors

• Map matching

• eHorizon map and dynamic data about the upcoming road details

• Passenger position and categorization

Powertrain and Driving HMI

• Changes in the position of accelerator and brake pedals including trend analysis

• Changes in the position of the steering wheel and other controls including trend analysis

• Dynamic engine and transmission status

• Fuel usage

Active Safety and Vehicle Sensor Systems

• Use of ACC, including specific system settings (set speed, adjustable following distance, and autonomous braking settings) and distance to object ahead

• Nearby objects detected, categorized, and tracked

• Lane markings and relative vehicle position

• Wheel slippage

• Street signs detected and signage information recognized and understood (construction zone speed limit, school zone active)

REMOTE DATA

Connectivity data from outside of the vehicle can also be leveraged:

• eHorizon and ADAS map data, particularly data from remote navigation and map servers

• Dynamic map information (additional data layers), such as real-time road segment speeds and construction zone data

• Local warnings and incident data such as disabled vehicles, road condition hazards, weather warnings, and approaching emergency vehicles

• Dynamically-changing speed limits

• Car-to-Car alerts (such as via RKE or DSRC)

• Other data via broadcast, multicast, or point-to-point (FM RDS, DARS/SDARS, cellular packet data)

IMPROVING TRANSPORTATION

Using this "sensor fusion" of shared information across vehicle subsystems, new use cases can be enabled to improve everyday driving experiences. In keeping with the goal of minimal cost, these enhanced features could be performed with the technologies and systems available to production vehicles in 2013 - no new hardware would need to be created. Of course, the specific use cases would differ, depending on the subsystems installed on each particular vehicle.

Improving Navigation

• Using a vehicle's LKA function, the vehicle can count the number of lanes traversed to determine in which lane the vehicle is driving, allowing the navigation system to provide a more accurate display of the vehicle on the road.

• Lane traversal may also be determined using a vehicle's backup camera or forward-facing camera.

• Local construction data received over-the-air can provide information on the number of open lanes.

• Sign recognition from the forward-facing sensors can provide updated information for the navigation display, including overriding the speed limit data in the map database or changing the road color to indicate a construction zone.

• Sign recognition indicating start of a construction zone can be correlated with local regulations for maximum speeds in construction zones (received over-the-air) to provide guidance to the driver.

• Approaching train information could be provided before the train crossing is visible, allowing the navigation system to calculate the fastest route around a long train.

Improving Fuel Efficiency and Safety with Map Data

• Prototype systems with OEMs have shown that eHorizon or ADAS map data, provided via the navigation system, connectivity system, or eHorizon device to the powertrain system, can reduce fuel usage by 3-5%. eHorizon systems are now shipping in commercial vehicles, and can save thousands of dollars per truck per year.

• Map data can be also be used by the vehicle safety systems. For example, knowledge of blind driveways, low-visibility intersections, and pedestrian crossings could be used to help optimize visibility and detection of objects (such as directing some light from the headlights toward a blind driveway to increase visibility, or toward the side of a country road where deer crossings are common).

• Speed limit changes, gathered from multiple sources including sign recognition, the on-board map database, and over-the-air data, can be used by an AFFP to more reliably prompt a driver to slow down, even if all data sources are not available.

• When approaching a neighborhood or school zone, an AFFP could prompt the driver to slow slightly before the signs are visible.

• A combination of engine management system and safety system could be used to bring a vehicle closer to the recommended speed when entering a dangerous curve.

Improving Active Safety

• By comparing digital map data with a camera (forward or rear) and/or lane detection sensors, the vehicle can better determine the position of the vehicle relative to the multiple lanes on the road.

• The vehicle could use lane position information to confirm whether an object detected on the side is another vehicle or a side barrier.

• Knowing whether the vehicle is beside the edge of the road or in the middle of a multi-lane roadway may be useful during evasive maneuvers, particularly where the vehicle provides assistance in swerving and braking.

• Information from forward-looking sensors can be combined with map data and dynamic incident data to better estimate whether an object directly ahead is just the center median on a curving road, or whether the object might be in the vehicle's lane.

Improving HMI

• The vehicle can act as a sixth sense for the driver, extending beyond the capabilities of human senses.

• The vehicle can automatically start to slow and provide notifications when approaching hazards or stopped traffic, before such hazards are visible, and take actions to alert the driver.

• Because modern vehicles are well-insulated for sound, making it harder to hear nearby sirens, the system can provide notification of approaching emergency vehicles and school buses, as well as information on local regulations when nearby such vehicles (for example, some regulations require drivers to stop under certain situations).

• The vehicle can adapt its HMI to the driver, depending on both the driving situation and what the driver is currently doing. For example, automatically slowing the vehicle earlier and providing spoken warnings, if the system detects the driver is adjusting the infotainment system in a risky situation.

• Using the infotainment HMI (cluster displays, center stack displays, audio system) as a coordinated part of the overall vehicle driving and safety HMI, the vehicle could reduce/blank the infotainment HMI during quick reaction situations, or use the infotainment system to provide active safety notifications and messages.

Improving Over-the-Air Data

• Roadway data has traditionally been sourced from roadside sensors and manually-activated user reports (including applications for which a user presses a button to submit a report).

• However, each vehicle that passes by a hazard, whether a disabled vehicle on the road, a large pothole, or a construction zone lane shift, can automatically provide this data as an exception report to a server (detection of swerving or departure from a lane, suspension force detection, RADAR target of immobile object, or camera image recognition of a stopped vehicle that is blocking a lane).

• As additional vehicles pass by the same hazard and provide corroborating reports using their own sensors, servers can process and release the information to other vehicles approaching the same location.

• The reports from the various vehicles could contain different but complementary information, depending on which sensors were installed on each vehicle.

These use cases, and many others, show features that could be improved or implemented by linking the vehicle sensors and actuation systems with the myriad data sources available to cars today.

By using a sensor fusion approach of merging data across multiple subsystems, situational awareness and understanding of driver intent are improved. For situational awareness, data from multiple sources could be correlated to increase the likelihood of the correct interpretation. For driver intent, the vehicle could determine whether a driver might be manipulating infotainment functions and thus not paying attention when a high-risk situation is imminent.

SUMMARY/CONCLUSIONS

When we began some years ago to analyze what could be performed by combining the different automotive domains of connectivity, infotainment, safety, and powertrain subsystems, we had not anticipated the number of additional and useful features that could be enabled. At that time, it was still a matter of waiting for many of the OEMs to install the next generation of safety and powertrain subsystems, and for there to be available data for the connectivity subsystem to receive.

Today, we see many of these subsystems being installed on vehicles, from RADAR to cameras to electronic braking systems. Connectivity capabilities are anticipated on over 50% of vehicles, and the percentage of vehicles that can access navigation and map data is increasing year-over-year. The externally-generated data sets are also improved, with availability of ADAS and eHorizon map data for many countries. A delaying factor has been the testing and deployment of each of these individual subsystems, but by 2013, OEMs will have had years of experience with these products.

Given the potential improvements to safety, powertrain efficiency, and comfort, now is the time to consider the implementation of these cross-subsystem vehicle features. Not only would it help the first OEMs to differentiate themselves as leaders in improving transportation, but the Four Corners framework suggests that the value to consumers could be very high, thus driving consumer interest for those vehicles so-equipped.

The last barrier, the open and active coordination across the planning and engineering teams of each of the different vehicle domains, is already starting to fall as the automotive industry strives to improve the parameters of cost, quality, and architectural modularity. What started as an exercise to improve each subsystem to benefit the overall vehicle is expanding as the discussion shifts to consider how cooperation across domains could create value by improving the vehicle's performance.

We therefore propose the next steps: working with each of the OEMs, but in joint technology sessions that combine safety, powertrain, and infotainment/connectivity organizations, in order to optimize what could be done with each OEM's system architecture, installed equipment, and service providers. We have already been meeting and working with some of the OEMs to help drive these topics. However, it is clear that there are many potential improvements to transportation that are now technologically feasible, so the goal would be to examine and redouble those activities in the quest to provide not only subsystems, but safe, efficient cars.

By redrawing the internal system boundaries, Fourth Generation Connected Vehicle Systems can help to make the vehicle much more than just the sum of its parts.

REFERENCES

1. National Highway Traffic Safety Administration, "Vehicle Weight, Fatality Risk and Crash Compatibility of Model Year 1991-99 Passenger Cars and Light Trucks," NHTSA Technical Report, DOT HS 809 662, Oct. 2003.

[2]. Insurance Institute for Highway Safety, "The Risk of Dying in One Vehicle or Another," Status Report, Vol. 40, No. 3, Mar. 2005.

[3]. Insurance Institute for Highway Safety, "Federal Proposal Would Unlink Fuel Economy Requirements from their Safety Consequences," Status Report, Vol. 41, No. 2, Feb. 2006.

[4]. Hu, P., and Reuscher, T., "Summary of Travel Trends, 2001 National Household Travel Survey," US Department of Transportation, Federal Highway Administration, Dec. 2004.

[5]. International Telecommunications Union, "The World in 2009: ICT Facts and Figures," http://www.itu.int/ITU-D/ict/material/Telecom09_flyer.pdf. Oct 2009.

[6]. Gartner Research, "Forecast: Mobile Internet Devices and Connected Portable Consumer Electronics, Worldwide, 2007-2013," http://www.gartner.com, Aug. 2009.

[7]. International Organization of Motor Vehicle Manufacturers (OICA), "2009 Production Statistics," http://oica.net/category/production-statistics, 2009.

[8]. Strategy Analytics, "Navigation Usage across Product Platforms," http://www.strategyanalytics.com, Sep. 2009.

[9]. US Department of Commerce Bureau of Economic Analysis, "Personal Saving Rate," http://www.bea.gov/briefrm/saving.htm, 2010.

[10]. National Institutes of Health, National Institute of Mental Health, "Teenage Brain: A Work in Progress (Fact Sheet)," http://www.nimh.nih.gov/health/publications/teenage-brain-a-work-in-progress-fact-sheet/index.shtml. 2001.

[11]. National Highway Traffic Safety Administration, "Teen Driver Crashes - A Report to Congress," http://ntl.bts.gov/lib/30000/30300/30301/811005.pdf, Jul. 2008.

[12]. Gee, R.A. and Gonsalves, S., "Resource Management for Third Generation Telematics Systems," SAE Technical Paper 2000-01-C021, 2000, doi:10.4271/2000-01-C021.

[13]. Strategy Analytics, "Global Automotive Vehicle-Device Connectivity Forecast 2008-2016," http://www.strategyanalytics.com, Feb. 2010.

[14]. Strategy Analytics, "Global Automotive OE Telematics Forecast 2008-2016," http://www.strategyanalytics.com. Mar. 2010.

[15]. General Motors, "OnStar Extends Partnership with the National Center for Missing and Exploited Children," Press Release, http://www.prdomain.com/companies/G/GeneralMotors/newsreleases/200782444833.htm, Aug. 2007.

CONTACT INFORMATION

Robert Gee is a Product Line Manager within Continental's Infotainment and Connectivity Business Unit. Gee works with automotive OEMs, partner companies, and Continental's Chassis & Safety and Powertrain Division's to develop the next generation of connected vehicle systems to better enable safe transportation, sustainable resource usage, and consumer information needs.

Gee's previous experience at IBM, Motorola, and other companies includes satellite mission control center architecture, military safety system process development, systems/software process definition and quality assurance to develop an SEI CMM® Level 5 organization, proposal negotiations and contract management, network communications architecture, and engineering on over 40 commercial and government product development efforts.

Gee is a television Emmy Award recipient for his work as a producer for CNN. He is based in Deer Park, Illinois, and may be reached at Robert.Gee@Continental-Corporation.com.

ACKNOWLEDGEMENTS

The author would like to extend his appreciation to Nancy Ann for her detailed reviews and providing a name for the previously untitled analysis framework.

DEFINITIONS/ABBREVIATIONS

ACC
: Adaptive Cruise Control

ADAS
: Advanced Driver Assistance Systems

AFFP
: Accelerator Force Feedback Pedal

C2X
: Car-to-Car and Car-to-infrastructure (Car-to-X), also known as Vehicle-to-X (V2X)

CAFÉ
: Corporate Average Fuel Economy

CAN
: Controller-Area Network, a vehicle data bus

CVS
: Connected Vehicle Systems

DARS/SDARS
: Digital Audio Radio Service, Satellite Digital Audio Radio Service

DSRC
: Dedicated Short Range Communications, a radio system designed to enable vehicle-to-vehicle and vehicle-to-infrastructure communications

E-Call
: Emergency Call

eHorizon
: Electronic Horizon, which is map-based data that provides driving-oriented information such as road curvature and slope, in order to provide a look-ahead capability to the vehicle

EMC
: Electromagnetic Compatibility

GPS
: Global Positioning System

HMI
: Human-Machine Interface

I-Call
: Information (or Concierge) Call

Infotainment
: Information and Entertainment

LIDAR
: Light Detection and Ranging, a technology similar in function to RADAR

LIN
: Local Interconnect Network, a vehicle data bus

LKA
: Lane Keep Assist

MOST
: Media Oriented Systems Transport, a vehicle data bus

NAD
: Network Access Device, embedded cellular communications

NHTSA
: National Highway Traffic Safety Administration

OEM
: Original Equipment Manufacturer

PAN
: Personal Area Network

PND
: Portable Navigation Device

POI
: Point of Interest

RADAR
: Radio Detection and Ranging

RDS
: Radio Data System

RKE
: Remote Keyless Entry

SEI CMM
: Software Engineering Institute Capability Maturity Model

TSP
: Telematics Service Provider

WAN
: Wide Area Network

WLAN
: Wireless Local Area Network

Telematics
: The combination of telecommunications and computing systems, and in particular for vehicles

Metrics for Evaluating Electronic Control System Architecture Alternatives

2010-01-0453
Published
04/12/2010

Arkadeb Ghosal, Paolo Giusto and Purnendu Sinha
General Motors

Massimo Osella
GM R & D

Joseph D'Ambrosio
General Motors R&D Ctr.

Haibo Zeng
General Motors

ABSTRACT

Current development processes for automotive Electronic Control System (ECS) architectures have certain limitations in evaluating and comparing different architecture design alternatives. The limitations entail the lack of systematic and quantitative exploration and evaluation approaches that enable objective comparison of architectures in the early phases of the design cycle. In addition, architecture design is a multi-stage process, and entails several stakeholders who typically use their own metrics to evaluate different architecture design alternatives. Hence, there is no comprehensive view of which metrics should be used, and how they should be defined. Finally, there are often conflicting forces pulling the architecture design toward short-term objectives such as immediate cost savings versus more flexible, scalable or reliable solutions. In this paper, we propose the usage of a set of metrics for comparing ECS architecture alternatives. We believe the set of metrics constitutes a relevant aspect to address the existing design gaps. We define the set of metrics based on non-functional requirements (reliability, vehicle availability, safety, monetary cost, and timing), the degree to accommodate changes (reusability, flexibility, scalability, and expandability), the customer requirements (integrity, maintainability, energy efficiency, and security), and the compatibility to legacy designs (complexity, organizational alignment, backwards compatibility, and packagability). The key pillars for the metric-based evaluation framework of ECS architecture alternatives are as follows: (a) incremental design knowledge (by a combination of prediction, measurement, and analysis of the architecture alternatives), (b) metric evaluations (ranging from qualitative to quantitative depending on the stage of the design), and (c) comparison of the alternatives (once the metrics have been evaluated). The proposed metric-based evaluation of architecture design alternatives is relative (i.e., the results are usable within the scope of a group of alternatives being considered). The comparison is potentially more qualitative in the early phases of the design cycle as fewer data are available; for a more quantitative evaluation, data sets and design details (which are usually available in the later phases of the design cycle) are essential.

INTRODUCTION

Electronic systems are being introduced at an exponential rate in automotive systems. The inclusion of electronic systems is causing a major shift in both cost and development of automotive products. While in the last decade the mechanical parts of a car accounted for the majority of the costs (mechanical: 76%, electrical: 13%, software: 2%, and others: 9%), electrical/software parts will be increasing in content, and consequently increasing in percentage cost in the next decade (mechanical: 55%, electrical: 24%, software: 13%, and others: 8%). The cost breakdown may not reflect the reality of the increase in content. In some cases, the number of the parts constituting an automotive embedded system is astounding. While the number of ECUs will almost triple in the following two decades, the software content will increase by about 100 times (1 million lines of code in the last decade to 100 million lines of code in the next decade).

Given the increasing importance of Electronic Control System (ECS) based features, there is an increasing need to select the "correct" architecture for implementing the features. The term "architecture" not only includes an hardware architecture but also encompasses functional, software, and physical architecture. A functional architecture defines the set of features, along with the partitioning and/or their allocation to hardware (e.g., CPU) and software resources (e.g., software mailbox). A software architecture sets the ground rules of software coding which may include interfaces, drivers, I/Os, and execution rules (e.g., of software tasks). A hardware architecture defines the set of computing resources, and components in the resources. The physical architecture consists of a topology architecture (distribution of resources), harness architecture (wiring and connections), sensor/actuator architecture (description of I/O devices), power supply architecture (power generation and transmission), and communication architecture (message transmission and related protocols). Finally, a product-line architecture [4, 6, 7] may be used for multiple markets across different products. Hence, an architecture is not constrained to only one product targeting one market.

Architecture design by an OEM usually entails several stakeholders; however there is no clear definition and understanding of how to evaluate and compare design alternatives resulting from alternative design decisions. Specifically, there is no comprehensive view of which metrics should be used for the evaluation, and what their definitions should be. In the early design phases, an objective evaluation of the alternative ECS architectures is essential in addition to relying on the experience of the designers. To provide more objectivism in the design decisions, we propose a set of metrics for comparing ECS architecture alternatives. This set, in our judgment, includes the most relevant and important metrics that enable objective design decisions during the early exploration phases of the ECS architecture cycle. This paper "builds on top" of the work that was started in [4, 5, 6, 7] in which a methodology for architecture exploration, a set of initial metrics (timing, cost, dependability), data modeling artifacts, and abstraction layers (functional, physical, logical) were defined. In this paper, we have leveraged the experience gathered in the earlier work, and extended the set of metrics, the methodology, and the modeling artifacts.

Previous work in this domain has focused on software architecture evaluation methods. Software architecture analysis methodologies such as Architecture Trade off Analysis (ATAM [22]), Architecture Level Modifiability Analysis (ALMA [24]), Software Architecture Analysis Method (SAAM [21]), and Family-architecture assessment method (FAAM [23]) have been discussed in detail in [20]. The underlying theme of these evaluation techniques is the concept of scenarios. Scenarios are brief narratives of expected or anticipated use of a system from both development and end user view points. The set of scenarios are used to assess quality attributes of architectures such as modifiability, usability, performance, maintainability, etc. Specifically, SAAM can be used for understanding modifiability of software architectures; ATAM is an improved version of SAAM which gives insight into the trade off among the quality attributes in addition to the assessment of the quality attributes; ALMA can predict the maintenance cost and the flexibility of software for modifications; FAAM is used for evaluating families of architectures for information systems.

METHODOLOGY

[4, 7] outline a methodology for quantitative evaluation of architecture alternatives with the following concepts:

- An iterative process in which design alternatives are produced.
- A meta-model to describe the system model, and the modeling artifacts.
- A set of degrees of freedom representing the exploration space for the design choices.
- A set of (quantitative) metrics and (qualitative) criteria by which each design alternative is scored.
- A federation of tools which are used to capture the different aspects of a design, and which are used to compute the metrics.

The steps outlined in [4, 7] have been enhanced by additional inputs, such as trend analysis and constraints from legacy designs, to complete the overall methodology (Figure 1).

- The use cases come from market demands and design experts.
- The requirements are typically generated by marketing teams that assess the need for a specific feature (e.g., autonomous driving) to achieve competitive advantage. The constraints are usually tied to legacy designs that are re-used (e.g., for cost reduction reasons), constraints from technologies, etc.
- The degrees of freedom are defined by the system level teams, and they are complementary to the constraints. Generating requirements would ideally entail a tool which formally translates the use cases to formal requirements; however developing requirements may need knowledge on

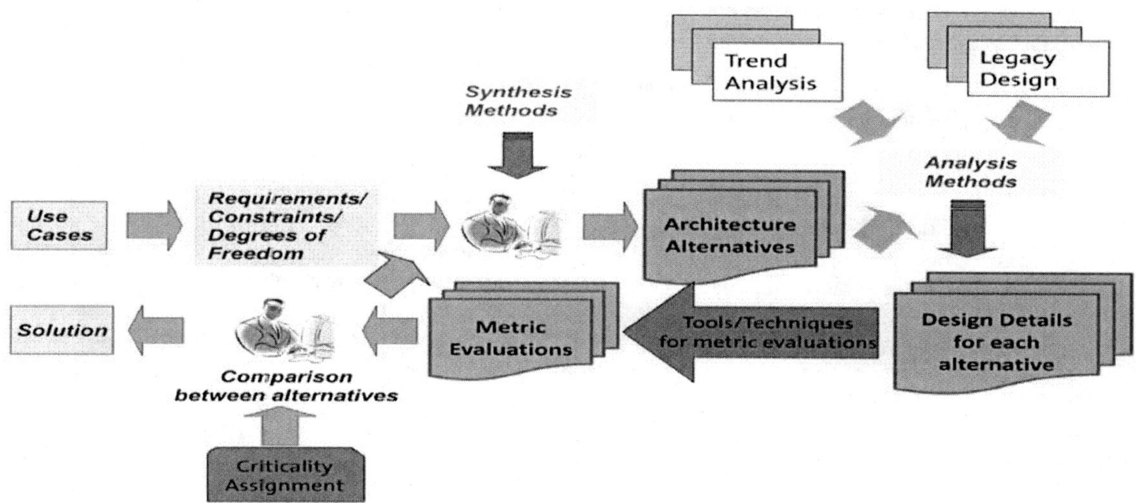

Figure 1. Metric-based Methodology for the Design and Selection of an Architecture Alternative

design expertise, technology trends, regulations, and market constraints.

<figure 1 here>

• Once the requirements are available, synthesis techniques are used to generate architecture alternatives. Traditionally, alternatives are manually generated, but if degrees of freedom and technology constraints are formally stated, architecture alternatives are generated automatically. For example, given the set of constraints on end-to-end latency for a data path, synthesis techniques may generate alternatives based on priority and offset assignment to tasks and messages in the path.

• In the initial steps of the design cycle, the alternatives may be specified at a very basic level; e.g., only the communication backbone, topology architecture, one software task per function, etc., may be available. If some design details for the analysis at hand are required (e.g., execution times for tasks, failure rates for hardware components, etc.), then they can be obtained from trend analysis, historical data bases, and legacy designs. Most of the design details may not be present during the early design phases, and the metric based evaluation will be qualitative or aimed at sensitivity-like studies as budgets may be assigned for parameters of uncertain values. The design process is iterative, and as design phases evolve, more design data may become available.

• Once the design data are available, the metrics can be computed. For different metrics, a multitude of techniques can be used. For example, for a timing metric, both corner case (e.g., worst case), and average case (e.g., probability distribution) analysis methods can be used: while corner case analysis provides an upper/lower bound on response times, probabilistic method provides a probability distribution of possible response times. The reader should refer to description of each metric for more on analysis tools/ techniques, and required design data/models.

• Different metrics may have different degrees of importance for different designs, and across the design life cycle. For example, the objective of one design may be higher security/ integrity, and for another one, it may be higher energy efficiency. Importance assignment of metrics for different designs is crucial. Once the designer decides the relative importance of metrics, the metric computations are fed into comparison tools (e.g., Pugh analysis [15] and Spider charts [17]) to evaluate the alternatives. Depending upon the phase of the design, the evaluations can be used to improve the architecture alternatives in the next design cycle, or an alternative is selected.

• Similar to Technical Cost Modeling [9], the proposed methodology does not compute the absolute value of the metric. For example, the evaluation of the cost metric focuses on relative differences in cost; factors such as taxes, tool-set up, office space, etc. are not be included in the evaluation as the contribution of these factors is independent from the architecture alternatives under consideration; in other words, we consider them "sunk" costs - costs which are not under control of ECS architect, and therefore invariant across the different alternatives.

METRICS

Our proposed set of metrics is chosen according to the following five criteria:

• Goal: The basic requirement is to implement a safety-critical hard-real-time system; the related metrics are *timing*, *safety*, *reliability*, and *vehicle availability*.

• Consumer: The current customer trends center on high fuel efficiency, low maintenance, and consumer features requiring secured systems (e.g., no hacking of the navigation software);

the metrics for these desired aspects are *energy efficiency*, *security*, *integrity*, and *maintainability*.

• Market: Automotive markets are evolving quickly with new demands in features and product varieties; the metrics *flexibility*, *scalability*, *expandability*, and *reusability* capture these new market demands.

• Legacy: While changes are desired, process/tools/design methods for new architectures may deviate significantly from the structure or legacy designs, adding significant challenges; the metrics *organizational alignment*, *backwards compatibility*, and *packagability* address the legacy concerns.

• Cost: Monetary *cost* is a critical measure to evaluate alternatives, and can be combined with any other metrics discussed above.

Our proposed set of metrics with their definitions is as follows:

• ***Timing***: The ability of the system to ensure timely computation and communication.

• ***Reliability***: The ability of the system to perform intended/ specified function satisfactorily for a prescribed time, and under stipulated environmental conditions.

• ***Vehicle Availability***: The ability to use a vehicle over a period of time.

• ***Safety***: The ability to manage risk to an acceptable level.

• ***Reusability***: The degree to which system components are adapted to support different products in the product-line.

• ***Flexibility***: The ability to accommodate changes in features without any change in hardware resources.

• ***Scalability***: The ability to accommodate "anticipated" changes in hardware resources.

• ***Expandability***: The ability to accommodate "unanticipated" changes in hardware resources.

• ***Security***: The ability to prevent unauthorized access to, or handling of, system state.

• ***Integrity***: The ability to prevent improper system alterations.

• ***Maintainability***: The ability to restore a failed system to an operationally effective condition within a given period of time through repair and corrective action.

• ***Energy Efficiency***: The net electrical power consumed when no system is running, when a partial network of resources is running, and when all resources are running (operational).

• ***Cost***: The net monetary cost for design/development, manufacturing, vehicle integration, and ownership (including repair/maintenance).

• ***Complexity***: An assessment of intricacies in a design.

• ***Organizational Alignment***: The alignment of the design to organization for design, development and integration.

• ***Backwards Compatibility***: The alignment of the design to historical data including past designs, data sets, code, supplier relations, tool availability, etc.

• ***Packagability***: A measure of the freedom to place/ redistribute hardware resources inside a vehicle.

Metrics are different from ***design choices***. Design choices outline the degrees of freedom in selecting alternatives. The metrics do not encompass design choices; rather, the metrics study the effect of the design choices. For example, the adoption of globally synchronized communication is a design choice, and timing is a metric to analyze between two alternative implementations (FlexRay vs. TTP/C) of the design choice.

The ***maturity of the tools/techniques*** available for respective metric evaluations is heterogeneous. We classify the metrics into the following four classes depending upon the maturity of the related tools/techniques:

• <u>Mature</u>: Techniques and tools for the evaluation of timing, reliability, availability, and safety metrics are the most mature ones. There exist standard techniques for measuring latency and jitter. Similarly, extensive work has been done in the research/tools community to compute reliability and fault coverage of control systems. In addition, standards such as ISO26262 will lay the foundation for the formal definition and acceptance of safety related metrics in the automotive domain.

• <u>Semi-mature</u>: While there are techniques/tools for estimating power, cost, integrity, and security, there has been little work on estimating them for automotive system development. There exists a significant need to develop tools for evaluating these metrics for automotive architectures design.

• <u>Early-stage</u>: We believe that the evaluation of metrics such as reusability, flexibility, scalability, expandability, and maintainability is in its infancy. While standardization of interfaces, timing slacks, or gateway separation can be counted and/or measured, there has not been extensive work to develop tools/techniques, due to the differences in how these metrics are defined in various application domains.

• <u>Qualitative</u>: To the best of our knowledge, metrics such as complexity, organizational alignment, backwards compatibility, and packagability have no techniques/tools available to evaluate them. Thus, their evaluation requires domain expertise. Techniques such as regression analysis are in the right direction, but much more work needs to be done to enable quantitative estimates of these metrics.

In the following sections, we will provide an overview of the definitions of the metrics, and refer to related methods/tools available to compute the mature set of metrics; for the rest of the metrics, we will provide a definition of the metric, and present possible methodologies to compute/evaluate them; a detailed discussion of the modeling abstractions used is outside of the scope of the paper.

TIMING ANALYSIS

Timing analysis entails the evaluation/computation of the following sub-metrics: (1) end-to-end latencies along computation paths, (2) maximum jitter requirements on activations (queuing and release jitters), and output generations (output jitter), (3) utilization of computing/ communication resources, and (4) time correlation between any signal pair originating (resp. terminating) from (resp. to) a functional block. Evaluation of timing metrics such as latencies, response times, jitter, and utilization is a complex topic that requires a deep understanding of the abstractions necessary to perform such analysis. In its simplest yet very powerful abstraction, timing analysis pertains to the definition of a set of agents (e.g., tasks and messages) requesting time for their execution on shared resources (e.g., CPUs and communication links, shared variables, mailboxes, etc.). As such, the agents are described as black boxes (no functionality) with an activation policy (e.g., sporadic, periodic, periodic with jitter, sporadically bursty, etc). Other parameters include the requested time (e.g., execution time for tasks), and blocking time for the access to shared variables. Shared variables have associated control policies (e.g., OSEK priority ceiling protocol) used to prevent deadlocks, unbounded blocking delays, and priority inversion. Within this simple model, several techniques are available to compute either a single value (e.g., best/worst case response time, end-to-end latency) or a probability distribution of the latencies, jitters, etc. Mature techniques include simulation, and best/worst case schedulability analysis. New approaches have been developed recently, such as stochastic analysis [19] to compute distribution of response times, and formal methods [13] to compute best/worst case response times. The other important dimension in these analysis techniques is their accuracy to handle implementation details such as multi-queue/multi-buffer CAN controllers, copy times of TX buffers, etc. In general, the analysis techniques of CAN and OSEK based systems are more mature than their counterpart standards (e.g., Ethernet and Wireless protocols, Autosar). As for FlexRay, the real analysis challenge is in the dynamic segment, as the schedule for the static part is time-triggered, and thus its timing behavior is deterministic.

In the following, we summarize the worst case resource-level (bus, CPU etc.) schedulability analysis; please refer to [4] for a detailed set of references on timing and schedulability analysis. The analyses enable the estimation of lower (best) and upper bounds (worst) of the task and message response times. The response times include latency effects due to jitter (output, queuing, and release), bursts ("dense" activation of tasks/messages), and periodic/sporadic activation of tasks/ messages. In addition, implications due to the usage of software and hardware shared resources (e.g., critical regions, shared memory locations) that affect response times with additional blocking delays are also considered in the analysis. The message and task response times are then used to compute end-to-end latencies (e.g., sensor/actuator). These results are useful both in the prediction/exploration phases of the architecture design and in the verification phase. Schedulability analysis dates back to 1973 [12]. Since then, a large number of sophisticated methods have been developed in the real-time systems community to analyze increasingly complex systems. [14] provided a guaranteed worst-case response time calculation. In this work, the main contribution is the computation of the task response time. Specifically, a task experiences its largest number of preemptions by higher-priority tasks when all tasks are activated simultaneously, a situation called the "critical instant of a task". The formula below (Equation 1) is an example of how to compute the response time of task assuming the task deadline is less than or equal to its period. The general case of arbitrary deadlines is much more complicated, and we refer the reader to [4]. The response time R_i of a task t_i (with period T_i, and deadline D_i equal to the period) is calculated as the sum of the task's worst-case execution time C_i, and its worst-case interference from execution of other tasks. The interference captures the preemption experienced by task t_i due to the tasks with higher priority than task t_i (denoted by the set $hp(i)$). While the priority and period are typically part of a functional description, the execution time is usually measured from actual traces, static code analysis, or estimates.

$$R_i = C_i + \sum_{j \in hp(i)} C_j \left\lceil \frac{R_i}{T_j} \right\rceil \leq D_i \leq T_i$$

Equation 1

RELIABILITY ANALYSIS

Reliability can be defined as the probability that a system will perform its intended/specified function satisfactorily for a prescribed time, and under stipulated environmental conditions. The crucial aspect of this definition is that reliability is dependent on the environment. Besides, the reliability of a system will vary according to its operational usage; for example, if the vehicle is being driven in different environment (or operating conditions) by different users, then it is likely that the reliability will be significantly different. A metric for reliability is the Mean Time to Failure (MTTF); MTTF is a basic measure of reliability for non-repairable systems, and is equivalent to the mean of its failure time distribution. The time to failure is modeled by an exponential distribution. If $R(t)$ is the reliability function for a system over time t, then the reciprocal of the failure rate parameter is equivalent to its distribution mean. Thus, $MTTF = \int_0^{\infty} R(t)dt = \frac{1}{\lambda}$, where λ is the constant failure rate. Since the definition is given as a probability, we can additionally make use of the following metrics:

- Probability of failure per operating life of the vehicle (e.g., 450 hrs/year for 10 years).
- Mean number of trips (or ignition cycles) for the likelihood of a failure as a result of a latent fault; latent faults lay dormant in a system (e.g., undetected maintenance errors which introduce faults).
- Incidents per Thousand Vehicles (IPTV) is computed through warranty calls; the simplest definition of IPTV at a given time (say, Months in Service) is :

$$Cum. IPTV = \frac{total\ claims}{total\ vehicles} \times 1000$$

SAFETY ANALYSIS

Safety is defined as the condition where risks are managed to acceptable levels, where risk is a combination of the probability of occurrence of harm, and the severity of that harm [8]. However, safety cannot be absolutely guaranteed. Accident prevention and risk management contribute to increasing safety. Safety and reliability usually entail design/ development tradeoff - increasing reliability might not increase safety, and increasing safety by means of redundancy will typically result in decreasing reliability (as more resources imply more potential failures). Mean Time to Critical Failure (MTTCF), defined similarly to MTTF, and normally expressed in hours, is a quantitative metric for safety.

The charter of the ISO26262 [8] committee is the definition an international standard for functional safety of E/E systems in road vehicles. The standard defines metrics for the hardware architecture, and addresses only safety-related aspects. Each fault occurring in a safety-related hardware element can be classified as one of the following:

- Safe fault is a fault which does not significantly increase the probability of violation of a safety goal.
- Residual fault is a fault which leads to the violation of a safety goal, occurring in a hardware element where a safety mechanism only covers some of the faults.
- Single-point fault is a fault of a hardware element where no fault is covered by any safety mechanism, and can lead to the violation of a safety goal.
- Multiple-point fault is a fault which leads to the violation of a safety goal as a result of a combination of several independent faults; this scenario includes detected, perceived, and latent faults.
- Latent fault is a fault which is neither detected by a safety mechanism nor perceived by the driver.

The standard also presents two hardware architectural metrics:

- Single Point Fault (SPF) Metric is the proportion of residual hardware single points of failure, and residual failures that can lead directly to a safety goal violation.
- Latent Fault (LF) Metric is the proportion of residual latent hardware failures that can contribute to a safety goal violation.

In addition to the above metrics, ISO 26262 also identifies another important metric, that is, the probability of violation of a safety goal. ISO 26262 provides criteria to demonstrate that the risk of a safety goal violation due to a random hardware failure of the item is sufficiently low; e.g., with regard to ASIL (Automotive Safety Integrity Level) D systems, ISO 26262 sets quantitative target values for maximum probability of violation of each safety goal due to hardware random failure to be less than 10^{-8} per hour. The metric can be computed by quantitative fault tree analysis.

VEHICLE AVAILABILITY ANALYSIS

Vehicle Availability is defined as "the vehicle's capability of being used over a period of time", and its measure is defined as "the period in which the vehicle is in a usable state". In layman words, availability is the degree to which a system is operable, and is in a usable state at the start of a mission, when the mission is called for at a random time. This aspect requires the definition of "acceptable performance". Acceptable performance is defined taking into account the consequences of failure or unsuccessful operation of the system, and the key requirements needed to restore the operation or performance. Common measures are Mean Time between Failures (MTBF), Mean Downtime (MDT), and Mean Time to Repair (MTTR). We define these concepts as follows:

- Mean Time between Failures, a key metric for evaluating a system that can be repaired or restored, is the expected value of the time between two consecutive failures of a system.
- Mean Downtime is the average time a system is unavailable for use due to failure. This time includes the actual repair time plus all the delay time associated with technician arriving with required replacement parts.
- Mean Time to Repair is the expected time to repair or restore the system after a failure occurs.

The following metrics are also closely linked to availability. Unavailability is the probability that the component/system is unavailable at any given time. Total Down Time is the total time the component/system is expected to be unavailable for the specified system lifetime. Total Up Time is the total time the component/system is expected to be available for the specified system lifetime.

DATA INTEGRITY ANALYSIS

We make use of the term Data Integrity to avoid any confusion with Safety Integrity Level (related to the safety metric). The expectations for data integrity of a system encompass the steps taken to ensure integrity in the case of an undesired event, such as the alteration/corruption of data (which is essential for the correct operation of the system) in databases, in memories, and during transmission. In some control functions where information integrity is absolutely necessary, if data becomes corrupt or incorrect or disappeared, then it could cause serious malfunctions in the system. Data failure detection and correction codes (such as Hamming codes or CRC) during transmissions, or error checking and correction (ECC) mechanisms in memories are possible methods to increase data integrity.

MAINTAINABILITY ANALYSIS

Maintainability is defined as the probability that a system that has failed can be restored to an operationally effective condition within a given period of time through repair action, or corrective maintenance action. Maintainability is measured by Mean Time to Repair (MTTR) which is the time to restore service since the last failure occurrence. Typically, it includes fault identification, removal, replacement, and repair of component(s) that has (have) failed, and final checking. One of the key objectives in maintenance is the system ability to detect the fault, and diagnose the problem correctly so that proper/accurate corrective actions can be taken. The ability to repair is proportional to the ease of availability of parts/ components, ease of replacing (if parts are inexpensive, and can be replaced), ease of repairing faulty parts, and/or ease of integrating new parts. Maintainability can also be measured in terms of service time and cost.

SECURITY ANALYSIS

Security has several connotations, and has been applied relatively to software, communication, infrastructure, transactions, and information. A unified definition of security is provided in [1] as "unauthorized access to, or handling of, a system state". In the context of data, security is composed of the following attributes: (1) Data Availability - the prevention of unauthorized withholding of information, (2) Data Confidentiality - the absence of unauthorized disclosure of information, and (3) Data Preservation - the prevention of unauthorized change, and/or deletion of information. Data confidentiality of ECS architectures may not be a critical factor for traditional designs; however as vehicles are increasingly connected (Vehicle-to-Vehicle/Vehicle-to-Infrastructure), confidentiality may become a critical factor in choosing different design alternatives. The issue of correct data availability is also critical: consider a malicious router sending valid-but-erroneous data to imply road traffic (while there is no traffic) to reroute vehicles. In those scenarios, ensuring that information is not withheld maliciously is critical.

FLEXIBILITY ANALYSIS

The flexibility metric is a measure of ability to accommodate changes (i.e., addition, removal or modification) in features without any changes in hardware resources that implement the architecture. A feature is modified if there is a change either in the interface or in its functional behavior. A feature interface is a communication protocol comprised of a set of methods used to communicate/transmit data to other features (e.g., send/receive or client-server methods), and a set of parameters to configure the feature. The feature behavior (the semantics of the defined functionality) is encapsulated, and is accessible via its interface. Flexibility can be measured by checking the standardization of the feature interface to company-wide adopted standards that enable reuse, measuring slack (in execution time or memory usage) in the sense that additional time and memory resources are available for future extensions making the design flexible, or computing the tolerance to uncertainties in design parameters (e.g. min-max analysis). Consider two software architectures: one based on proprietary standards, and the other one based on AUTOSAR; the second design would be more flexible as AUTOSAR standardized interfaces would allow many different software components to be interconnected as long as their interfaces adhere to the AUTOSAR standard, while the semantics is compatible to the intended behavior.

SCALABILITY ANALYSIS

The scalability metric measures the ability to accommodate "anticipated" changes in hardware resources; in other words, scalability accounts for the changes in usage of hardware resources without the addition of hardware resources. Although the definition may sound contradictory, an example is provided for illustration purposes. Consider two architecture alternatives: one where each node has only CAN controllers, while the other one has nodes with both CAN and FlexRay network controllers. Suppose both the alternatives use CAN for communication. While the second architecture is scalable for FlexRay communication, the first one is not. Similarly an extra processor or memory in hardware nodes makes an architecture design more scalable for a future increase in processing loads or memory size. Obviously a system with higher scalability has higher unused resources (e.g., the FlexRay controllers in the last example). The trade-off between the give-away cost and the cost of changing the architecture to accommodate new design (e.g. FlexRay channel) must be evaluated. Typically, when no information is available, designers tend to make the architecture as scalable as possible to accommodate anticipated changes at the price of overly designing the system. The scalability metric also may help to evaluate product-line dispersion. For example, the ECU (with FlexRay and CAN connectors) may be connected to FlexRay for a high-end vehicle (with active safety and autonomous features), and connected through CAN for basic chassis control. The tradeoff is then the extra cost of unused hardware in the basic platforms with respect to the economy of scale of a single standardized unit.

EXPANDABILITY ANALYSIS

The expandability metric measures the ability to accommodate "unanticipated" changes in hardware resources. Unanticipated changes may imply integrating new hardware resources (e.g., adding new processors/channels/ gateways/ sensors/actuators) and/or modification of topology. Consider two alternatives: one using a larger number of nodes, while the other using a smaller number of nodes; given that any communication system restricts the maximum number of nodes that can be connected over the network, the second alternative is more expandable than the first one as more nodes can be added.

REUSABILITY ANALYSIS

The reusability metric measures the ability to reuse components (hardware and/or software). The evaluation can be within the same ECS architecture, and across ECS architecture instances within a product-line architecture. A possible evaluation [4] for component reusability is to use the number of instances of the component within each architecture instance; a component may denote a CPU, a software task, or a communication controller used in an architecture.

COST ANALYSIS

Cost analysis computes the effect of design decisions on the cost of an architecture across its life-cycle. [5] presents a cost model based on four stages in the life-cycle: (1) the design/ development stage includes the planning for the architecture, and research/prototype development of the new hardware/ software components, (2) the fabrication stage encompasses the manufacturing of hardware parts, (3) the assembly stage includes the integration of (hardware and software) parts during production, and (4) the in-service stage accounts for the post-sale life time of a car. The life-cycle phases described above are not comprehensive, and can be further refined; however, the model is sufficient to study the effect of architecture design decisions on cost [5, 6]. The cost model accounts for costs incurred in each of the above phases; different parties may be responsible for cost in different phases. While the design phase is usually performed jointly by the suppliers and OEMs, fabrication is carried out by suppliers, and assembly is done by the OEMs. The in-service cost accounts for the cost from repair and/or maintenance. The focus of the model is relative comparison; hence the model disregards the effect of decisions outside the scope of the designer (e.g. cost of IT infrastructure, software licenses, office space, test benches, and prototype set up). The major components of the cost model are:

- The ***software design cost*** accounts for the software design effort which can be computed by using methods like COCOMO [3]; the design effort is computed based on the characteristics of software modules (e.g., interfaces, lines of code etc), and the development constraints (like project and organization expertise).

- The ***hardware design cost*** accounts for OEM engineering activities (like system specification and validation), summed with supplier engineering activities (such as development, redesign, and validation of the hardware modules). An accurate mathematical model for such costs may not be possible, and may need to be predicted based on historical trends.

- The ***fabrication cost*** accounts for the cost of manufacturing the physical parts. The current supply chain for parts relies almost exclusively on suppliers for fabrication; so, if the cost evaluation is done from the OEM side, the model will include the best guess estimate (from market data and trends). The cost includes the piece cost (accounting for fabrication of individual hardware modules), and the cost of interconnects (e.g. cut-leads); the cost may also include some one-time costs like investment in tools.

- The ***assembly cost*** accounts for the cost associated with placement (of individual modules in the car), harness, part maintenance (engineering and production effort to keep track of part numbers of new hardware, and calibration sets for new software), flashing (for software), and end-of-line verification.

- The ***in-service cost*** accounts for the cost of ownership, and is incurred by the vehicle owner (resulting from repair and/or maintenance not covered by warranty), and the OEM (resulting from repair under warranty).

ENERGY EFFICIENCY ANALYSIS

The energy efficiency metric compares architectures on the basis of the net electrical power consumed. Three different operation modes are considered: (1) none of the systems is running (i.e. energy spent on parasitic current), (2) a partial network (set) of resources is running (e.g., energy spent on driving mode such as parking is different from the energy spent while cruising), and (3) all systems are running (e.g., energy spent in operation). The currently available methods of estimation are ***benchmarking*** [16], ***mathematical modeling*** of power usage by individual components (e.g. hardware [18] or software [2] components), ***analyzing trends***

[11] to provide estimations on possible future bounds on energy usage, and ***black box modeling*** (testing power usage of an electronic module).

COMPLEXITY ANALYSIS

Classical definitions of complexity tend to express critical elements in a design and their relationships; e.g. algorithmic complexity relates the number of inputs to execution time. In line with this concept, we define complexity to be a set of critical elements of a design (e.g., communication protocol types, calibration parameters, software size etc), and the inter-relationship (e.g., relation of communication protocol to calibration parameter and software size); however assignment of lower/higher complexity may be relative/qualitative and vary over design projects. There are also ongoing works in automotive domain [8] to address complexity by identifying design properties such as hierarchy, modularity, encapsulation, interfaces, cohesion, and testability.

BACKWARD COMPATIBILITY ANALYSIS

The backward compatibility metric analyzes the compatibility of an architecture to legacy designs and data. The term compatibility may have different connotations depending on whether functional, hardware, software, or a communication architecture is being compared to. A functional architecture is compatible to another functional architecture if both implement identical applications and/or have similar structure/partition/allocation. A software architecture is compatible to another software architecture if both have similar interfaces and provide similar execution environment. A communication architecture is compatible to another communication architecture if any message mapped on the former one can be mapped to the latter one and vice versa. A hardware architecture is compatible to another hardware architecture if both have similar components and/or have similar functional properties (e.g. speed, reliability etc). The comparison can either be qualitative or quantitative.

ORGANIZATIONAL ALIGNMENT ANALYSIS

The organizational alignment metric analyzes the alignment of an architecture to the organizational structure of the OEM. The analysis evaluates how different organizations within the OEM are responsible for different parts of the design (e.g. requirement capture, software development, hardware development etc). In case of new technologies, separate organizational roles need to be defined; e.g. introduction of FlexRay in the next generation ECS architectures requires different groups for FlexRay related development and implementation.

PACKAGABILITY ANALYSIS

The packagability metric is a measure of the freedom to place/redistribute hardware modules inside the car; e.g. an architecture based on wireless communication allows more styling opportunities than wire based communication (which requires harness and related hardware connections). A possible evaluation is a measure of space requirement/ constraints for introducing new architectural elements; e.g., a radar should not be placed in a position where it is more susceptible to damage which increases packaging/warranty cost.

COMPARISON TECHNIQUES

While evaluating metrics is an important step of the process, comparing the relative values of the metrics, and selecting the best choice available is equally important. We will focus on two common techniques for selection:

• ***Pugh analysis*** [15] which consists of the following basic steps: (a) selecting a list of criterion (to which a set of alternatives are to be analyzed), (b) selecting an alternative as a baseline, (c) evaluating each design concept against a scale for each of the criteria, (d) assigning a value (better/same/ worst) for each criterion of each alternative to represent how the alternative fairs with respect to the baseline, (e) summing the assigned values of the criterion for each alternative, and (f) selecting the alternative with the best score OR analyzing the scores to check for possible room for improvement. In our case, the criteria are the metrics; and each alternative is assigned a number in relation to quality of the metric with respect to a baseline alternative.

• ***Spider charts*** [17] which "… consist of a sequence of equi-angular spokes, called radii, with each spoke representing one of the variables. The data length of a spoke is proportional to the magnitude of the variable for the data point relative to the maximum magnitude of the variable across all data points. A line is drawn connecting the data values for each spoke." In our case, each spoke represent one metric. In a mathematical sense, expressive power of Pugh analysis and Spider charts are identical.

SUMMARY/CONCLUSIONS

This paper attempts to fill a gap in the current automotive ECS architecture development process. As architecture design by an automotive OEM entails several stakeholders, there is no clear definition and understanding of how to compare design alternatives. Specifically, there is no comprehensive view of which metrics should be used, and what their definitions are. As such, we address this gap by proposing a set of the metrics for comparing ECS architecture alternatives, a related methodology, and two techniques for selecting the best alternative.

REFERENCES

1. Avizienis A., Laprie J.-C., Randell B., and Landwehr C.. *Basic Concepts and Taxonomy of Dependable and Secure Computing*. In IEEE Transactions on Dependable and Secure Computing. Vol. 1(1). pp. 11-33. 2004.

2. Chatzigeorgiou A., and Stephanides G.. *Energy Metric for Software Systems.* In Software Quality Control. Vol. 10 (4). pp. 355-371. 2002.

3. Boehm B., Clark B., Horowitz E., Westland C., Madachy R., and Selby R.. *Cost Model for future Software Life Cycle Processes: Cocomo 2.0.* In Special Volume of Software Processes and Product Measurement, Annals of Software Engineering. 1995.

4. Di Natale M., Giusto P., Pinello C., Popp P., and Kanajan S.. *Architecture Exploration for time-critical and cost-sensitive systems.* In SAE Transactions Journal of Passenger Cars: Electronic and Electrical Systems. 2007. pp 381-392.

5. Ghosal A., Kanajan S., Sangiovanni-Vincentelli A., and Urbance R.. *An Initial Study on Monetary Cost Evaluation for the Design of Automotive Electronic Architectures.* In SAE Transactions Journal of Passenger Cars: Electronic and Electrical Systems. 2007. pp 362-371.

6. Ghosal A., Sangiovanni-Vincentelli A., and Kanajan S.. *A Study on Monetary Cost Analysis for Product Line Architectures.* In SAE Transactions Journal of Passenger Cars: Electronic and Electrical Systems. 2008.

7. Giusto P., Kanajan S., Pinello C., and Chiodo M.. *A Conceptual Data Model for the Architecture Exploration of Automotive Distributed Embedded Architectures.* In Proceedings of IEEE International Conference on Information Reuse and Integration. 2007. pp. 582-587.

8. ISO 26262 Road Vehicles - Functional safety.

9. Kirchain R., and Field F.. *Process-Based Cost Modeling: Understanding the Economics of Technical Decisions.* In Encyclopedia of Material Science and Engineering. 2001.

10. Laprie J.C.. *Dependable Computing and Fault Tolerance: Concepts and Terminology.* In Proceedings of IEEE International Symposium of Fault-Tolerant Computing. 1985. pp. 2-11.

11. Leen G., and Heffernan D.. *Expanding Automotive Electronic Systems*. In IEEE Computer. Vol. 35 (1). pp.88-93. Jan 2002

12. Liu C. L., and Layland J. W.. *Scheduling algorithms for multiprogramming in a hard-real-time environment.* Journal of the ACM. 20(1). pp. 46-61. 1973.

13. Mohalik S., Rajeev A. C., Dixit M. G., Ramesh S., Suman P. V., Pandya P. K., and Jiang S.. *Model checking based analysis of end-to-end latency in embedded, real-time systems with clock drifts.* In Annual ACM IEEE Design Automation Conference. pp 296-299. 2008.

14. Joseph M., and Pandya P.. *Finding response times in a real-time system.* In The Computer Journal. Vol. 29(5). pp 390-395. 1986.

15. Pugh Analysis. http://thequalityportal.com/q_pugh.htm

16. Rivoire S., Shah M. A., Ranganathan P., Kozyrakis C., and Meza J.. *Models and Metrics to Enable Energy-Efficiency Optimizations.* In IEEE Computer. Vol. 40(12). pp 39-48. 2007.

17. Spider Charts. http://www.itl.nist.gov/div898/handbook/eda/section3/starplot.htm

18. Wang A. Y., and Sodini C. G.. *On the Energy Efficiency of Wireless Transceivers.* In IEEE International Conference on Communications. Vol. 8. pp 3783-3788. 2006.

19. Zeng H., Di Natale M., Giusto P., and Sangiovanni-Vincentelli A.. *Stochastic analysis of CAN-based Real-time Automotive Systems.* In IEEE Transactions on Industrial Informatics. Special Section on In-Vehicle Embedded Systems. Vol. 5(4). 2009.

20. Bass L., Clements P., and Kazman R.. *Software Architecture in Practice.* Pearson Education. 2003.

21. Kazman R., Bass L., Abows G., and Webb M.. *SAAM: A method for analyzing the properties of software architecture.* In Proceedings of the International Conference on Software Engineering. pp 81-90. 1994.

22. Kazman R., Klein M., Barbacci M., Longstaff T., Lipson H., and Carriere J.. *The architecture tradeoff analysis method (ATAM).* ICECCS, 1998.

23. Dolan T. J.. *Architecture assessment of information-system families.* Ph.D. Thesis. Department of Technology Management. Eindhoven University of Technology. SEI, 2002.

24. Bengtsson P., Lassing N., Bosch J., and Vliet H.. *Architecture-level modifiability analysis (ALMA).* In Journal of Systems and Software. Vol. 69(1-2). pp 129-147, 2004.

Connected Vehicle Accelerates Green Driving

2010-01-2315
Published
10/19/2010

Tsuguo Nobe
Nissan Motor Co., Ltd.

ABSTRACT

After the turn of the century, growing social attention has been paid to environmental concerns, especially the reduction of greenhouse gas emissions and it comes down to a personal daily life concern which will affect the purchasing decision of vehicles in the future.

Among all the sources of greenhouse gas emissions, the transportation industry is the primary target of reduction and almost every automotive company pours unprecedented amounts of money to reengineer the vehicle technologies for better fuel efficiency and reduced CO2 emission.

Besides those efforts paid for sheer improvements of genuine vehicle technologies, NISSAN testified that "connectivity" with outside servers contributed a lot to reduce fuel consumption, thus the less emission of GHG, with two major factors; 1. detouring the traffic congestions with the support of probe-based real-time traffic information and 2. providing Eco-driving advices for the better driving behavior to prompt the better usage of energy.

This article explains how the connected vehicle via network realized the reduction of fuel consumption and, thus, CO2 emission in real-life deployments in Japan and China.

INTRODUCTION

Nissan is pursuing technical developments and commercial deployments of various kinds of Telematics solutions for the improvement of environmental concerns toward ecological use of energy and less emission of CO2.

This article refers to the effectiveness of Dynamic Route Guidance (DRG) and Eco Drive Advice (EDA) deployed for CARWINGS, a Japanese commercial implementation of Telematics, and DRG for the world's highest density real time traffic information based on taxis in Beijing, China.

CASE STUDY #1: CARWINGS IMPLEMENTATION IN JAPAN

In Japan, governmental agencies gather traffic data mostly from infrastructure-based sensors and provide traffic information through VICS[1] for the use of general vehicles. In addition to this governmental traffic information, with consents from vehicle owners, major Japanese vehicle companies gather, mostly via either customers' carry-in cellular phones or embedded cellular phone modules, and analyze anonymized probe data, which contains location-related data with time stamps, to obtain broader, detailed, and nearly real-time traffic information.

With using probe data in conjunction with VICS data, NISSAN started commercial implementation of DRG, with statistical analysis to forecast the near-future traffic congestions in 2006. Based on our experiments in Kanagawa Prefecture in Japan, we obtained the data which showed that a vehicle with a DRG-enabled Navigation System increased the average speed by 25% and **reduced the emission of CO2 by 17%**, when compared with a plain vanilla Navigation System without any traffic information as shown in Figure 1. (Red bar v.s. Green bar)

In addition to these DRG effects, NISSAN provides Eco-Driving Advice (EDA), as shown on Figure 2 and Figure 3, for drivers and proved that further improvement of fuel consumption would be attained. This is an implementation based on our findings that the differences of customer's driving habit yield diversified levels of fuel efficiency.

For the reference of drivers, we report the up-to-date status of EDA result on drivers' Web site as well as on Navigation

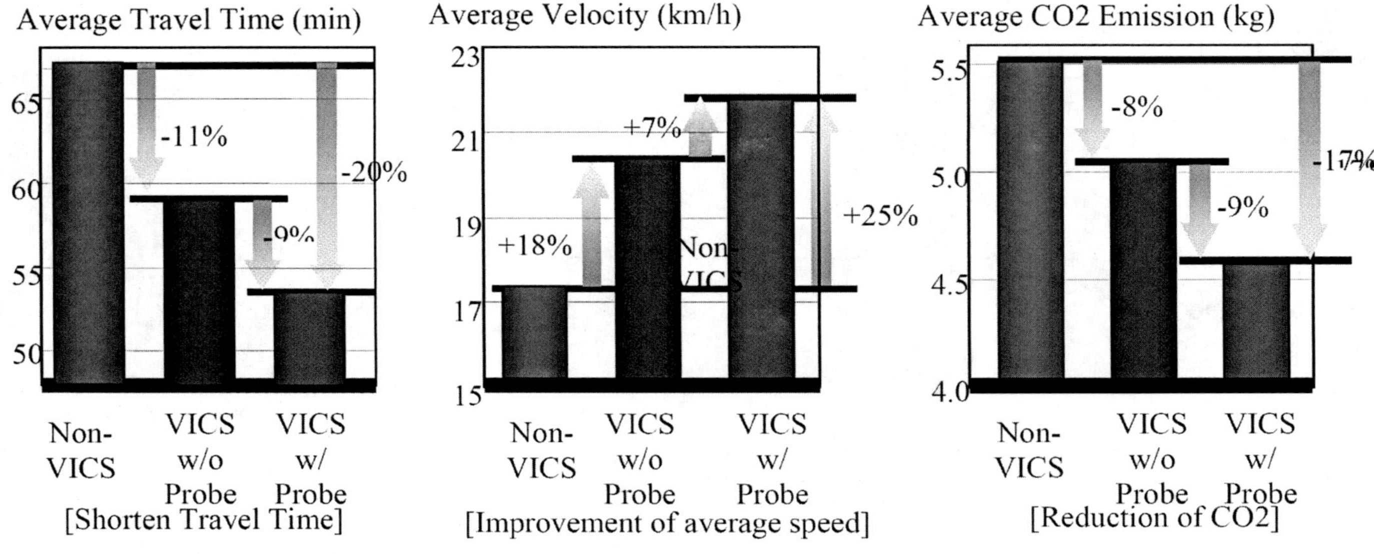

Figure 1.

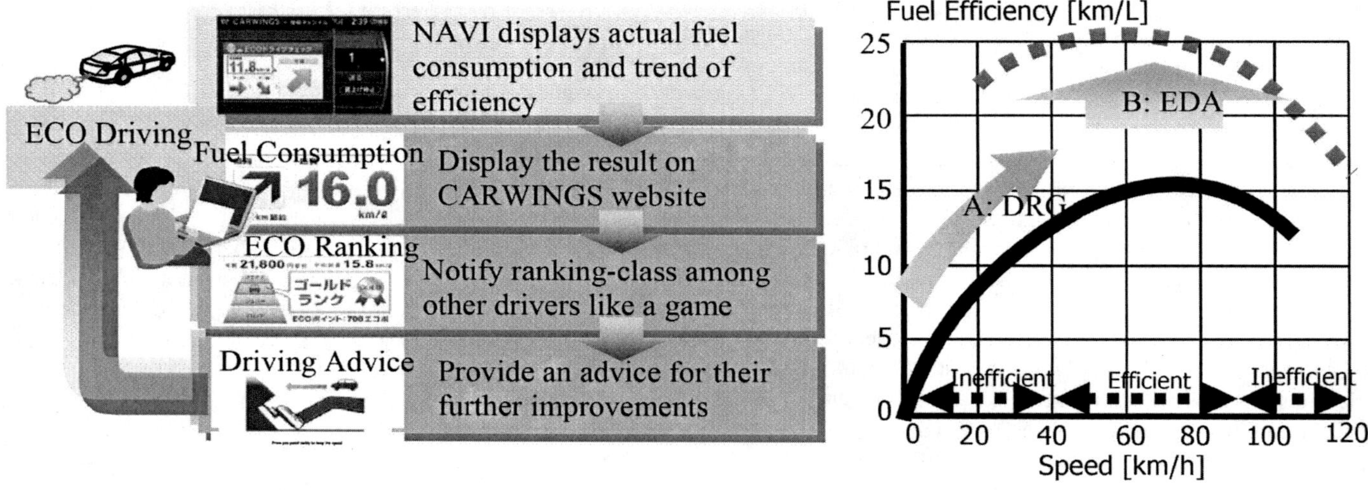

Figure 2.

since July '07 in Japan as a packaged service of the Eco Management System (EMS).

Such a daily awareness through Eco-related activities will contribute a lot for the improvements of real life Eco Management. As a result of those EMS activities, we testified through our examinations that **EDA would bring in additional 18% of improvements, even after DRG,** on Fuel Efficiency among controlled group.

As shown on the right side of Figure 2, NISSAN proved the effective solutions toward Green Driving with those combination of A: DRG and B: EDA and was awarded with two major ECO prizes from Ministry of Land, Infrastructure, Transport, and Tourism in 2007 and from Ministry of International Trade and Industry consecutively in 2008, 2009, and 2010.

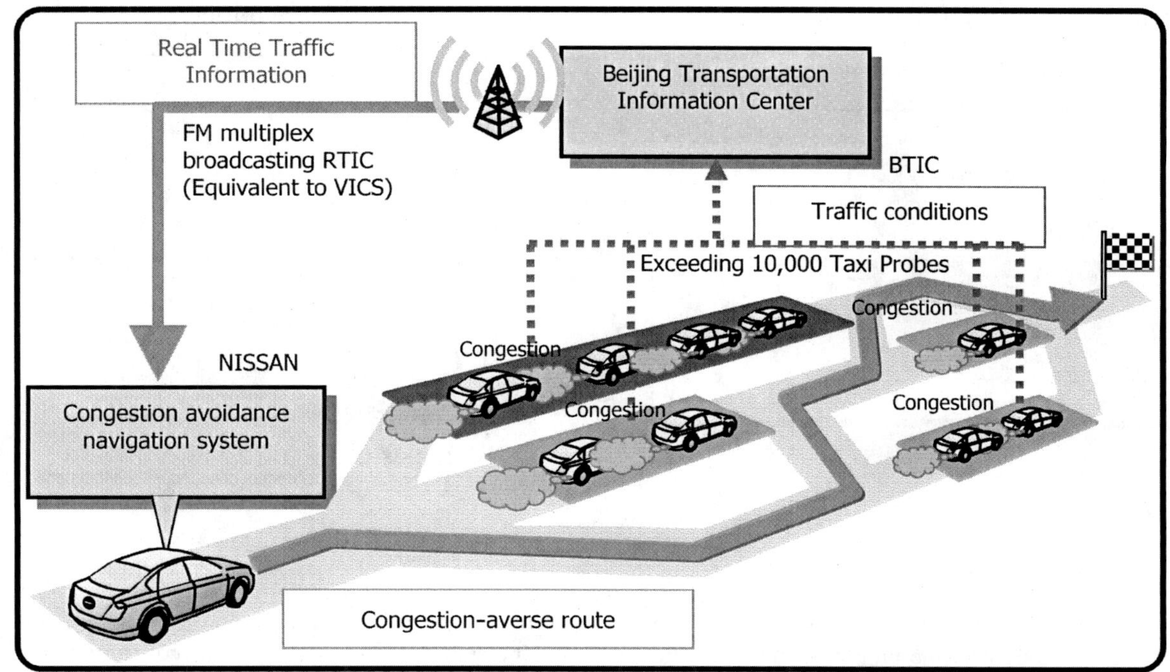

Figure 3.

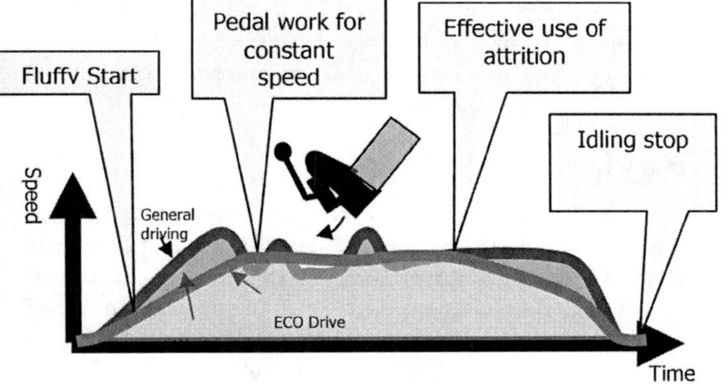

Figure 3.

CASE STUDY #2: STAR WINGS IMPLEMENTATION IN BEIJING, CHINA

In the year of 2006, NISSAN started to discuss with Beijing Traffic Information Center (BTIC) regarding the use of taxi-based real-time traffic information to improve the traffic and environmental conditions in Beijing. Within this collaboration, BTIC provided their traffic information processing technology with the world's highest density traffic information[2] gathered from taxis, while NISSAN provided Traffic Forecast Technology based on our cutting-edge Navigation & Telematics Technology.

Through this collaboration, NISSAN testified that **18 percent of travel time was reduced** on average in the real traffic conditions in Beijing in July 2007 and demonstrated this total solution, named as STAR WINGS, at Beijing ITS-World Congress in Oct. 2007.

At the Beijing Motor Show in Apr. 2008, NISSAN announced the Navigation System with STAR WINGS, which utilized the BTIC's Traffic Information received via FM multiplex broadcasting, with a capability of DRG to avoid congestions, and then commercially launched NISSAN's new passenger vehicle called TEANA with the DRG-enabled the Navigation System in China. (Figure 3)

In addition, in Sep. 2009, NISSAN explored the possibility to apply **EDA in the Chinese market** and implemented an educational driving course for trial in Shanghai to find **15%, on average, of improvements** of fuel efficiency among those who took the course.

TRAFFIC FLOW SIMULATION FOR 2020 IN BEIJING

In the real market, penetration of Navigation Systems with DRG capability takes a long time[3], because the take rate of those devices at the point of new car sales would be just a small percentage, especially in a non-Japanese market, and

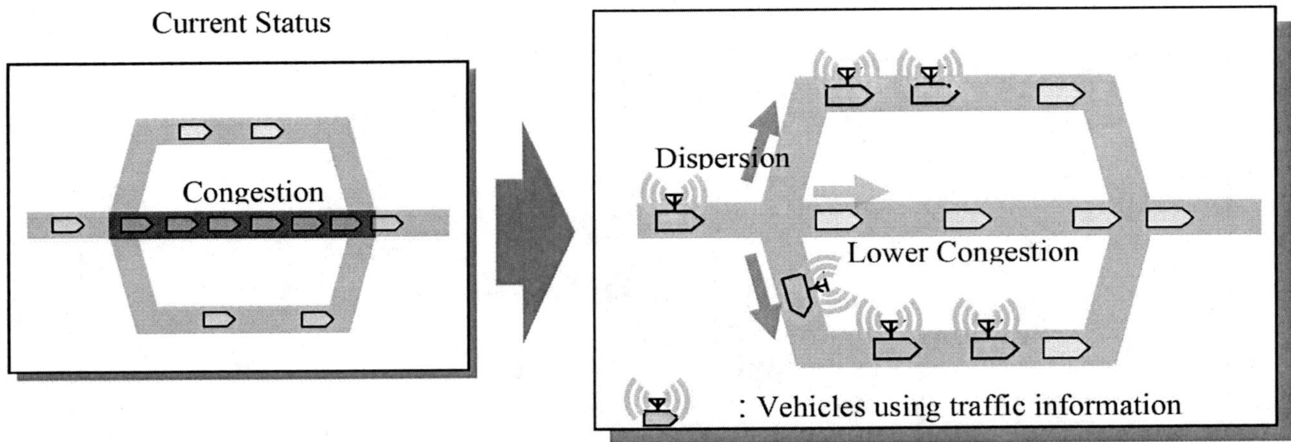

Figure 4.

any vehicle stays in the market for about a decade. So, in order to evaluate the long-term effectiveness of DRG, we simulated the effects of DRG for the improvements of traffic flow in Beijing by analyzing the situation of **available road network, congestion rate of vehicles, driver's route selection behavior**, and **market penetration rate of DRG-enabled Navigation Systems** to receive traffic information such as STAR WINGS toward 2020.

We defined the effect of improved traffic conditions as:

[Effect of improved traffic conditions] = ***[available road network]*** x ***[congestion rate of street]*** x ***[driver's route selection behavior]*** x ***[penetration rate of DRG-enabled Navigation Systems]***

Here;

- ***Available road network*** has been forecasted based on the Beijing's Future Road Plan in 2020. The improved road infrastructure provides more available alternative routes for the better traffic conditions.

- ***Congestion rate of street*** was assumed from the increasing rate of 5% per year considering Beijing's policy to shift for public transport system.

- ***Route selection behavior*** was analyzed from the interviews[4] of two thousand professional taxi drivers in Beijing.

- ***Penetration rate of DRG-enabled Navigation Systems*** is set as variable.

As shown in Figure 4, we simulated how the traffic information will disperse the traffic congestions depending on the market penetration rate of DRG-enabled Navigation Systems among total vehicles.

We conducted the simulations in four selected areas in Beijing as shown in Figure 5. For this simulation, Beijing University of Technology and Hiroshima University are joining in STAR WINGS project, within which Beijing University of Technology, in collaboration with BTIC, provided us with traffic flow analysis and its forecast; NISSAN provided our expertise in DRG technology using real-time traffic information; and Hiroshima University provided us with the analysis of driver behavior and its modeling.

The in-vehicle system is rather simple and shown below in Figure 6. Most of the data processing and analysis would be done on servers on the network.

Result of the simulation

Based on our simulation in the area of CBD (Figure 7), those vehicles using DRG experienced the improvement of vehicle speed by 12km/h on average. This improvement level was almost constant and independent from the penetration rate of DRG-enabled Navigation Systems. On the other hand, the average speed of vehicles even without traffic information will also be improved due to the broader optimization of traffic flow. As for the average, if the penetration rate of DRG-enabled Navigation Systems exceeds 30%, overall traffic speed will be improved from 9km/h to 15-20km/h.

A similar effectiveness was found with the simulation in the area of Zhong Guan Cun as shown in Figure 8. Improved vehicle speed was 7km/h for the vehicles using DRG-enabled Navigation System and also would improve the vehicle speed for the vehicles not using traffic information as the penetration rate of the devices goes higher. 30% penetration of rate of DRG-enabled Navigation Systems contributes effectively to improve vehicle speed from 10 to 16km/h in overall.

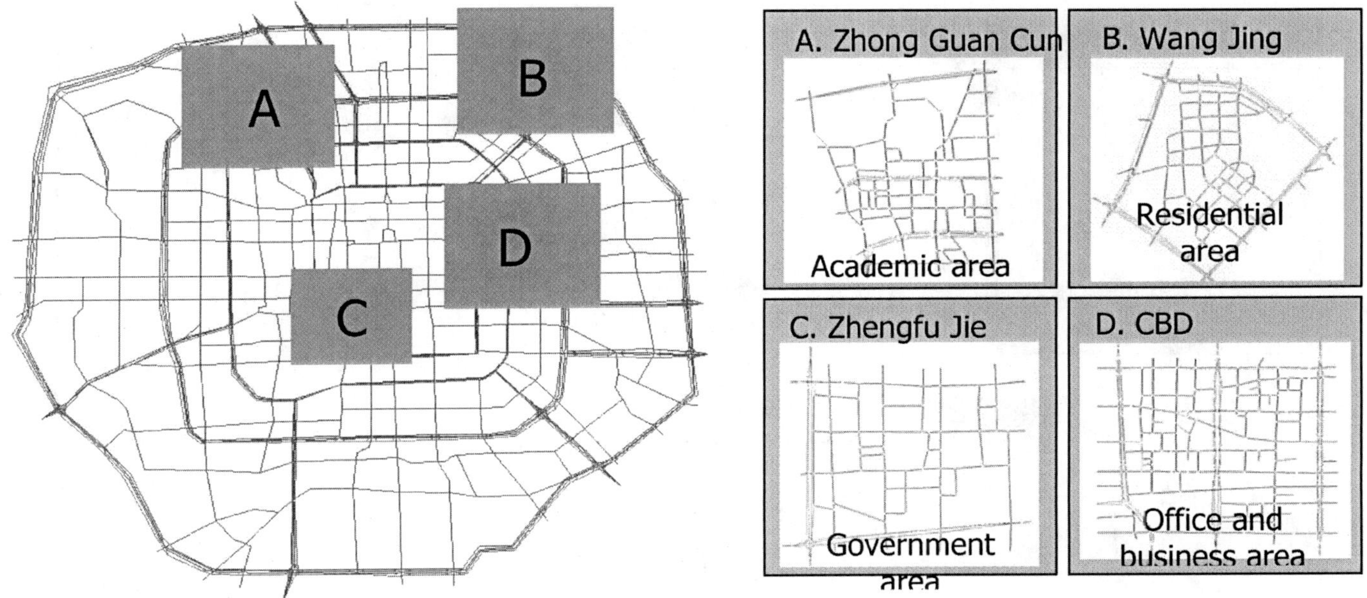

Figure 5.

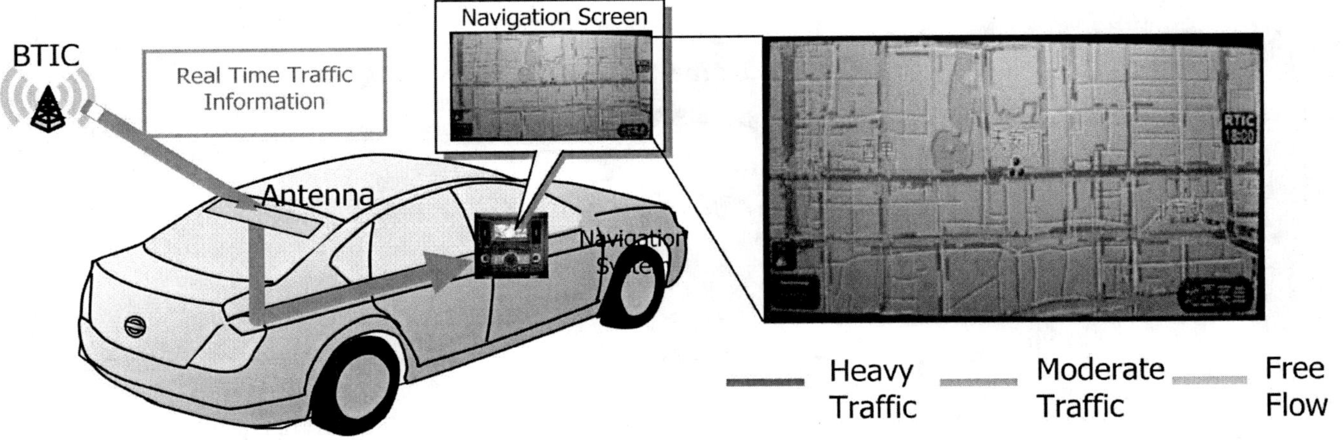

Figure 6.

Summery of the simulation

We verified that STAR WINGS is effective to improve traffic flow and, with 30% of penetration rate of DRG Navigation Systems, overall traffic flow would reach an optimized level.

SUMMARY/CONCLUSIONS

- Dynamic Route Guidance (DRG) helps drivers to avoid traffic congestions and results in shorter travel time up to around 20%[6] with better fuel usage and, thus, reduction of CO2 emission.
- If the penetration rate of DRG-enabled Navigation Systems exceeds 30% among the total number of vehicles, overall traffic flow will be optimized.
- Eco-driving Advice (EDA), which changes the human factor of vehicle driving, has been proved efficient for the better fuel efficiency and ecologically friendly driving in Japan and China. With more penetration of navigation systems and communication devices worldwide, further implementation of Green Driving would be pursued.

REFERENCES

1. VICS: Vehicle Information and Communication System. "VICS is an innovative information and communication system, enables you to receive real-time road traffic

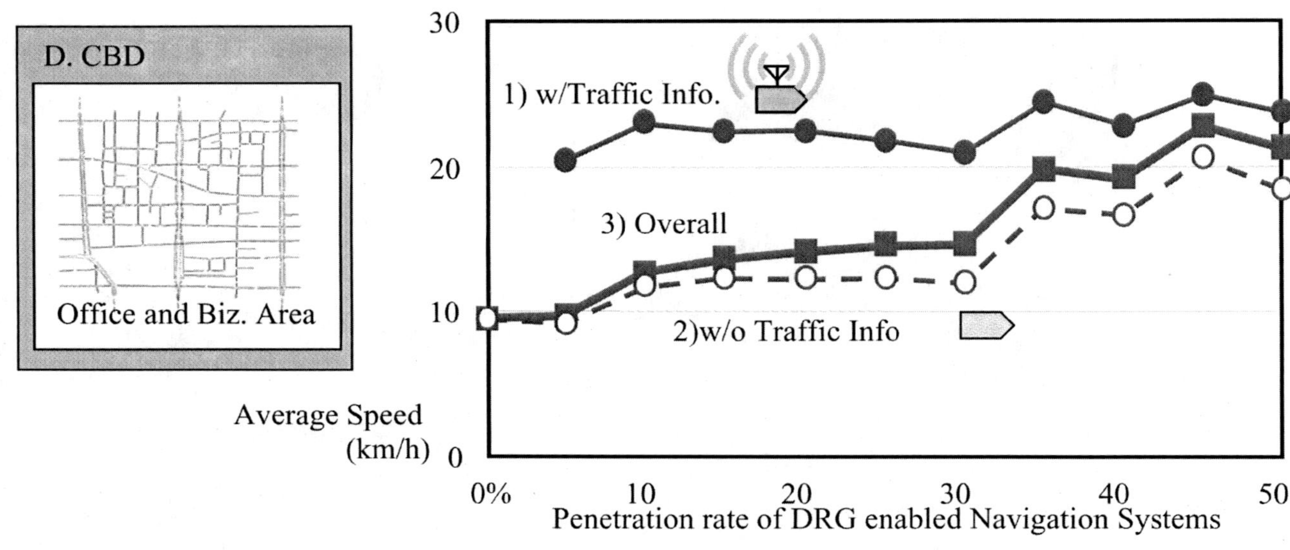

Figure 7.

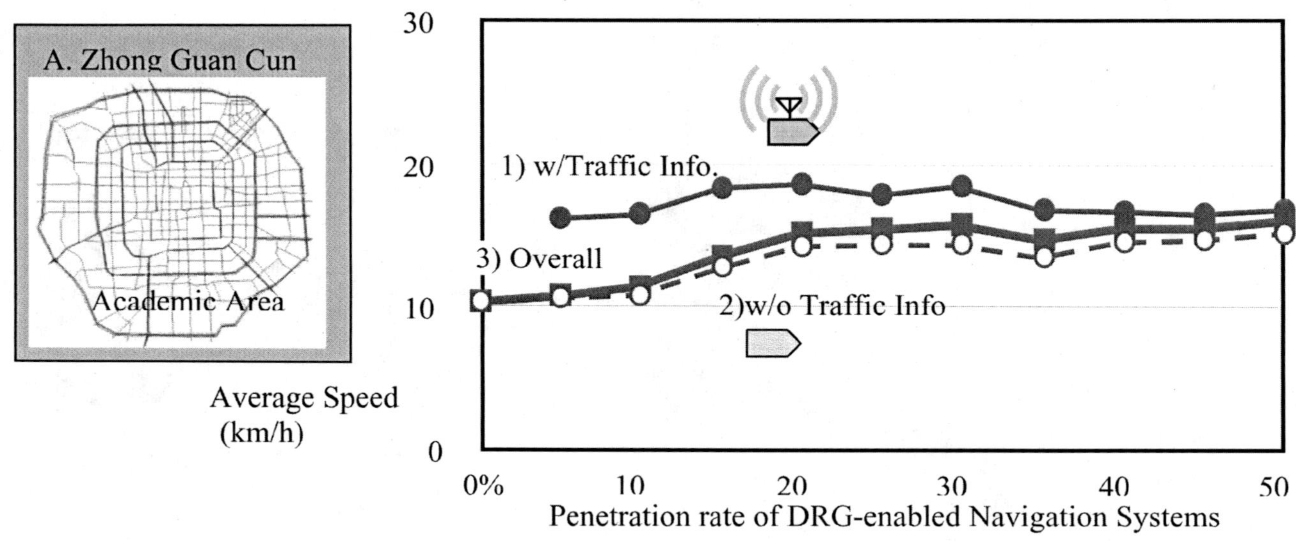

Figure 8.

information about congestion and regulation. This information in edited and processed by Vehicle Information and Communication System Center, and shown on the navigation screen by text or graphical form. You can receive information 24 hours a day, everyday. (from VICS home page: http://www.vics.or.jp/english/vics/index.html)"

2. The density of traffic information availability is 70% in Beijing, while 24% in Tokyo and 23% in Paris.

3. Japanese traffic information service, VICS, took 12 years to reach the penetration rate of 27%.

4. Through the interviews, we found that, if the travel time forecast suggests 10-15mins of trip time reduction, drivers will use traffic information and take alternative route. My additional calculation resulted in as follows; ***Y=8.56 × log(X)***, where X is a number of minutes to make a trip on a given route and Y is a number of expected minutes when alternative route would be taken. If Y is smaller than 8.56 × logX, drivers may take the alternative route.

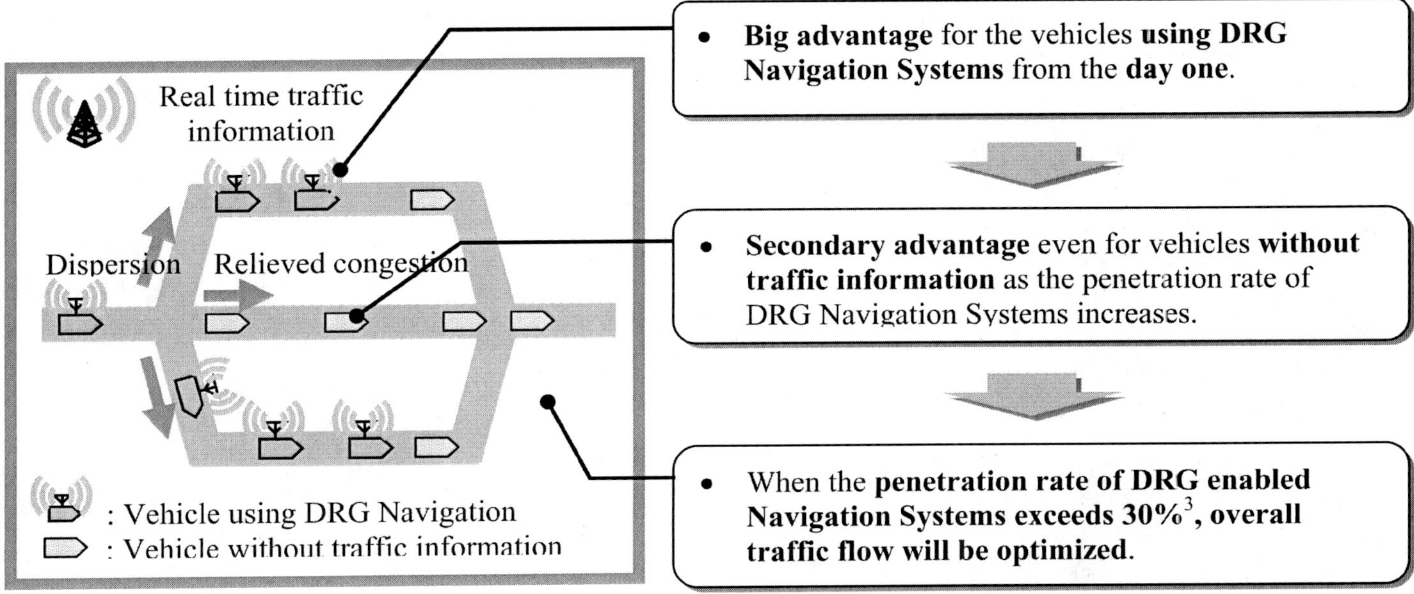

Figure 9.

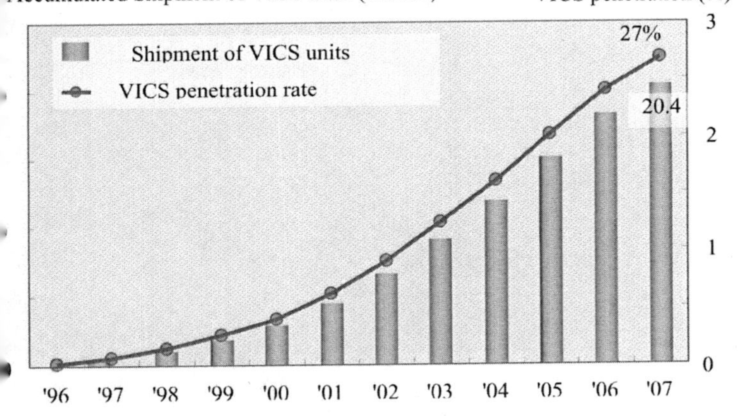

5. 30% of penetration of traffic information receivers is equivalent to shipments of three million units.

6. NISSAN made similar experiment in the US and obtained the reduction of travel time by 16% and fuel consumption by 8% on average. The experiments were done in Michigan and Virginia in July and August in 2008 and California in September 2009.

CONTACT INFORMATION

Tsuguo Nobe
Program Director
Vehicle Information Technology Business Unit
NISSAN MOTOR CO., LTD.
Kanagawa 243-0192, Japan
Phone: +81-50-2029
Fax: +81-46-270-3434
nobe@mail.nissan.co.jp

ACKNOWLEDGMENTS

Tsuguo Nobe would like to thank members of NISSAN Research Strategy Office and IT&ITS Technology Planning Group for their support while writing this paper.

DEFINITIONS/ABBREVIATIONS

CARWINGS

NISSAN'S Telematics service started in 2002 in Japan. CARWINGS service is available for NISSAN customers who purchased Make-option Navigation Systems for free of charge, except for Operator Service which are free only for the first three years.

BTIC

Beijing Traffic Information Center

CBD

Central Business District

DRG

Dynamic Route Guidance

EDA

Eco-Drive Advice

EMS

Eco-Management System

GHG

Green House Gas

VICS

Vehicle Information and Communication System

Communications - Vehicular Safety

Enabling Safety and Mobility through Connectivity

2010-01-2318
Published
10/19/2010

Chris Domin
Ricardo Inc.

ABSTRACT

Vehicle-to-Vehicle (V2V) and Vehicle-to-Infrastructure (V2I) networks within the Intelligent Transportation System (ITS) lead to safety and mobility improvements in vehicle road traffic. This paper presents case studies that support the realization of the ITS architecture as an evolutionary process, beginning with driver information systems for enhancing feedback to the users, semi-autonomous control systems for improved vehicle system management, and fully autonomous control for improving vehicle cooperation and management. The paper will also demonstrate how the automotive, telecom, and data and service providers are working together to develop new ITS technologies.

INTRODUCTION

A primary goal of ITS is to provide substantial benefits in real world fuel economy, road congestion, and general road safety. ITS has its roots in leveraging leading edge technologies, beginning with driver-focused applications, building towards semi-automatic operations, and ultimately arriving at autonomous operations.

FROM SIMPLE FEEDBACK TO AUTONOMOUS SYSTEMS

A number of passive information systems are available to drivers today. Nowadays it is common for basic driver information systems to provide some kind of vehicle status relative to the environment. For instance, a basic collision warning system can alert the driver if there is an impending rear-end collision or if it detects a pedestrian obstacle. Similarly, a travel information system can alert the driver for upcoming road obstructions. By combining information from the immediate environment with longer-range environment, higher degrees of fuel economy and safety are achieved.

As an example of fuel-economy and safety improvements, Ricardo UK Limited with its academic, business, and ITS committee partners, have developed in-vehicle applications to provide feedback to the driver about fuel-economy and safety conditions. Currently available technologies, such as GPS, cell phone, back-office systems, are used. Two recent programs of note are

• Foot-LITE - A smart electronic co-pilot provides fuel economy information and impending economy-changing situations to the user. A small portable display unit is connected to ITS infrastructure and provides the vehicle driver real-time feedback about actual driving behaviour vs. ideal fuel-efficient behaviour. A web-based service provides historical trends for the driver and allows information sharing with other users.

• Co-Driver - An electronic hazard warning system provides situational awareness of road safety conditions and impending hazards. An in-vehicle unit processes hazard information such as steep grade, sharp curves, obstructions, etc., and provides advanced warning of the potential hazard. Co-Driver also indicates the degree of urgency the hazard presents to the user, for instance, a fallen tree across road versus routine road construction markers. Vehicle passengers can easily enter information back to the system in order to alert other drivers of transient hazards, such as obstacles in the road or accidents.

Applications such as these help the driver make decisions about driving habits and navigation. Reduced fuel costs, reduced CO_2 emissions, and safer driving are direct benefits of these technologies. In addition, even further advances are possible by utilizing semi-autonomous and autonomous technology.

CASE STUDY #1: SEMI-AUTONOMOUS "SENTIENCE"

Fuel efficiency can be improved by integrating topographical and geophysical data with automatic vehicle control subsystems. *Sentience* is a recently completed 2 ½ year collaborative R&D program that was co-funded by innovITS[1]. It was jointly developed with six European partners: Ricardo, innovITS, Jaguar/Land Rover/Ford, Ordnance Survey, Orange, and TRL. The overall achievement of this program is the identification and development of a system to improve the fuel efficiency of vehicles using "electronic horizon" data collected with V2I communications. *Sentience* performs intelligent speed adaptation based on situational awareness.

Sentience is built using a web-based server and mobile client application. The server environment includes the telecommunications infrastructure, GPS satellites, weather data, ITS traffic data, historical traffic trend data, and the *Sentience* application web-server. The server translates data from the environment, categorizes them, and communicates them to the client using V2I.

The *Sentience* client application resides in a smart-phone mobile device that is part of the *Sentience* on-board system. The *Sentience* on-board system includes the GPS receiver, the mobile device (cell phone), and real-time supervisory controller unit (SCU). The vehicle interface software and supervisory control algorithms execute on the SCU. The SCU software communicates directly with the acceleration/braking subsystem electronics and over dedicated Ethernet with the client software on the mobile. The SCU software is responsible for optimization of regenerative-braking, air-condition boosting, and EV mode operations.

The team selected a Ford Hybrid Escape as the target vehicle for the prototype system. A hybrid vehicle presents several opportunities not available on conventional vehicles. The Ford Escape is a full hybrid vehicle and can operate in several modes: full electric, conventional combustion, and mixed (parallel) mode. It also utilizes regenerative braking.

SENTIENCE REQUIREMENTS AND SIMULATION

Phase one of the project focused on simulation and requirements specification. Ordnance Survey, Orange, and TRL focused on defining the vehicle routes and supporting data. Ricardo focused on simulation, control strategies, and prototype architecture. The team created and validated a vehicle model to assess baseline vehicle performance.

The primary opportunities found for energy savings are regenerative braking, EV usage during acceleration, and air-conditioning usage. Through measurement and analysis, the team found when it was best to run the electrical motor and when it was best to charge the battery based on road conditions and vehicle characteristics. This analysis allowed selection of optimum tradeoff points between electrical drive and conventional drive for vehicle speed and wheel torque.

U.S. EPA cycles were used to validate the model. Subsequent simulations included different route profiles and varying drive conditions, such as level or hilly routes, constant or varying speeds, with/without air-conditioning, head/tail wind, etc.

- Flat 12.4 mi route, 60mph
- 12.4 mi route (0.6mi flat, 3.7 mi uphill, 7.5 mi downhill, 0.6 mi flat), 60mph
- Flat 12.4 mi route, 30mph, air-conditioning turned on
- Flat 12.4 mi route, variable speed (multiple discrete target speeds) with average of 30mph
- 12.4 mi route (0.6 mi flat, 11.2 mi of repeated alternate 0.6 mi uphill, 0.6 mi downhill gradients, 0.6 mi flat to end), 30mph

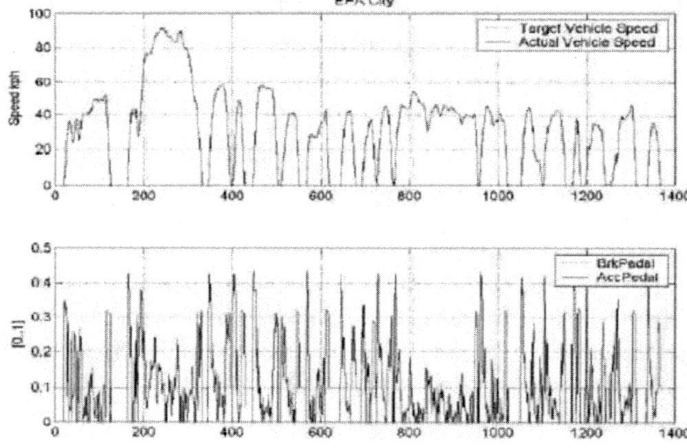

Figure 1. EPA Test Cycle

The on-board *Sentience* architecture incorporates V2I communications to access electronic horizon data, such as topographical, geographical, and traffic data, from the *Sentience* web server. The team performed a sensitivity analysis of the look-ahead algorithm to characterize how deep the queue of traffic/map data must be in order to maximize efficiency of the algorithms and to account for temporary interruptions of service. As a result, the on-board *Sentience*

[1, 2] InnovITS is the UK Centre of Excellence for sustainable mobility and intelligent transport systems. See http://www.innovITS.co.uk Simulink® is a registered trademark of The MathWorks™.

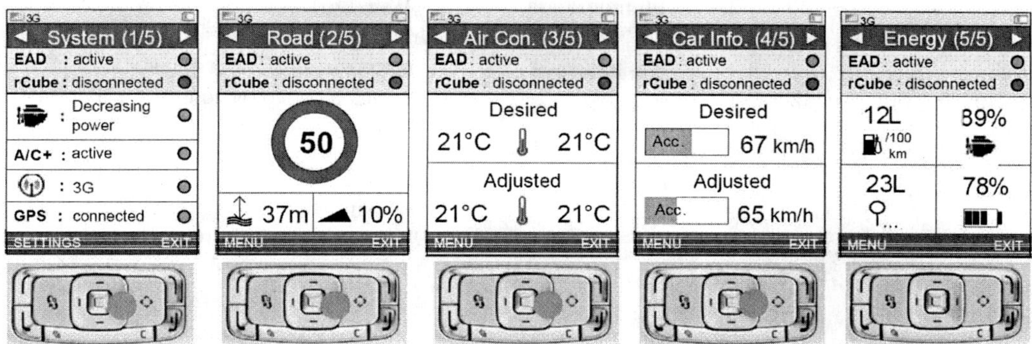

Figure 2. Sentience HMI sample

system views an electronic horizon of up to 3 miles to calculate optimum acceleration/regeneration potential.

A key requirement of the on-board system was a safe implementation with minimal cost. For this reason, a Ricardo rapid prototyping system was selected for the SCU. The SCU intercepts and overrides controls for cruise control and air conditioning. It communicates with the powertrain controller for hybrid, engine, and safety functions. A safety cut-off function for the enhanced acceleration and deceleration was identified as being required during the preliminary safety analysis.

The team, with input from Ford, assessed vehicle systems for suitability, and concluded that a small amount of additional hardware was required to ensure the vehicle system did not raise faults against the cruise control or air conditioning switchgear.

SENTIENCE DEVELOPMENT AND ASSESSMENT

Phase two of the project focused on development, integration, and assessment.

A Nokia N95 cellular phone served as the mobile communications device and human-machine interface (HMI) for the on-board system. For convenience, an external GPS was connected to the phone to provide location information. Ricardo and Orange defined and implemented a telecommunications protocol for communication between the phone and the SCU. Ordnance Survey data provided the historical traffic data. For future use, the *Sentience* architecture supports the of real-time traffic data from ITS infrastructure sources. *Sentience* focuses on three main areas of system operation to optimize energy storage and transfers: engine loading, air-conditioning, and acceleration/braking. See discussions on OEL, EAC, and EAD below.

The *Sentience* HMI on the mobile device displays road information as well as *Sentience* status information, e.g,

- Road speed limit, height and gradient
- Enhanced Air-Conditioning level desired and adjusted temperature set-point
- Enhanced Acceleration/ Deceleration level desired and adjusted vehicle speed
- Energy status information such as battery state of charge

Sentience detects when the vehicle is approaching significant changes in driving conditions due to traffic or geography, and displays pop-ups on the HMI. *A* configuration screen allows the user to select the desired features. The user can selectively enable and disable both pop-ups and audio messages by feature.

Sentience subsystem components are discretely installed in the vehicle under the passenger seat and in the dash. *Sentience* components include a custom harness, a modified A/C control unit, a custom cruise-control unit, the SCU, a wireless router & GPS receiver, and a CAN data logger device.

Sentience Optimized Engine Loading (OEL) executes on the SCU and optimizes the efficiency of the hybrid powertrain through intelligent management of electric, mixed and combustion modes of operation. The OEL algorithm communicates to the powertrain controller via the CAN network and provides supervisory control. Advanced knowledge of opportunities to recharge the battery system allows more flexibility in EV use, e.g., the battery state of charge limits are adjusted with the vehicle operation utilizing these limits. Ricardo developed supervisory control strategies in Simulink® for execution on the *Sentience* SCU. With the current OEL strategies, a 4-9% improvement in fuel consumption is realized.

Sentience Enhanced Air Conditioning (EAC) executes on the SCU and optimizes the A/C operation in order to reduce the CO_2 emissions from the vehicle. Because the combustion engine drives the A/C compressor directly, a specialized strategy was developed to keep the passengers comfortable during extended vehicle stops. EAC overrides the A/C switchgear signals and "pre-cools" the interior 1 or 2 degrees cooler whenever extended stops are predicted. By minimizing the amount of occurrences when the engine runs exclusively for cooling the vehicle interior, a result of a 2-10% improvement in fuel consumption is realized.

Sentience Enhanced Acceleration / Deceleration (EAD) is a form of adaptive cruise control, where vehicle speed as well as acceleration and deceleration profiles are controlled with the knowledge of future traffic and geography features. EAD augments the existing cruise control strategy. *Sentience* automatically controls the speed at a more optimal rate than might be expected through normal driving, allowing potentially significant savings in fuel. Speed set points are a combination of fixed-feature speed limits and probabilistic-feature speed limits. Fixed-features include actual speed limits, bends, roundabouts, speed bumps and stop signs. Probabilistic features include traffic lights, junctions, traffic conditions and pedestrian crossings. EAD slows or accelerates the vehicle at an optimum rate to match legal or safe speeds. The driver can manually override EAD at any time for safety or convenience. Depending on traffic conditions, EAD may have an impact on journey time; the driver therefore could make an informed decision as to whether the trade-off with increased comfort and fuel efficiency is acceptable on that occasion.

Figure 3. OEL Hilly Terrain Optimization

SENTIENCE VALIDATION AND CONCLUSIONS

1. Three new control systems were added to those on a production hybrid vehicle. OEL for enhanced hybrid system efficiency is useful under any mode of operation. EAC can be used whenever air conditioning is turned on. EAD can be used whenever cruise control is active.

2. Track and road testing results indicated significant savings. Improvements in fuel consumption on the order of 5%-10% can be obtained for a low implementation cost.

- With OEL, simulations predicted savings of 4-9%; Initial track test measurements show savings of approximately 2%, with speculation of higher percentage savings on specific routes. Further analysis and testing continues.

- With EAC, dynamometer-testing using an NEDC cycle with simulated sunlight loading has been performed. Over 9% improvement in fuel consumption was seen on an NEDC cycle under moderate mild summer weather conditions.

- With EAD, initial measurements show an average saving ranging between 5% and 24%, with an average of 12% during track testing. Scaling this data to average vehicle usage on real roads gives a total estimated fuel saving of nearly 14%. Initial real world road tests have already shown a fuel savings of over 5%.

3. Implementation costs for *Sentience* using a production system is not restrictive. Typical 3G mobile phones come with GPS capabilities and can be acquired for low cost. Memory and CPU requirements for OEL, EAC, and EAD functions do not prevent those features from being co-resident with software in production ECUs. Suitable sources of traffic data are required, but these traffic data can be easily supported by future ITS infrastructure.

CASE STUDY #2: AUTONOMOUS "SARTRE"

Both fuel efficiency and safety are improved by integrating V2V communications and automatic vehicle control subsystems. *SARTRE* is a Ricardo-led program that shares situational awareness data between vehicles using V2I and V2V communications, thus enabling autonomous vehicle coordination and the creation of "road trains".

The *SARTRE* project began in September 2009 and is scheduled to complete in August 2012. It is being jointly developed with seven European partners in the UK, Sweden, Spain, and Germany: Ricardo, IDIADA Automotive Technology, Institute for Automotive Engineering (ika) of RWTH Aachen University, SP Sveriges Tekniska Forskningsinstitut, TECNALIA Robotiker, Volvo Car Company, and Volvo Technology. The concept behind *SARTRE* is that vehicle platoons improve fuel consumption, increase safety, and reduce congestion on freeways.

Since human driver errors contribute to well over 85% of road fatalities, it is expected that safety will improve dramatically by using autonomous control to remove distractions and errors in judgment. Because autonomous systems can process data much more quickly than a human can, congestion can be reduced automatically by optimizing gaps between vehicles, minimizing traffic dynamics and delaying traffic collapse. Fuel economy can be improved by

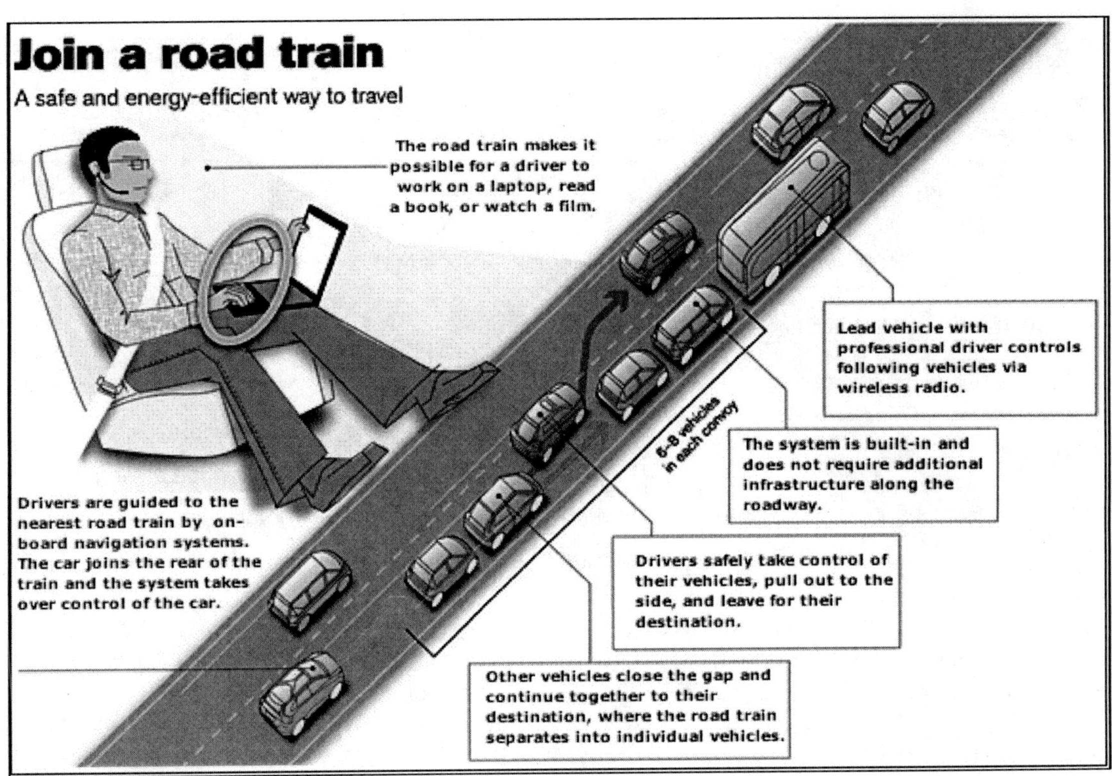

Figure 4. The SARTRE concept

reducing aerodynamic drag, due to drafting in each vehicle's slipstream.

Each road train will consist of up to eight vehicles. Each road train or platoon has a lead vehicle that drives exactly as normal, with human control over the various functions. This lead vehicle is controlled by an experienced driver who is familiar with the route. For instance, the lead may be taken by a taxi, a bus or a truck. A driver approaching the convoy requests entry into the convoy using a human-machine interface. The convoy accepts the vehicle and the vehicle automatically enters the convoy, after which it is completely under autonomous control. A driver approaching his destination leaves the convoy by exiting off to the side and then continues on his own to his destination under his own control. The other vehicles in the road train automatically close the gap and continue on their way until the convoy splits up.

The advantage of such road trains is that all the other drivers in the convoy have time to perform other business while on the road, e.g., talking on the phone, eating, working on a computer, etc. The road trains increase safety and reduce environmental impact thanks to lower fuel consumption compared with cars being driven individually. The reason is that the cars in the train are close to each other, exploiting the resultant lower air drag. Simulation results show the energy saving to be in the region of 20 %: Road capacity is utilized more efficiently by minimizing distance between vehicles.

Researchers see road trains primarily as a major benefit to commuters who cover long distances by motorway every day, but they will also be of potential benefit to trucks, buses, coaches, vans and other commercial vehicle types. As the participants meet, each vehicle's navigation system is used to join the convoy, after which the autonomous driving program then takes over. As the road train approaches its final destination, the various participants can each disconnect from the convoy and continue to drive as usual to their individual destinations.

SARTRE REQUIREMENTS AND SIMULATION

Phase one of the project considered scenarios and constraints during interaction with other road users. A use-case analysis was performed with an emphasis on the human factors. Modeling of the use cases focused on creating a combination of vehicle and traffic specific models, taking into consideration all interchanges occurring between driver, vehicle and other traffic.

An important constraint for *SARTRE* is that the architecture and implementation has to be feasible and use available production components and subsystems. So the team performed additional analysis to understand business

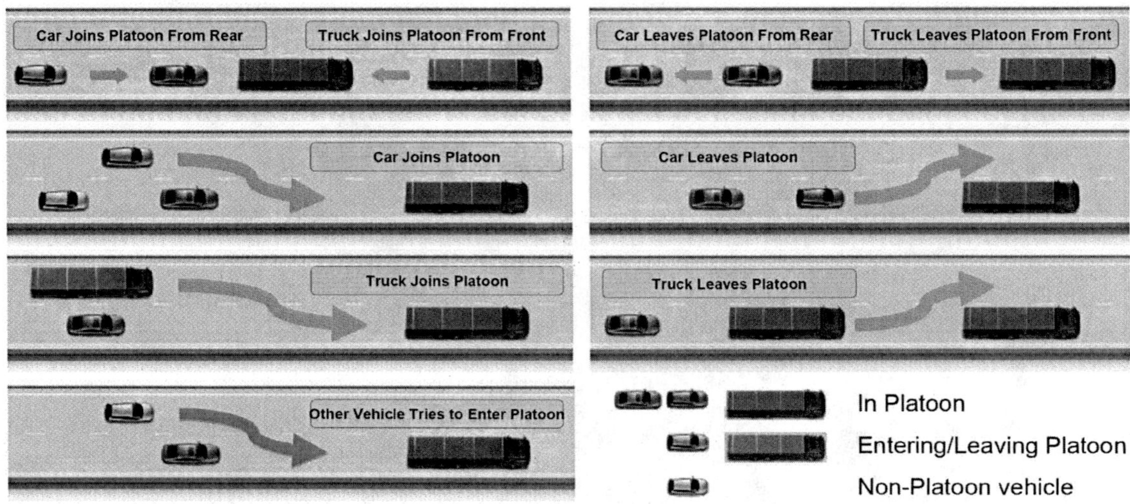

Figure 5. SARTRE Platoon Use-Cases

requirements, usability, risk, and safety, as well as the system itself. As the concept solutions were balanced against available technology, they were rationalized against draft ISO/DIS 26262 using InnovITS[2] Framework Architecture and Classification for ITS (FACITS) process.

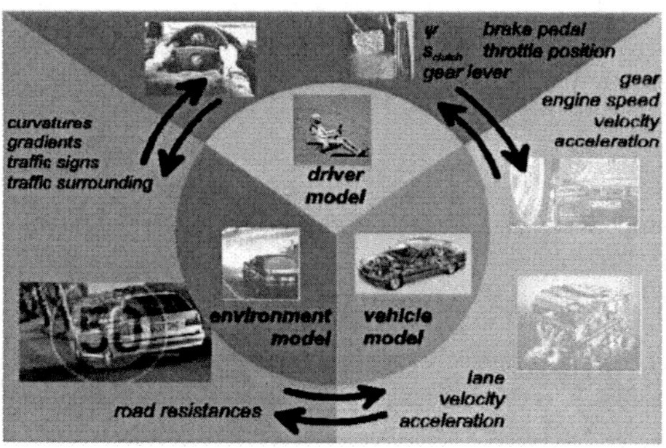

Figure 6. Modeling Process

Use-cases (see Figure 5) needed to take into account a significant number of factors, including, performance/failure of vehicles, braking/acceleration/turning procedures, other vehicles, platoon size, and gap length, and human behaviors, among others. Example use cases are:

• authorized car/truck enters platoon from rear or joins middle of platoon

• unauthorized other car/truck enters platoon or leaves platoon from middle

• authorized car/truck leaves from rear or middle

• authorized leader joins or leaves from front

After all the primary modeling, analysis, and concept generation were complete, the team focused on concept implementation.

SARTRE CONCEPT IMPLEMENTATION

Phase two of the project involves concept selection and implementation. Since intellectual property is being developed by partners to support the implementation of *SARTRE*, only a general discussion of the architecture is given here. Each vehicle is equipped with a dedicated short-range communications (DSRC) radio, an active safety control module, short-range radar, vision systems, active cruise control system, actuators, and supervisory control unit (SCU). DSRC is used to communicate platoon information among all vehicles in the platoon. Once in the platoon, V2V communications, V2I communications, and other active subsystems in each vehicle support autonomous behavior.

To date, the project partners have reached agreements on the factors necessary to proceed with implementation and the concept implementation is underway. Transport behaviour modelling and platoon strategies continue in parallel with human behaviour studies and safety studies. Development of lead vehicles and following vehicles has started. Track studies will soon be performed with the *SARTRE* road-train using three cars and two trucks.

Some of the areas for continued research and refinement are in the areas of

[1, 2] InnovITS is the UK Centre of Excellence for sustainable mobility and intelligent transport systems. See http://www.innovITS.co.uk Simulink® is a registered trademark of The MathWorks™.

• Number of vehicles in a *SARTRE* platoon and the mix of vehicles (cars/trucks)

• Specification and architecture updates.

• Safety requirements and analysis

• Updates to V2V Communications (DSRC)

• Inputs regarding V2I findings to infrastructure organizations

• Sensor Fusion Systems

• Actuator Systems

• Human Machine Interfaces

• Autonomous Control System

• Platoon Management System

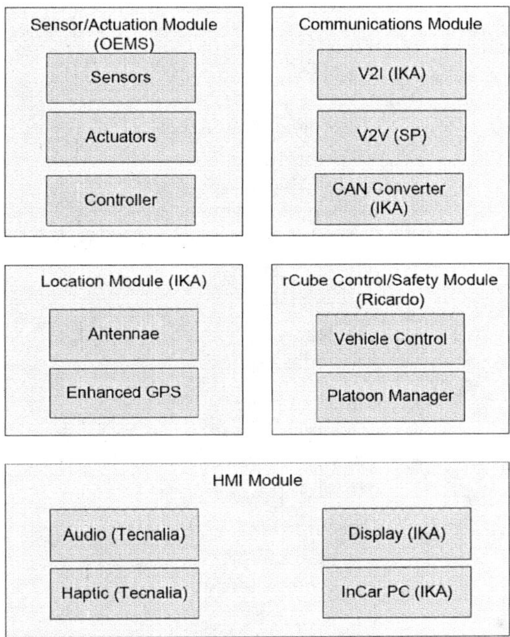

Figure 7. SARTRE Concept Architecture

VALIDATION AND ASSESSMENTS

At the time of the writing of this paper, validation of the systems has not been completed. The plan is to validate the on-vehicle systems, the remote systems, end-to end systems, and fuel consumption claims. Once validation is completed, results of studies will be made available that include assessments of the commercial viability of *SARTRE*, the net impact on infrastructure and vehicles, and potential policy impacts.

SUMMARY/CONCLUSIONS

V2V and V2I communications are changing the ways that people interact with their vehicles. Driver assistance systems are making way for semiautonomous mobility improvements in fuel economy and safety. Future automotive systems will leverage V2I and V2V in order to allow drivers to select semiautonomous and autonomous behaviors, with net gains in safety and mobility. Partnerships between science researchers, policy makers, academia, infrastructure manufacturers, and automotive manufacturers will change the landscape of automotive transportation to a more efficient and safer experience for drivers and passengers.

THE PARTNERSHIPS

The *Sentience* program included each of the following organizations.

Organization	Organization Overview	Role
Ricardo	Automotive Engineering Consultants with expertise in Control Systems, Vehicle Systems, and ITS	• Project Management • Rapid Prototyping Electonics • Vehicle control algorithms
innovITS	UK Dept of Business Enterprise Reform (formerly DTI)	• Promotes UK Telematics/ITS • Funding/Coordination • Program goals
Jaguar/LandRover/ Ford	Prestige UK Vehicle OEM	• Base Hybrid Vehicle • Vehicle Data and Interfaces
Ordnance Survey	UK mapping organization	• Enhanced map data • Traffic congestion data
Orange	Telecoms company	• Telecoms engineering expertise and equipment
TRL	Transport Research and testing Lab	• Vehicle system testing • Test facilities

The *SARTE* program included each of the following organizations.

Organization	Organization Overview	Role
Ricardo (UK)	High value engineering services to the automotive, ITS and clean energy communities	• Coordinator and Management WP leader • Safety Analysis • Platoon Management & Autonomous Control
IDIADA Automotive Technology (Spain)	World-leading company for automotive testing and demonstration	• Validation/Assessment work package leader • Test lead • Road trial lead
Institute for Automotive Engineering (ika) of RWTH Aachen University (Germany)	Leading university in automotive technology	• Concept definition WP leader • Modelling lead • Back office and organisation assistant
SP Sveriges Tekniska Forskningsinstitut (Sweden)	Research institute experienced in automotive safety and communication	• Dissemination WP leader • Use case lead • V2V communications
TECNALIA Robotiker (Spain)	Expert technology centre specialising in ICT	• Human factors assessment • HMI design and implementation
Volvo Car Corporation (Sweden)	Major passengar car OEM	• Implementation WP leader • Following vehicle lead • Following vehicle (car) sensor fusion
Volvo Technology (Sweden)	Major trucks, buses and construction equipment OEM	• Lead vehicle lead • Lead/following vehicle (truck) sensor fusion

CONTACT INFORMATION

Chris Domin
Intelligent & Autonomous Systems
Ricardo, Inc.
40000 Ricardo Drive
Van Buren Twp., MI 48111 USA
Phone: (734) 394-4155
Fax: (734) 397-6677
chris.domin@ricardo.com

ACKNOWLEDGMENTS

Ricardo, Inc. and Ricardo UK Limited would like to thank its partners for support of Foot-LITE, Co-Driver, *Sentience*, and *SARTRE* programs. Chris Domin would like to thank Tom Robinson and Jonathon Hunt from the UK facility for their support while writing this paper.

DEFINITIONS/ABBREVIATIONS

A/C
: Air-conditioning unit

CPU
: Central Processing Unit, a micro controller

DSRC
: Digital Short Range Communications

EAC
: Enhanced air conditioning

EAD
: Enhanced acceleration/deceleration

ECU
: Embedded Control Unit

EV
: Electric Vehicle

HMI
: Human-machine interface or display

ITS
: Intelligent transportation systems

OEL
: Optimised engine loading

SARTRE
: EU Program: Safe Road Trains for the Environment

SCU
: Supervisory control unit

Sentience
: EU Program: Using Electronic Horizon Data to Improve Vehicle Efficiency

V2V
: Vehicle to vehicle

V2I
: Vehicle to infrastructure

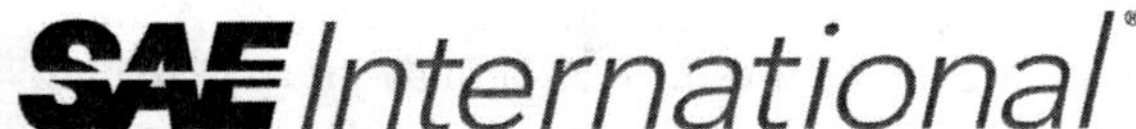

Vehicle Safety Communications - Applications: System Design & Objective Testing Results

2011-01-0575
Published
04/12/2011

Farid Ahmed-Zaid
Ford Motor Company

Hariharan Krishnan
General Motors Company

Michael Maile
Mercedes Benz REDNA

Lorenzo Caminiti
Toyota Motor Engineering & Mfg NA Inc.

Sue Bai
Honda R&D Americas Inc.

Steve VanSickle
Danlaw, Inc.

doi:10.4271/2011-01-0575

ABSTRACT

The USDOT and the Crash Avoidance Metrics Partnership-Vehicle Safety Communications 2 (CAMP-VSC2) Consortium (Ford, GM, Honda, Mercedes, and Toyota) initiated, in December 2006, a three-year collaborative effort in the area of wireless-based safety applications under the Vehicle Safety Communications-Applications (VSC-A) Project. The VSC-A Project developed and tested communications-based vehicle safety systems to determine if Dedicated Short Range Communications (DSRC) at 5.9 GHz, in combination with vehicle positioning, would improve upon autonomous vehicle-based safety systems and/or enable new communications-based safety applications. The project addressed the following objectives:

- Assess how previously identified crash-imminent safety scenarios in autonomous systems could be addressed and improved by DSRC+Positioning systems
- Define a set of DSRC+Positioning based vehicle safety applications and application specifications including minimum system performance requirements
- Develop scalable, common vehicle safety communication architecture, protocols, and messaging framework (interfaces) necessary to achieve interoperability and cohesiveness among different vehicle manufacturers. Standardize this messaging framework and the communication protocols (including message sets) to facilitate future deployment.
- Develop requirements for accurate and affordable vehicle positioning technology needed, in conjunction with the 5.9 GHz DSRC, to support most of the safety applications with high-potential benefits
- Develop and verify a set of objective test procedures for the vehicle safety communications applications

In this paper, we summarize the work that took place in the VSC-A Project in the areas of system design and objective testing. We first introduce the VSC-A system framework. We then list the crash imminent scenarios addressed by the VSC-A Project and the safety applications selected to potentially address them. Next we describe the VSC-A test bed system development. This test bed was ultimately used to verify Vehicle-to-Vehicle (V2V) communication interoperability between Ford, GM, Honda, Mercedes-Benz, and Toyota vehicles. Public demonstrations of V2V interoperability were held in New York City at the 2008 Intelligent Transport Systems (ITS) World Congress. The test bed also served to validate the system and minimum performance specifications that were developed as part of this project. We discuss one of the most important achievements of the project in the communication area, i.e., implementation, testing, verification, and standardization of a safety message that supports all of the VSC-A safety applications. The result is the Basic Safety Message (BSM) as defined in the SAE J2735 Message Set Dictionary standard. Details of the objective test procedures are presented next and are followed by a summary of the performed test scenarios (test descriptions, speeds, number of runs for each test, type of test, etc.) with the corresponding objective testing results. We conclude the paper with a section summarizing the accomplishments of the project and also identify potential next steps and recommendations based on the technical results and engineering experience gained throughout the execution of the VSC-A Project.

INTRODUCTION

Vehicle-to-Vehicle (V2V) safety communications can play a major role in addressing vehicle crashes where multiple vehicles are involved. According to [1], this technology can reduce, mitigate, or prevent 82 percent of crashes by unimpaired drivers. The communications technology for V2V is 5.9 GHz Dedicated Short Range Communications (DSRC). This wireless communications technology has a very low latency and is considered to be the technology of choice for the types of crash avoidance applications that were prototyped in the Vehicle Safety Communications-Applications (VSC-A) Project [2]. The major objectives of the VSC-A development activities were the:

- Selection of high-value safety applications
- Development of a test bed that allowed interoperability between different car manufacturers
- Development and standardization of a message set for vehicle safety communications
- Development of an accurate relative positioning system
- Prototyping of safety applications
- Objective testing of the safety applications

A primary goal of the VSC-A Project was to determine whether systems that utilized DSRC-based V2V communications and positioning can help overcome limitations of autonomous systems and enhance the overall performance of safety systems. One potential advantage of V2V safety communications is that it may provide significant, additional information about the driving situation and expand the awareness horizon of the vehicle well beyond the capabilities of vehicle-autonomous sensors. Another advantage of V2V systems is that it may be possible to integrate such systems on vehicles in which the system was not original equipment, including retrofit of existing vehicles.

In order to gauge the feasibility of such systems, a reference system and applications to address crash imminent scenarios were implemented. This reference system ("test bed") combined communications, accurate relative positioning and security and was integrated with the vehicles from the five Original Equipment Manufacturers (OEMs) that participated in the VSC-A Project. A fundamental aspect of the project was the establishment of interoperability between different OEMs. This interoperability requirement led to the development of the V2V message set, which was standardized in SAE J2735 as the Basic Safety Message (BSM) [3]. The development of the test bed and the applications followed a systems engineering process and the resulting minimum performance requirements formed the basis for the development and the testing of the applications. To test the performance of the test bed and the applications, objective test procedures were developed together with the United States Department of Transportation (USDOT) and the testing was performed at the Transportation Research Center (TRC) in East Liberty, Ohio with the aid of the National Highway Traffic Safety Administration's (NHTSA) Vehicle Research and Test Center (VRTC).

CRASH SCENARIOS AND APPLICATION SELECTION

To provide a foundation for the VSC-A Project, the USDOT evaluated pre-crash scenarios based on the 2004 General Estimated Systems (GES) crash database. This list served as the basis for the selection of the safety applications to be prototyped under the VSC-A Project. Each crash scenario was assigned a composite crash ranking determined by taking the average of the crash rankings by frequency, cost, and functional years lost for each scenario. The crash scenarios were then sorted based on the composite ranking and were analyzed to evaluate whether autonomous safety systems and/or vehicle safety communications would offer the best opportunity to adequately address the scenarios.

From this ranked list of crash scenarios (based on crash frequency, crash cost and functional years lost) the top seven (7) crash scenarios to be addressed by the VSC-A Project were selected. The selected crash-imminent scenarios were analyzed and potential, DSRC-based, safety application concepts capable of addressing them were developed. The crash imminent scenarios and the applications selected to be part of the VSC-A safety system is shown in Table 1. The VSC-A team together with the USDOT analyzed the scenarios in Table 1 and developed concepts for safety applications that could potentially address them through vehicle safety communications. This analysis resulted in the identification of the following safety applications as part of the VSC-A system:

Emergency Electronic Brake Lights (EEBL), defined as follows

The EEBL application enables a host vehicle (HV) to broadcast a self-generated emergency brake event to surrounding remote vehicles (RVs). Upon receiving the event information, the RV determines the relevance of the event and issues a warning to the driver, if appropriate. This application is particularly useful if the drivers' line of sight is obstructed by other vehicles or bad weather conditions (e.g., fog, heavy rain)

Forward Collision Warning (FCW), defined as follows

The FCW application is intended to warn the driver of the HV of an impending rear-end collision with an RV ahead in traffic in the same lane and direction of travel. FCW is intended to help drivers in avoiding or mitigating rear-end vehicle collisions in the forward path of travel.

Blind Spot Warning+Lane Change Warning (BSW+LCW), defined as follows

The BSW+LCW application is intended to warn the driver during a lane change attempt if the blind-spot zone into which the HV intends to switch is, or will soon be, occupied by another vehicle traveling in the same direction. Moreover, the application provides advisory information that is intended to inform the driver of the HV that a vehicle in an adjacent lane is positioned in a blind-spot zone of the HV when a lane change is not being attempted.

Do Not Pass Warning (DNPW), defined as follows

The DNPW application is intended to warn the driver of the HV during a passing maneuver attempt when a slower moving vehicle, ahead and in the same lane, cannot be safely passed using a passing zone which is occupied by vehicles in the opposite direction of travel. In addition, the application provides advisory information that is intended to inform the driver of the HV that the passing zone is occupied when a vehicle is ahead and in the same lane and a passing maneuver is not being attempted.

Intersection Movement Assist (IMA), defined as follows

The IMA application is intended to warn the driver of a HV when it is not safe to enter an intersection due to high collision probability with other RVs. Initially, IMA is intended to help drivers avoid or mitigate vehicle collisions at stop sign-controlled and uncontrolled intersections.

Control Loss Warning (CLW), defined as follows

The CLW application enables a HV to broadcast a self-generated, control-loss event to surrounding RVs. Upon receiving such event notification, the RV determines the relevance of the event and provides a warning to the driver, if appropriate.

Table 1 illustrates the mapping between the crash imminent scenarios and the safety applications defined above.

DEVELOPMENT OF THE TEST BED

Each OEM in the VSC-A Project developed a vehicle test bed to serve as a prototype platform for the V2V communications system. The OEMs jointly developed system specifications and performance requirements that served as the basis for the system and application developments. The test bed was based on a common prototype platform referred to as the On-Board Equipment (OBE). The selected OBE allowed development flexibility and was representative of current (or future) automotive grade processing power. The OBE contained a DSRC radio, a processor and various interfaces (e.g., for vehicle data, Global Positioning System (GPS) data, etc.). The test bed was an effective tool for validating safety application concepts, system test procedures and for answering critical research questions regarding V2V communications. Those issues included relative lane-level positioning, time synchronization, communications scalability and practical security and anonymity.

SOFTWARE ARCHITECTURE

In order to support the functionality of the safety applications described earlier and their development, the activities initially focused on the development of a system architecture based on various modules that could be upgraded independently from each other, if necessary. This approach allowed for fast and efficient prototyping throughout the development phase of the project. This architecture was used during the test bed design stage for the definition of the Hardware (HW) and Software (SW) architectures and required interfaces. The various modules forming the system test bed were categorized into the following major groups: Interface, Positioning & Security, Core, Safety Applications, Threat Process and Reporting, and Data Analysis. The system block diagram (Figure 1) shows the breakdown of the individual modules that make up each of the major module groupings. This provided a good framework for a comprehensive V2V safety system.

The focus of the system design activities was the core modules (Target Classification, Host Vehicle Path Prediction and Path History) and the positioning, security and safety application modules. The system design was based on the preliminary requirement specifications developed for each of the modules. Testing of the system resulted in updates to the modules throughout the project, culminating in the final test bed implementation. In the next section the software modules are described briefly.

SOFTWARE MODULES

The VSC-A software modules are composed of support and application functions. The support functions provide the interface to any external equipment and they calculate the necessary parameters to support the application modules and the Engineering Driver-Vehicle Interface (DVI). The primary software modules are:

- Threat Arbitration (TA)
- Driver-Vehicle Interface Notifier (DVIN)
- Target Classification (TC)
- Host Vehicle Path Prediction (HVPP)
- Path History (PH)
- Data Logger (DL)
- Engineering Graphical User Interface (EGUI)
- Sensor Data Handler (SDH)
- Wireless Message Handler (WMH)

The application modules evaluate potential safety threats based on the data and inputs from the support modules. The application modules contain the warning algorithms for the safety applications shown in Table 1. The SDH and WMH are basic, functional blocks necessary for parsing inputs from and submitting data to the software services of the system platform and those in use by the other support and application elements. The SDH interfaces to the vehicle Controller Area Network (CAN) gateway device to transmit and receive CAN messages and detect communication errors. It also connects to the GPS receiver to obtain National Marine Electronics Association (NMEA) data including Universal Coordinated Time (UTC) time, position, speed and heading, as well as raw GPS data. The SDH also interfaces to the external computing platform that executes the Real Time Kinematic (RTK) software to obtain accurate relative positioning of the neighboring vehicles. The WMH interfaces to the DSRC radio and to the Security Module (SM) software. It transmits and receives WAVE Safety Messages (WSM) using the SM to generate and verify message signatures. The TC categorizes the surrounding vehicles according to their position and heading relative to the HV, using the HVPP and the PH of the HV and RV. The TA arbitrates between multiple threats and chooses the one with the highest crash energy as the one to display to the driver and sends the respective request to the DVIN, which activates the corresponding alert in the EGUI.

The VSC-A team decided to use the shared memory interface concept. This allows for data in memory to be accessed by multiple modules for inter-process communication. This is advantageous, because there are many cases of one module supplying data to other functional blocks. For example, consecutive host and remote GPS time and position data points may be used by HVPP, PH, TC and the warning algorithms at the same time. The shared memory scheme used in the architecture fulfills the requirements for support of the VSC-A functionality while allowing for extensibility of the architecture.

ENGINEERING GUI

The EGUI is an "engineering-type" graphical user interface with the purpose to provide a simple engineering tool that could be used to understand, evaluate, and configure the VSC-A platform. It allows representation of visual and auditory vehicle driver warnings as a result of the application module processes. The touch-screen interface also allows the user to view and control parameters necessary for the operation of the VSC-A safety applications. Figure 2 shows examples of the graphical interface as depicted on a Video Graphics Array (VGA) touch screen.

This allowed the EGUI to display the warning states of a particular threat (e.g., DNPW in Figure 2). Only one of the warning screens is visible at any particular time. In order to ensure that the most important warning was shown on the DVI screen, the TA uses the threat level, relative speed, and location of the threat from each of the application modules to assess the severity and determine the highest priority request to be used by DVIN.

IN-VEHICLE HARDWARE INTEGRATION

The in-vehicle HW integration involved the selection, purchasing, installation and integration of all the HW and SW required for completion of the test bed. Table 2 identifies the model and manufacturer of the equipment installed on the VSC-A test bed vehicles.

MESSAGING STANDARDS

A major goal of the VSC-A Project was to define a single Over-the-Air (OTA) message whose contents could support all of the VSC-A safety applications as well as other safety applications that are likely to be developed in the future. That goal was achieved with the standardization of the SAE J2735 BSM [3]. An internal version of the OTA message was defined and implemented in the test bed with the objective testing verifying that this message supports all of the VSC-A applications. The BSM consists of Parts I and II. A proposal was prepared and presented for SAE to redefine both Parts I and II of the BSM. Part I consists of vehicle state data that is so critical for safety applications that it must be included in every BSM. Part II consists of data that is either required by applications at regular intervals (potentially at a reduced frequency), required to notify applications of a given event or optional for applications. Figure 3 shows the components and format of the BSM in SAE J2735.

The SAE J2735 conformant BSM uses the Distinguished Encoding Rules (DER) to encode the message for OTA transmission. In addition to the effort to develop and standardize the BSM, the VSC-A team also initiated a new SAE DSRC standards project (SAE J2945) for BSM minimum performance requirements. This standard will augment SAE J2735 to define rules necessary for effective V2V safety communications interoperability (e.g., minimum message rate, minimum data accuracy, etc.).

OBJECTIVE TESTING

OVERVIEW

The objective testing activity included the development of the Objective Test Procedures (OTPs) and test plan, conducting the objective tests, and analyzing the test results. The purpose of the objective testing was to ascertain that:

- The performance of the VSC-A system test bed was sufficient to enable the safety applications in the project
- The safety applications satisfied the minimum performance requirements developed in the system design activity of the project

The OTPs were developed for each application and were designed to include the most common scenarios that the application would encounter. The procedures included the following:

- True positive tests, where the objective is to get a warning
- False positive tests, where the objective is to suppress a warning because it is not needed

The outcomes of the objective tests were used by the Volpe National Transportation Systems Center (Volpe) to estimate the safety benefits opportunity for V2V communications based safety applications. In total, 33 test procedures were developed, 22 true positive tests and 11 false positive tests. For the benefits estimate, only the true positive tests which all had successful/unsuccessful criteria associated with them were evaluated. The OTPs were discussed with NHTSA and Volpe and agreed upon by all the participants. Following the OTP development, the test plan was written. It included the number of runs for each test, test speeds, validation criteria for each test (allowable speed ranges, etc.) and detailed setup procedures to make the OTPs as repeatable as possible. The test plan was also agreed upon by Volpe and NHTSA prior to the start of testing. The objective testing took place from June 1, 2009 to June 3 2009 at TRC in East Liberty, Ohio.

The data that was collected during the testing was recorded in a data logging and visualization tool called CANape [4]. CANape is a SW tool developed by Vector CANtech, Inc. and is used for the development, calibration and diagnostics of Electronic Control Units (ECUs) as well as data acquisition and analysis. The CANape software was customized by Vector for the VSC-A Project. Figure 4 shows an example of the primary screen that was used for the objective testing. The screen is divided into four quadrants as follows:

- Quadrant 1 contains a birds-eye view, which is a graphical representation of the location of the HV, centered at (0,0) and the RVs that the HV is in communication with

- Quadrant 2 contains the camera data which will consist of a single image, as shown below, or up to four images multiplexed together

- Quadrant 3 contains the HV's sensor data and GPS data

- Quadrant 4 contains the RV track data as determined by the TC core module

OBJECTIVE TEST RESULTS

The complete list of tests, the speed for the runs, the number of runs for each test and the test outcome is shown in Table 3. As can be seen from Table 3, all the applications passed the objective tests.

SUMMARY/CONCLUSIONS

The major accomplishments of the project are:

- Defined a set of high-priority, potential crash scenarios that could be addressed by V2V communication

- Selected and developed a set of V2V applications to address the above set of potential crash scenarios

- Defined efficient system architecture for V2V safety system where all VSC-A safety applications are enabled at the same time

- Successfully implemented a test bed with all the safety applications on a platform running an automotive grade processor (400 MHz)

- Successfully incorporated and evaluated in the test bed two relative positioning approaches (RTK and Single Point (SP))

- Successfully incorporated in the test bed the necessary OTA communication protocol (SAE J2735) and security protocol (IEEE 1609.2 Elliptic Curve Digital Signature Algorithm (ECDSA) [5] with Verify-on-Demand (VoD) [6])

- Defined OTPs for all the VSC-A safety applications, including true positive and false positive tests

- Successfully executed and passed all objective tests for all the VSC-A safety applications

• Refined, with field data, the required OTA message set for V2V safety (BSM within SAE J2735) which led to the recently published version of the standard [3]

• Conducted a study to quantify availability and accuracy of GPS-based relative positioning by using RTK and SP methods for V2V

• Confirmed that IEEE 1609.2 ECDSA with VoD functioned properly under all test conditions for the VSC-A safety applications

• Performed and analyzed initial scalability with up to 60 radios [8] to characterize channel behavior under IEEE 1609.4 [7] and under dedicated full time use of channel 172

Another outcome of the technical work was the identification of technical questions and topics that still need to be answered for any successful deployment:

• How does the system perform with large numbers of communicating nodes?

• How can security certificates be managed and privacy preserved?

• Are the standards sufficient for interoperability?

• What are requirements for data reliability and integrity?

• What are technical solutions for acceleration of market penetration?

• How to enhance the safety applications and system design?

• How to enhance relative vehicle positioning?

Those questions and topics are being addressed under the current NHTSA V2V safety roadmap [1] which outlines the next set of activities needed to support a NHTSA decision regarding V2V safety in 2013.

REFERENCES

1. DOT-Sponsored Research Activities: V2V Communications for Safety, http://www.intellidriveusa.org/research/v2v.php.

2. Vehicle Safety Communications - Applications (VSC-A) - First Annual Report, http://www.intellidriveusa.org/documents/2009/05/09042008-vsc-a-report.pdf

3. SAE International Surface Vehicle Standard, "Dedicated Short Range Communications (DSRC) Message Set Dictionary," SAE Standard J2735, Rev. Nov. 2009.

4. CANape, A Versatile Tool for Measurement, Calibration and Diagnostics of ECUs, Vector, http://www.vector.com/vi_canape_en.html?quickfinder=1.

5. IEEE Trial-use Standard 1609.2TM-2006, WAVE - Security Services for Applications and Management Messages, 2006.

6. Krishnan, H., Technical Disclosure, "Verify-on-Demand" - A Practical and Scalable Approach for Broadcast Authentication in Vehicle Safety Communication, IP.com number: IPCOM000175512D, IP.com Electronic Publication: October 10, 2008.

7. IEEE P1609.4TMD6.0, Draft Standard for Wireless Access in Vehicular Environments - Multi-channel Operation, IEEE Vehicular Technology Society, March 2010.

8. Ahmed-Zaid, F., Krishnan, H., Maile, M., Caminiti, L. et al., "Vehicle Safety Communications - Applications: Multiple On-Board Equipment (OBE) Testing," *SAE Int. J. Passeng. Cars – Mech. Syst.* **4**(1):547-561, 2011, doi: 10.4271/2011-01-0586.

CONTACT INFORMATION

Farid Ahmed-Zaid
Ford Motor Company
fahmedza@ford.com

ACKNOWLEDGMENTS

The CAMP VSC2 Participants would like to acknowledge the following USDOT personnel for their invaluable project support; Art Carter, Ray Resendes, and Mike Schagrin. The Participants would also like to thank VRTC personnel, especially Garrick Forkenbrock, for their outstanding support during the execution of the objective tests. Finally the Participants would like to express their appreciation to the following Volpe personnel; Wassim Najm, Bruce Wilson, and Jonathan Koopman for their support with the development and execution of the objective tests.

DEFINITIONS/ABBREVIATIONS

BSM
Basic Safety Message

BSW/LCW
Blind Spot Warning, Lane Change Warning

CAMP
Crash Avoidance Metrics Partnership

CAN
Controller Area Network

CLW
Control Loss Warning

DER
Distinguished Encoding Rules

DL
Data Logger

DNPW
Do Not Pass Warning

DSRC
Dedicated Short Range Communications

DVI
Driver-Vehicle Interface

DVIN
Driver-Vehicle Interface Notifier

ECDSA
Elliptic Curve Digital Signature Algorithm

ECU
Electronic Control Unit

EEBL
Emergency Electronic Brake Lights

EGUI
Engineering Graphical User Interface

FCW
Forward Collision Warning

GES
General Estimated Systems

GPS
Global Positioning System

HV
Host Vehicle

HVPP
Host Vehicle Path Prediction

HW
Hardware

IMA
Intersection Movement Assist

ITS
Intelligent Transport Systems

NHTSA
National Highway Traffic Safety Administration

NMEA
National Maritime Electronics Association

OBE
On-Board Equipment

OEM
Original Equipment Manufacturer

OTA
Over-the-Air

OTP
Objective Test Procedure

PH
Path History

RTK
Real-Time Kinematic

RV
Remote Vehicle

SDH
Sensor Data Handler

SM
Security Module

SP
Single Point (positioning)

SW
Software

TA
Threat Arbitration

TC
Target Classification

TRC
Transportation Research Center

USDOT
United States Department of Transportation

UTC
Universal Coordinated Time

V2V
Vehicle-to-Vehicle

VGA
Video Graphics Array

VoD
Verify-on-Demand

VRTC
Vehicle Research and Test Center

VSC2
Vehicle Safety Communications 2 (Consortium)

VSC-A
Vehicle Safety Communications - Applications

WMH
Wireless Message Handler

WSM
Wave Short Message

DISCLAIMER

This material is based upon work supported by the National Highway Traffic Safety Administration under Cooperative Agreement No. DTNH22-05-H-01277. Any opinions, findings, and conclusions or recommendations expressed in this publication are those of the Author(s) and do not necessarily reflect the view of the National Highway Traffic Safety Administration.

APPENDIX A

OBJECTIVE TEST PROCEDURE EXAMPLE AND TEST RESULTS

In this appendix we provide an example of the test plan and OTP together with the results of the testing. The chosen example is the FCW, Test 1.

FCW OBJECTIVE TEST PROCEDURES

FCW is a V2V, communication-based, safety feature that issues a warning to the driver of the HV in case of an impending rear-end collision with a vehicle ahead in traffic in the same lane and direction of travel. FCW is designed to help drivers in avoiding or mitigating rear-end vehicle collisions in the forward path of travel.

FCW-T1: HV Travel at a Constant Speed to a Stopped RV

Background

This test begins with the HV traveling on a straight, flat road at 50 mph. Ahead of the HV, in the same lane, is a single RV stopped in the lane of travel. The test determines whether the countermeasure's required collision alert occurs at the expected range. This test especially explores the ability of the countermeasure to accurately identify stationary in-path targets on a flat, straight road.

Test Setup

Figure 5 shows the vehicle positions and test setup for Test 1.

Cones with flags are placed so the driver of the HV is aware of the vehicle's location in reference to the required maneuvers. These flags are located by their distance from the starting point for the HV. It is assumed that flags will be placed using an accurate GPS handheld receiver. Alternate methods of flag location can be used. Flag locations are:

- A red flag is placed at the starting point where the HV begins its maneuver (cone not shown)
- A yellow flag is placed at the point where the HV reaches the target speed (cone HV-A), at least 650 meters from the red flag
- A white flag is placed at the earliest valid (from the driver's perspective) WARN point (cone HV-B)

A checkered flag is placed where the HV will make an evasive maneuver by changing lanes if the WARN has failed to occur (cone HV-C) which is positioned at 90 percent of the allowable alert range. At the test speed of 50 mph, this is 9 meters from HV-B cone

A green flag is placed at the stopping position for the RV (cone RV-A), at least 800 meters from the red flag

Driving Instructions

- The RV begins at the starting point and stops with its front bumper at the green flag
- The HV starts accelerating at least 800 meters behind the RV in the same lane to reach a speed of 50 mph
- The HV Cruise Control is set at the required speed of 50 mph
- The HV Cruise Control shall be engaged at least 150 meters behind the RV
- The warning will be given at around the nominal warn range (cone HV-B) after which the HV will change lane

[Note: If the warning is not given when the HV reaches the checkered flag (cone HV-C), the HV shall make an evasive maneuver by changing lanes and come to a safe stop in the adjacent lane.]

Successful Criteria

- The collision alert shall occur within the ranges specified in Table 4 in order to pass the run
- If at least six runs out of eight runs pass, then the test is successful

Unsuccessful Criteria

- A run is unsuccessful if any of the conditions below occur:
  - Collision alert occurrence outside the range calculated in Table 4 using run-specific variables
  - The warning is missed such that the HV passes cone HV-C and no alert is triggered
  - If at least three runs out of eight runs fail, the test is unsuccessful

Table 4. Alert Range for FCW Test 1

	Collision Alert Test
Maximum Range	93.7
Nominal Range	85.2
Minimum Range	76.7

Evaluation Criteria

Number of Valid Test Runs	HV Speed (mph)	RV Speed (mph)	Number of Successful Test Runs
8	50	0	≥ 6

FCW OBJECTEIVE TEST 1 RESULTS

For the FCW application to pass, the warning had to come between the maximum and minimum alert range that was calculated for each run. As can be seen from the test results table (Table 5), the application was successful in all the runs for the test.

APPENDIX B

TABLES AND FIGURES

Table 1. Mapping of VSC-A Program Applications to Crash Imminent Scenarios

	V2V Safety Applications/ Crash Scenarios	EEBL	FCW	BSW	LCW	DNPW	IMA	CLW
1	Lead Vehicle Stopped		✓					
2	Control Loss without Prior Vehicle Action							✓
3	Vehicle(s) Turning at Non-Signalized Junctions						✓	
4	Straight Crossing Paths at Non-Signalized Junctions						✓	
5	Lead Vehicle Decelerating	✓	✓					
6	Vehicle(s) Not Making a Maneuver – Opposite Direction					✓		
7	Vehicle(s) Changing Lanes – Same Direction			✓	✓			

Table 2. VSCA Test Bed Hardware List

Item Description	Manufacturer	Model
GPS Receiver	NovAtel®	OEMV® Flexpak V1-RT20A
GPS Antenna	NovAtel®	GPS-701-GG
LCD VGA Monitor	Xenarc	700TSV-B
USB CCD Monochrome Camera	The Imaging Source	DMK 21BU04
Car PC	Logic Supply	Voom PC-2
Inertial Measurement Unit	Silicon Sensing	DMU
OBE Vehicle CAN interface	Smart Engineering Tools	Netway 6
DSRC Antenna	Nippon Antenna	DEN-HA001-001
Ethernet Switch	Netgear	GS105

Table 3. Objective Test Scenarios and Results

Test Scenario	Description	Speeds	Number of Runs	Type of Test	Result
EEBL-T1	HV at constant speed with decelerating RV in same lane	50	8	True Positive	Successful
EEBL-T2	HV at constant speed with decelerating RV in left lane on curve	50	8	True Positive	Successful
EEBL-T3	HV at constant speed with decelerating RV in same lane and obstructing vehicle in between	50	8	True Positive	Successful
EEBL-T4	HV at constant speed with mild-decelerating RV in same lane	50	2	False Positive	N/A

Test Scenario	Description	Speeds	Number of Runs	Type of Test	Result
EEBL-T5	HV at constant speed with decelerating RV in 2nd right lane	50	2	False Positive	N/A
FCW-T1	HV travel at a constant speed\RV stopped	50	10	True Positive	Successful
FCW-T2	HV travel behind RV1\RV1 travel behind RV2\RV2 stopped	50	10	True Positive	Successful
FCW-T3	HV drive on a curve\RV stopped at the curve	50	8	True Positive	Successful
FCW-T4	HV tailgate RV	50	2	False Positive	N/A
FCW-T5	HV follows RV\RV brakes hard	40	10	True Positive	Successful
FCW-T6	HV driving into a curved right lane\RV stopped in the left curved lane	50	2	False Positive	N/A
FCW-T7	HV travels behind a slower RV	50	10	True Positive	Successful
FCW-T8	HV changes lanes behind a stopped RV	50	8	True Positive	Successful
FCW-T9	HV approaches two RVs in left and right adjacent lanes and passes between them	50	2	False Positive	N/A
BSW/LCW-T1	LCW Warning, Left	50	8	True Positive	Successful
BSW/LCW-T2	LCW Warning, Right	50	8	True Positive	Successful
BSW/LCW-T3	LCW Warning, Right with Left BSW Advisory	50	9	True Positive	Successful
BSW/LCW-T4	BSW Advisory Alert, Left	50	8	True Positive	Successful
BSW/LCW-T5	BSW Advisory Alert, Right	50	8	True Positive	Successful
BSW/LCW-T6	No Warning or Advisory for RV behind	50	2	False Positive	N/A

Test Scenario	Description	Speeds	Number of Runs	Type of Test	Result
BSW/LCW-T7	No Warning or Advisory for RV far Right	50	2	False Positive	N/A
BSW/LCW T8	LCW Warning in Curve, Right	35	8	True Positive	Successful
DNPW-T1	Attempt to pass with oncoming RV in adjacent lane	25/35	10	True Positive	Successful
DNPW-T2	Attempt to pass with stopped RV in adjacent lane	30/40	10	True Positive	Successful
DNPW-T3	Attempt to pass with oncoming RV not in adjacent lane	45	2	False Positive	N/A
IMA-T1	Variable speed approaches with stopped HV/moving RV/open intersection	20/30/40/50	12	True Positive	Successful
IMA-T2	Stopped HV/moving RV/open intersection	35/50	4	False Positive	N/A
IMA-T3	Variable speed approaches with moving HV/moving RV/open intersection	15/25/35/45	16	True Positive	Successful
IMA-T4	Moving HV/moving RV/open intersection	25	4	False Positive	N/A
IMA-T5	Stopped HV/moving RV/open intersection/parked vehicle	20/30/40/50	12	True Positive	Successful
CLW-T1	HV at constant speed with CLW RV in same lane ahead in same travel direction	40	8	True Positive	Successful
CLW-T2	HV at constant speed with CLW RV in 2nd right lane	30	2	False Positive	N/A
CLW-T3	HV at constant speed with CLW RV in adjacent lane ahead in opposite travel direction	30	12	True Positive	Successful

Table 5. FCW Test 1 Results

Run	Actual Values at Alert Onset				Calculated Run-Specific Alert Ranges					
	HV Speed	HV Accel	RV Speed	RV Accel	Maximum (meters)	Nominal (meters)	Minimum (meters)	Actual Alert Range (meters)	Pass/Fail	Headway
FCW T1	50 mph 22.35 m/s	0 g	0 mph	0 g	93.7	85.2	76.7			6.71
1	49.70 mph 22.21 m/s	0 g	0 mph	0 g	92.8	84.4	75.9	84	Pass	6.80
2	50.0 mph 22.35 m/s	0 g	0 mph	0 g	93.7	85.2	77.7	86	Pass	6.83
3	49.68 mph 22.21 m/s	0 g	0 mph	0 g	91.7	83.4	75.0	85	Pass	6.76
4	50.31 mph 22.49 m/s	0 g	0 mph	0 g	94.1	85.5	77.0	85	Pass	6.80
5	49.69 mph 22.21 m/s	0 g	0 mph	0 g	92.8	84.4	75.9	85	Pass	6.80
6	49.50 mph 22.13 m/s	0 g	0 mph	0 g	93.1	84.6	76.2	80	Pass (A # of communication outages from the RV)	6.82
7	49.84 mph 22.28	0 g	0 mph	0 g	93.9	85.4	76.8	83	Pass	6.73

	Actual Values at Alert Onset				Calculated Run-Specific Alert Ranges					
Run	HV Speed	HV Accel	RV Speed	RV Accel	Maximum (meters)	Nominal (meters)	Minimum (meters)	Actual Alert Range (meters)	Pass/Fail	Headway
	m/s									
8	49.75 mph 22.24 m/s	0 g	0 mph	0 g	92.7	84.3	75.9	82	Pass (A # of communication outages from the RV)	6.79
9	50.35 mph 22.51 m/s	0 g	0 mph	0 g	93.2	84.7	76.3	85	Pass	6.66
10	49.71 mph 22.22 m/s	0 g	0 mph	0 g	91.7	83.3	75.0	84	Pass	6.75

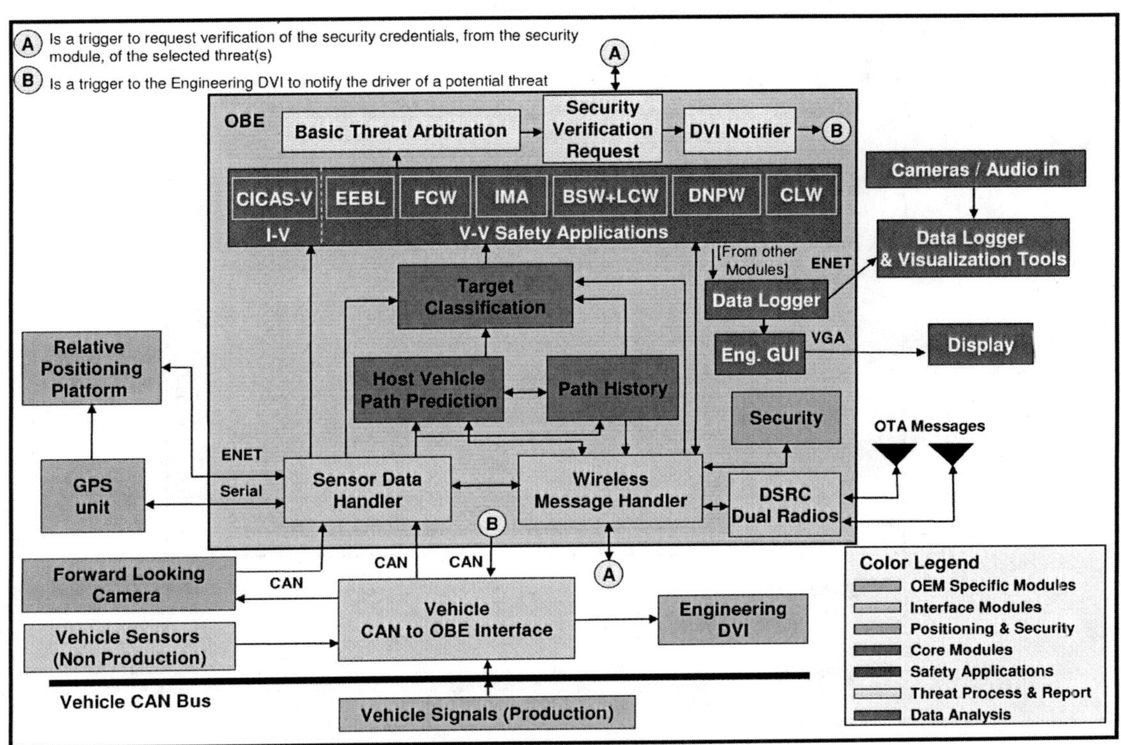

Figure 1. VSC-A System Block Diagram

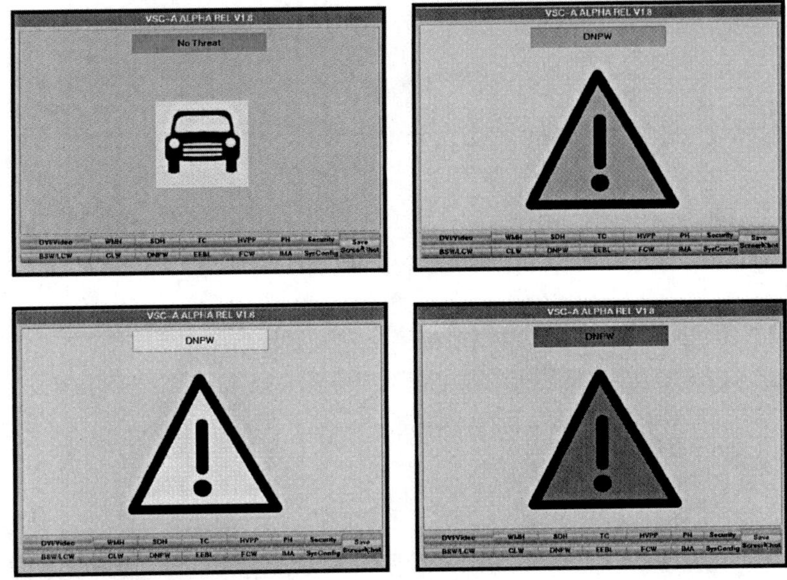

Figure 2. DVIN Stages (left to right, top to bottom) No Threat, Threat Detected, Inform Driver, Warn Driver

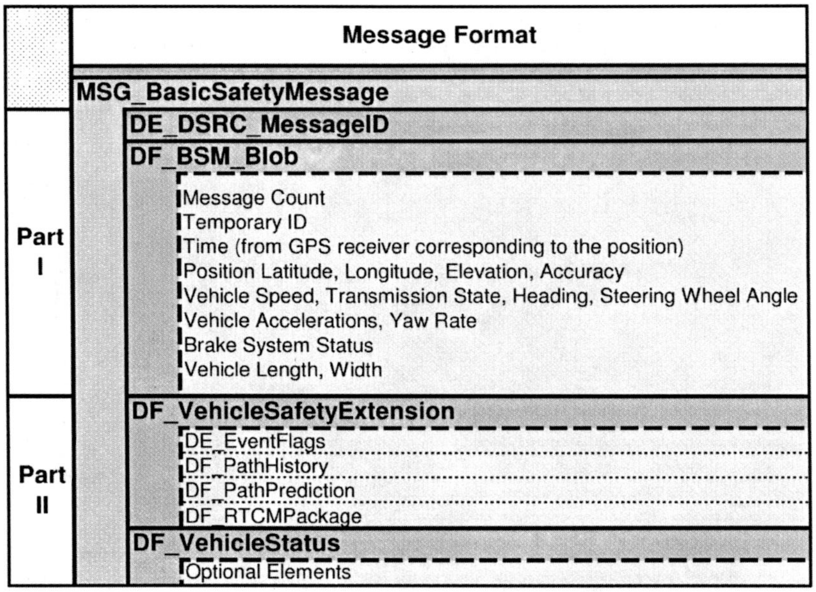

Message Format

	MSG_BasicSafetyMessage
Part I	DE_DSRC_MessageID
	DF_BSM_Blob Message Count Temporary ID Time (from GPS receiver corresponding to the position) Position Latitude, Longitude, Elevation, Accuracy Vehicle Speed, Transmission State, Heading, Steering Wheel Angle Vehicle Accelerations, Yaw Rate Brake System Status Vehicle Length, Width
Part II	DF_VehicleSafetyExtension DE_EventFlags DF_PathHistory DF_PathPrediction DF_RTCMPackage
	DF_VehicleStatus Optional Elements

Figure 3. SAE J2735 Rev 35 Basic Safety Message Format

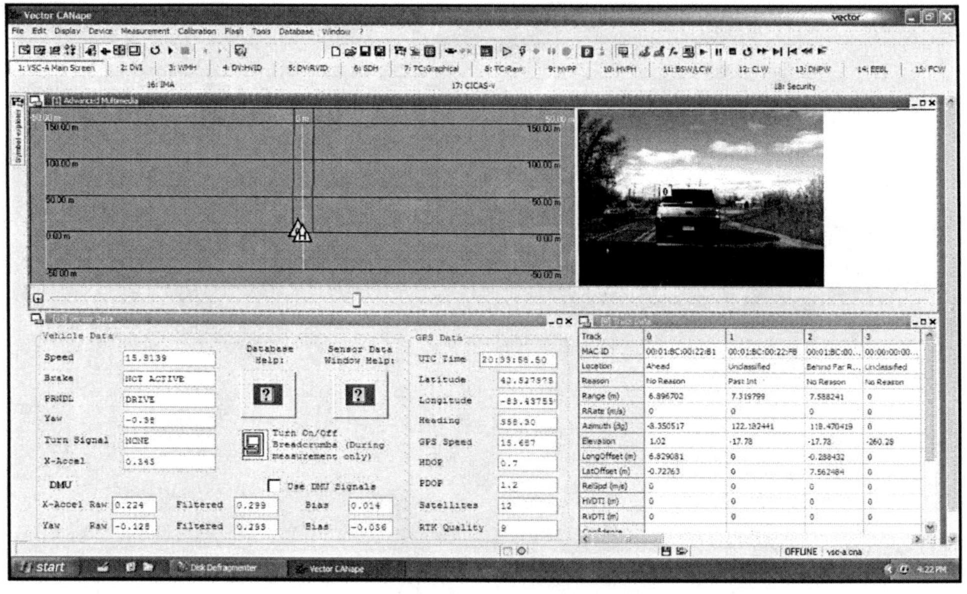

Figure 4. Example Layout Screen for OTP Testing

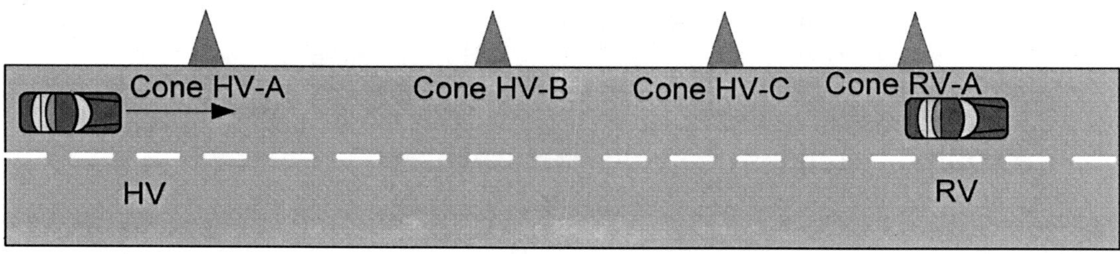

Figure 5. FCW Test 1 Test Setup - RV in Same Lane

2009-01-0165

Prioritized CSMA Protocol for Roadside-to-Vehicle and Vehicle-to-Vehicle Communication Systems

Jun Kosai, Shugo Kato, Toshiya Saito, Kazuoki Matsugatani and Hideaki Nanba
DENSO CORPORATION

ABSTRACT

This paper proposes Prioritized-CSMA (Carrier Sense Multiple Access) protocol for Japanese vehicle safety communications (VSC). To realize Japanese VSC, we have studied a protocol to carry out Roadside-to-Vehicle (R2V) and Vehicle-to-Vehicle (V2V) communications on single channel because a single 10MHz bandwidth channel on UHF band is allocated for VSC in Japan. In this case, R2V communication requires higher quality than V2V communication, so we have developed a protocol to prevent interference between R2V and V2V communications. The proposed protocol has been evaluated by field experiments and a simulation. The results confirm that the proposed protocol prevents the interference effectively and it has capability to achieve high quality R2V communication in actual case.

INTRODUCTION

Vehicle safety communications (VSC) including Roadside-to-Vehicle (R2V) and Vehicle-to-Vehicle (V2V) have been studied to improve vehicle safety. To realize VSC, frequency channels are allocated, and communication standards have been developed all over the world. In US and EU, multiple frequency channels on 5.9GHz band are allocated to VSC. From the viewpoint of communication standard, IEEE802.11p and IEEE1609 are used in US, and the same or similar standard would be used in EU[1-4]. On the other hand in Japan, a single 10MHz bandwidth channel on UHF band is allocated to VSC, but the standard for these communications is still being discussed. Japanese VSC environment differs from US and EU, so a communication protocol suitable for Japanese VSC is required.

Japanese VSC has an important issue to develop communication protocol. The issue is that both R2V and V2V communications have to be carried out on allocated single channel because allocated channel is just 10MHz single channel. From the viewpoint of communication requirements, R2V communication takes priority over V2V communication because roadside units (RSU) can send packet including crucial information to avoid traffic accidents especially at the intersection. However, in this case where both of communications are on single channel, interference between R2V and V2V should occur and affect the performance each other. Especially interference from V2V to R2V should be critical problem.

To prevent the interference, this paper proposes Prioritized-CSMA protocol, which is a CSMA-based medium access control (MAC) protocol. We assume that CSMA-based protocol is suitable for VSC because on-board units (OBU) can send their own packets without centralized scheme. To adapt CSMA protocol to Japanese VSC, we introduce a concept of time division into R2V and V2V time slots. RSUs play as a master that controls R2V and V2V slot timings; OBUs refrain from sending their packets during R2V slot and transmit packets during V2V slot by CSMA scheme. To evaluate the proposed protocol, we have developed UHF band wireless communication units and performed field experiments. Furthermore, we have evaluated the protocol in a scenario where large amount of vehicles exist by simulation.

This paper is organized as follows: Second section explains Japanese VSC system we assume. Third section presents our proposed protocol for Japanese VSC. Fourth section evaluates the proposed protocol by

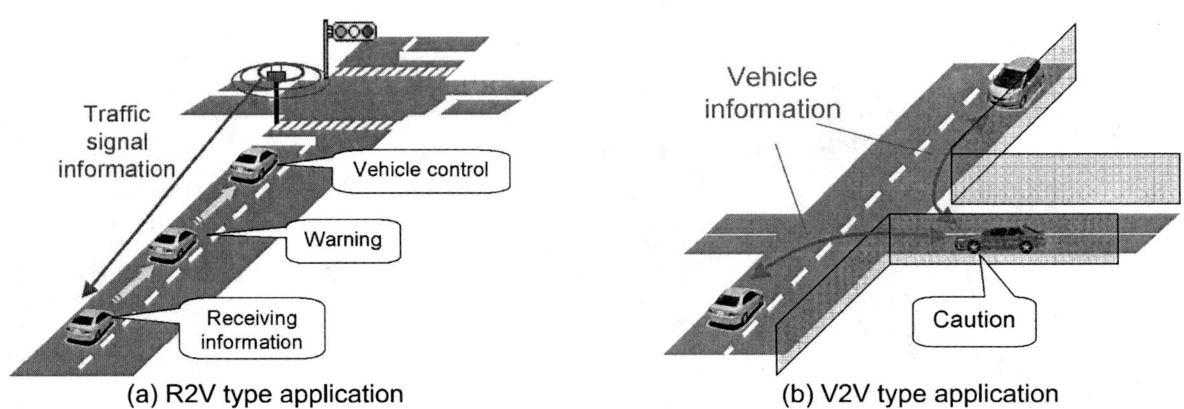

(a) R2V type application (b) V2V type application

Figure 1 Examples of Japanese VSC application

field experiments and a simulation. The last section describes the conclusion.

JAPANESE VSC SYSTEM

Vehicle safety applications and VSC systems have been discussed and developed by various organizations[5-7], and it is difficult to define Japanese VSC clearly. Therefore, this section introduces our assumed Japanese VSC system from the viewpoint of applications and requirements.

APPLICATIONS - Japanese VSC contains applications using both R2V and V2V communications. Figure 1 shows the examples. Figure 1 (a) depicts an example of driving support system based on roadside information. RSU sends traffic signal information and OBU, which receives the information, performs warning or vehicle control according to the situation. To realize vehicle control by roadside information, high communication quality is required. Figure 1 (b) illustrates an example of driving support system based on V2V communication. Each vehicle broadcasts the own data including the position, speed and heading. A receiving vehicle provides the driver with approaching vehicle information.

REQUIREMENTS - We assume that following four requirements are important to develop communication protocol for Japanese VSC.

1. Both RSU and OBU send packets on the same channel.
2. R2V communication has priority over V2V.
3. Both RSU and OBU broadcast their packets.
4. R2V communication occupancy is adaptable to R2V data amount.

Firstly, as explained above, a single channel on UHF band is allocated for VSC in Japan; RSU and OBU have to share the same channel.

Secondly, we assume that RSU has much more information than OBU because RSU would be connect with some facilities as traffic information center, database and so on via wired or wireless connection. Therefore RSU would transmit packets including more precise and crucial information to avoid traffic accidents especially at the intersection. That's why RSU transmission should not be interfered from OBU transmission. This paper aims to obtain 99% or more packet arrival rate for R2V communication.

Thirdly, every communication units send the information by broadcast. Unicast transmission is not suitable for safety application because a sending node cannot specify the receiving node in advance. We focus on broadcast transmission only. In addition in this paper, R2V communication means that RSU only sends packets and gives information to OBUs unilaterally.

Lastly, flexibility of R2V communication occupancy is required. The size of data, which is transmitted from RSU, may be changed depending on the situation such as supplying application, RSU located position and so on. Moreover, if there is no RSU, all communication resources should be assigned for OBU. Therefore, this is important requisite for Japanese VSC.

PRIORITIZED-CSMA PROTOCOL

This section proposes Prioritized-CSMA (P-CSMA) protocol. As mentioned above, we have developed a new MAC protocol based on CSMA because CSMA protocol does not need centralized scheme. In the case where CSMA protocol is applied to Japanese VSC, however, it is difficult to achieve 99% packet arrival rate for R2V communication caused by interference from V2V communication. This section states an issue, which is caused by applying CSMA protocol and presents P-CSMA as the solution.

INTERFERENCE FROM V2V - CSMA protocol has well-known problem, hidden node problem[8]. Figure 2 shows a situation where the hidden node problem occurs between R2V and V2V communications. While RSU sends a packet to OBUs within the communication range, the outside OBUs would send a packet simultaneously. As the result, packet collision happens between R2V and V2V. This collision leads to degradation of packet arrival rate for R2V communication.

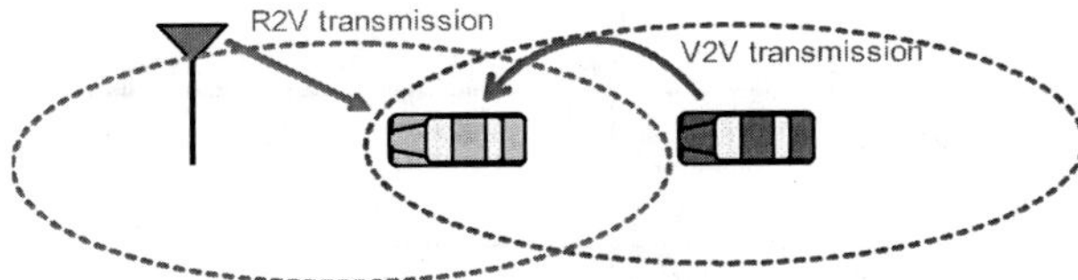

Figure 2 Hidden node problem

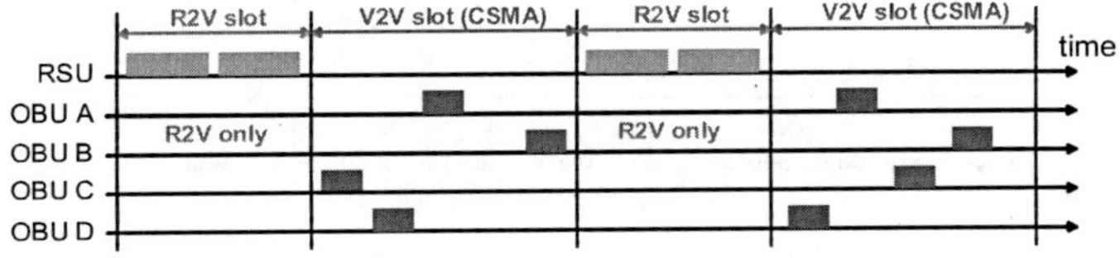

Figure 3 Division of R2V slot and V2V slot

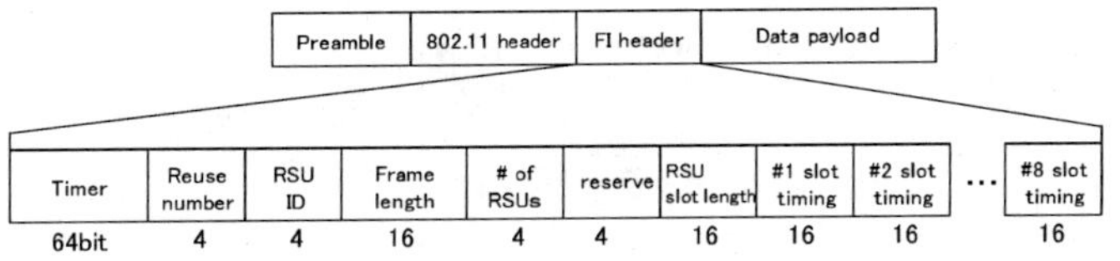

Figure 4 Frame format

PROTOCOL DESIGN

Concept - To avoid interference between R2V and V2V, we introduce time division scheme into ordinary CSMA. Time period is divided into two parts, R2V slot and V2V slot. Here, slot means a period that is allocated to RSU or OBU. Figure 3 shows the concept of P-CSMA. RSU works as a master to control R2V and V2V slots, and the slots are allocated by RSU. RSU sends packets only in R2V slot and OBU transmission is prohibited during the R2V slot. On the other hand in V2V slot, only OBU sends packets using conventional CSMA scheme. In this way, R2V communication is protected against the interference from V2V. In addition, R2V slot length can be adapted to R2V data amount because RSU assigns the slot itself.

To realize the concept, following two requirements have to be satisfied.

1. RSU informs OBU of slot information (SI).
2. OBU, which is hidden node from RSU, has to know the SI.

Hereinafter, this section describes frame format and SI propagation to meet the requirements.

Frame format - RSU sends SI to OBU to assign R2V and V2V slots. The SI includes *Timer*, *Slot length*, *Slot timing*. Figure 4 illustrates frame format and Table 1 shows the contents. SI is inserted between IEEE802.11 header and data payload. *Timer* is cyclic time in microsecond, and it is used for time synchronization between RSU and OBU. *Reuse Number (RN)* indicates freshness of the SI. The *RN* is utilized for updating SI on OBU and managing SI propagation area.

Table 1 Field contents of slot information

Field	Contents
Timer	Reference time which RSU keeps and generate using GPS.
Reuse Number	Indicator for freshness of slot information.
RSU ID	Identification of RSU.
Slot cycle	Slot cycle in which RSU send packets.
# of RSUs	The number of RSU slot timing in the slot information.
RSU slot length	RSU slot length calculated from RSU data amount.
#x slot timing	Start timings of RSU slot. (This field can contain up to 8 slot timings)

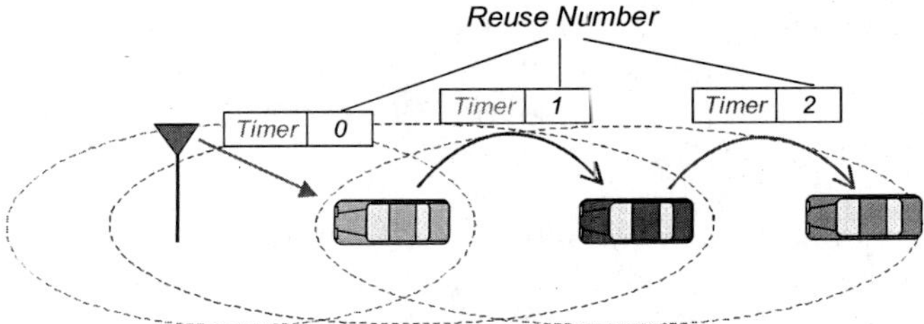

Figure 5 Propagation of slot information

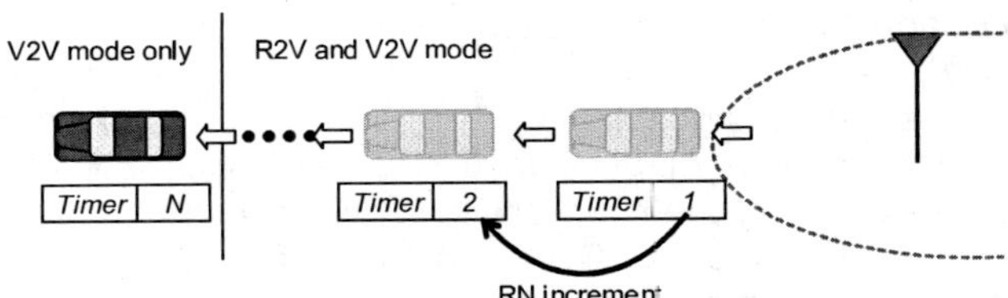

Figure 6 RN increment by elapsed time

SI is capable of including multiple *slot timings* (up to 8). As a result, P-CSMA works properly by containing all slot timing of neighbor RSUs even if multiple RSUs exist in the neighborhood.

Propagation of slot information - SI have to reach OBUs which are hidden nodes from RSU. A RSU generates SI and sends the SI to OBUs within the RSU communication range; however, the SI cannot reach to the hidden nodes. Therefore, OBU retransmits the received SI to surrounding OBUs.

Figure 5 shows the SI propagation. OBU A, which have received SI from RSU directly, synchronizes own timer with RSU using *Timer* field and controls to refrain from sending its own packets during R2V slot referring to *RSU slot length* and *slot timing* fields. When OBU A sends its own data packet, OBU A sends the packet in the same format as Figure 4. At this moment, OBU A stamps its own timer value on *Timer* field, and the other field are copied from received SI. In the same way, OBU B retransmits SI after receiving a packet from OBU A. In this way, SI is delivered to hidden nodes from RSU.

However, SI propagation should be limited within appropriate area. If OBUs, which do not interfere with

Figure 7 Experimental communication unit

R2V communication, receive SI, the prohibition of transmission causes unnecessary restrictions on V2V communication. So, P-CSMA controls the SI propagation area by *RN,* which indicates proximity to RSU and freshness of SI. SI is discarded when *RN* is over threshold. A couple of *RN* control methods are explained as follows.

SI retransmission - *RN* in SI is incremented at every SI retransmission as shown in Figure 5. When RSU sends SI, RN is 0, and the *RN* is incremented when OBU receives the SI. For example, *RN* is 3 on OBU C in Figure 5. If the *RN* threshold is 3, OBU C does not retransmit SI to surrounding OBUs. In this way, SI propagation area is controlled.

Elapsed time since latest SI - The other method to control *RN* uses elapsed time since receiving latest SI. Figure 6 shows the mechanism. OBU cannot receive SI when OBU get out of RSU communication range, but if OBU continues to keep the SI, V2V communication is restricted uselessly. To avoid this problem, *RN* is incremented every time definite period is elapsed. When *RN* is over the threshold, SI is invalidated. After that, OBU shifts to V2V only mode, in which OBU can send packets anytime.

P-CSMA uses both of methods to control SI propagation area.

EXPERIMENTAL COMMUNICATION UNIT - We have developed experimental communication units for performance evaluation. Figure 7 shows the external appearance, and Table 2 shows the specification. Slot cycle means intervals at which RSU sends packets. Now slot cycle is 100 ms; if 20 % are allocated for R2V, 20 ms are used for R2V and 80ms are used for V2V. To realize high accuracy time synchronization, the synchronization function is implemented in hardware logical circuit. As the result, the synchronization accuracy is less than 4us, which is sufficient for P-CSMA.

EVALUATION

This section describes performance evaluation of P-CSMA. The evaluation consists of two field experiments and a simulation analysis. Filed experiments are conducted to confirm that P-CSMA is capable of preventing interference from V2V to R2V. In addition, the simulation is carried out for performance analysis in the case where more than hundreds of OBUs exist.

FIELD EXPERIMENT IN STATIC ENVIRONMENT - This experiment is intended to evaluate the performance depending on interference from hidden node. Figure 8 depicts the experimental arrangement. Three

Table 2 Specification of experimental unit

Center frequency	792.5MHz
Maximum TX power	20dBm
Modulation	BPSK, QPSK, 16QAM
MAC protocol	Prioritized-CSMA
Slot cycle	100ms
Synchronization accuracy	less than 4us
Size	330(W) × 200(D) × 80(H)

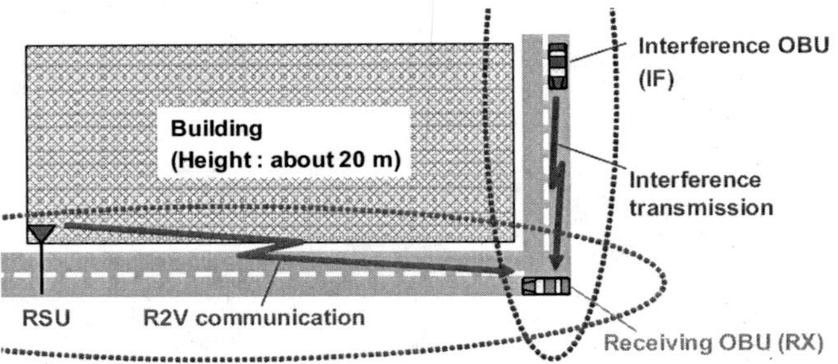

Figure 8 Experimental arrangement in static environment

Table 3 Experimental parameters

Bandwidth		12 Mbps
Slot allocation		R2V:20%, V2V:80%
RSU	Packet size	1500 byte
	Transmission rate	100 packets / 100ms
Receiving OBU	Packet size	100 byte
	Transmission rate	1 packet / 100ms
Interference OBU	Packet size	100 byte
	Transmission rate	0 ~ 100 packets / 100ms

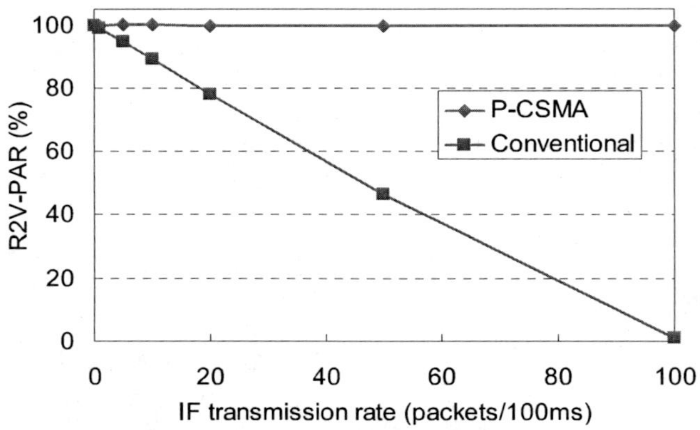

Figure 9 R2V-PAR vs. IF transmissions

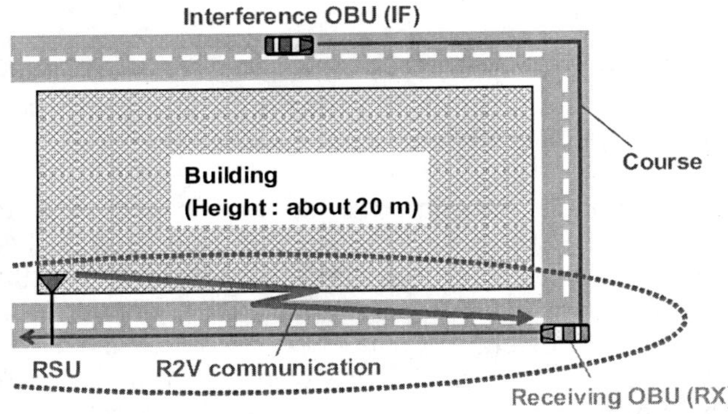

Figure 10 Experimental arrangement in dynamic environment.

communication units are located as RSU, receiving OBU (RX) and interference OBU (IF). Although RX can communicate with both RSU and IF, RSU and IF cannot communicate each other. In other words, RSU and IF are mutually hidden nodes. Therefore, if RSU and IF generate sending packets simultaneously, packet collision happens at RX because of incapable of carrier sense. In this arrangement, we compared R2V packet arrival rate (R2V-PAR) of P-CSMA with conventional CSMA's. Here R2V-PAR means communication successful rate from RSU to RX. Table 3 shows the experimental parameters. Slot allocations for R2V and V2V are 20 % and 80 % respectively. The packet size of RSU and OBU comes from supposed applications. We assume that OBU sends a packet in every 100 ms, so RX sends a packet per 100 ms. IF Transmission rate changes from 0 to 100 packets per 100ms. The number of packets per 100 ms from IF correspondents to the number of hidden nodes from RSU.

Figure 9 presents the experimental result, which is relation between the number of IF packets and R2V-PAR. The result shows that R2V-PAR of conventional method decreases as IF transmission rate increases. The reason of the R2V-PAR degradation is interference from IF to R2V communication. IF cannot sense R2V communication, so R2V-PAR degrades linearly as IF transmission rate increases. On the other hand, P-CSMA keeps almost 100% R2V-PAR regardless IF transmission rate. This result confirms that P-CSMA works properly to prevent hidden node problem on R2V communication in static environment.

FIELD EXPERIMENT IN DYNAMIC ENVIRONMENT - This experiment evaluates the performance in dynamic environment, where OBUs come into RSU communication range from outside. Figure 10 depicts the experimental arrangement. RSU and RX are placed in the same positions as the previous experiment. IF starts from outside of RX communication range and runs along the course shown in Figure 10 toward RSU. The vehicle speed is 20 km/h. IF does not have valid SI initially because IF is placed on outside of RX communication range. Considering this point, this experiment measures an effect of IF movement at the moment when IF receives SI. IF transmission rate is 20 packets per 100 ms.

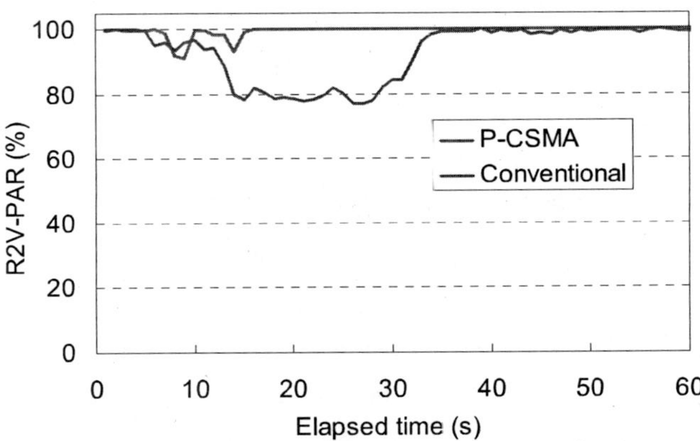

Figure 11 R2V-PAR vs. elapsed time

Figure 11 shows the experimental result. The horizontal axis indicates elapsed time from IF starting. The vertical axis indicates R2V-PAR at every second. In the graph, IF enters RX communication range at around 5 s, and IF enters RSU communication range at around 35 s. In other words, IF is hidden node from RSU between 5 s and 35 s. During the period, R2V-PAR of conventional method degrades to 80% because of hidden node effect. On the contrary, P-CSMA keeps higher R2V-PAR compared with conventional method. This result shows that P-CSMA functions effectively to keep R2V-PAR even in dynamic environment. However, instantaneous R2V-PAR drops at around 10s. Hereinafter, we consider why R2V-PAR drops at the timing.

The R2V-PAR degradation is cased by two reasons. First, a relation between communication range and radio propagation range affects R2V-PAR. Generally, radio propagation range is greater than communication range, where two nodes are able to send and receive packets each other. In this experiment, IF transmission interferes with receiving RSU packets at RX before IF receives SI from RX. Second, the degradation depends on experimental scenario. In this experiment, RX is the only node which can receive RSU packet and retransmit SI. In contrast, the number of hidden nodes from RSU corresponds to 20 because IF sends 20 packets per 100 ms. Therefore, this experimental scenario is the hardest case for P-CSMA, where delivering SI to hidden node is difficult.

When the actual operability of P-CSMA is considered, the result suggests that how long SI can spread is the most important for P-CSMA. So, we calculate the time by following simulation.

SIMULATION - We calculated the time SI spread in traffic flow by using traffic and network simulator. Figure 12 shows the simulation model and Table 4 shows the parameters. The duration SI spreads depends on vehicle density; this simulation evaluates the duration with multiple vehicle densities. From the communication ranges of RSU and OBU, hidden nodes appear in shaded area of Figure 12. So, the percentage of OBUs,

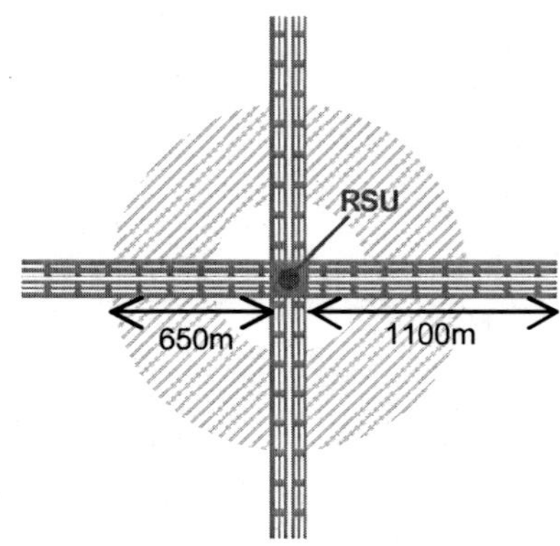

Figure 12 Simulation model

Table 4 Simulation parameters

Lane	Three lanes each way
Vehicle density	5, 10, 20, 30 vehicles/km
RSU communication range	200 m
OBU communication range	450 m
R2V slot rate	5 %

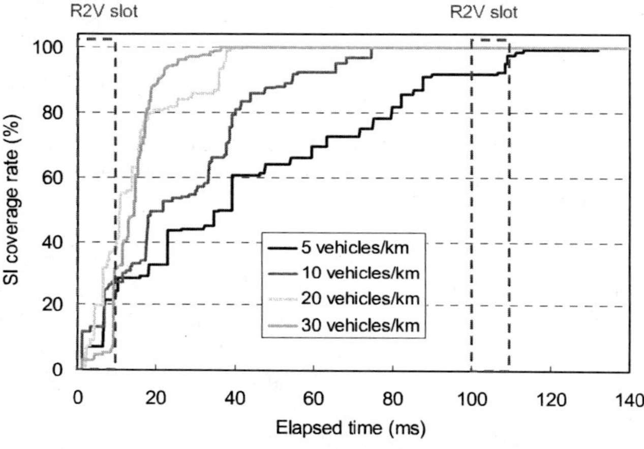

Figure 13 SI coverage rate vs. elapsed time

which have received SI in 650 m radius from RSU, is defined as SI coverage rate.

Figure 13 shows a relation between elapsed time and SI coverage rate. In the case of high vehicle density, SI coverage rate rises rapidly and it reaches 100 % in 40 ms. Also, SI coverage rate increases to 100 % in approximately 100 ms even if vehicle density is 5 vehicles per km.

This result demonstrates that SI can spread out over the area necessary to prevent hidden node problem in short period. Therefore, P-CSMA is effective to realize reliable R2V communication in actual traffic flow.

CONCLUSION

This paper proposed Prioritized-CSMA protocol to realize Japanese VSC, in which both R2V and V2V communications are carried out on single channel. We assumed that R2V communication takes priority over V2V communication, so we developed P-CSMA to prevent interference from V2V to R2V. P-CSMA introduced time slots of R2V and V2V to conventional CSMA by sending SI from RSU.

We developed experimental communication unit to evaluate P-CSMA and conducted field experiments. The experimental results show that P-CSMA improves the R2V-PAR sufficiently compared with conventional method. In addition, the simulation result demonstrates that P-CSMA has capability to achieve high quality R2V communication in actual case because SI coverage rate reaches 100 % in short period.

REFERENCES

1. Wireless LAN Medium Access Control (MAC) and Physical Layer (PHY) Specifications, IEEE Std 802.11-2007, 2007
2. Wireless LAN Medium Access Control (MAC) and Physical Layer (PHY) Specifications Wireless Access in Vehicular Environments, IEEE 802.11p/D3.0, 2007
3. Wireless Access in Vehicular Environments (WAVE) – Networking Services, IEEE Std 1609.3-2007, 2007
4. Wireless Access in Vehicular Environments (WAVE) – Multi-channel Operation, IEEE Std 1609.4-2006, 2006
5. O. Maeshima, S. Cai, T. Honda, and H. Urayama, A roadside to vehicle communication system for vehicle safety using dual frequency channels, presented at IEEE ITSC 2007, pp.349-354, 2007
6. Y. Tadokoro, K. Ito, J. Imai, N. Suzuki, and N. Itoh, A New Approach for Evaluation of Vehicle Safety Communications with Decentralized TDMA-based MAC Protocol, presented at V2VCOM2008, 2008
7. K. Tokuda, M. Akiyama, H. Fujii, DOLPHIN for inter-vehicle communications system, presented at IV2000, pp.504-509, 2000
8. F. A. Tobagi and L. Kleinrock, (1975). Packet Switching in Radio Channels: Part II – the hidden terminal problem in carrier sense multiple-access modes and the busy tone solution, IEEE Trans. in Communications, COM-23 (12), pp.1417-1433

Communications - Vehicle Networks

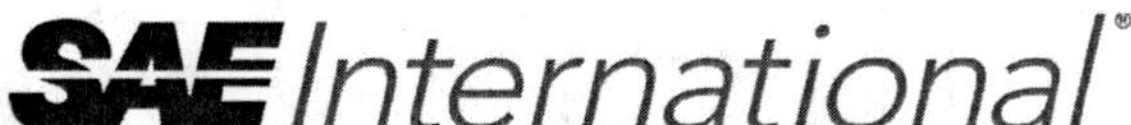

Vehicular Networks for Collision Avoidance at Intersections

2011-01-0573
Published
04/12/2011

Seyed Reza Azimi, Gaurav Bhatia and Ragunathan (Raj) Rajkumar
Carnegie Mellon University

Priyantha Mudalige
GM Technical Center

doi:10.4271/2011-01-0573

ABSTRACT

A substantial fraction of automotive collisions occur at intersections. Statistics collected by the Federal Highway Administration (FHWA) show that more than 2.8 million intersection-related crashes occur in the United States each year, with such crashes constituting more than 44 percent of all reported crashes [12]. In addition, there is a desire to increase throughput at intersections by reducing the delay introduced by stop signs and traffic signals. In the future, when dealing with autonomous vehicles, some form of co-operative driving is also necessary at intersections to address safety and throughput concerns.

In this paper, we investigate the use of vehicle-to-vehicle (V2V) communications to enable the navigation of traffic intersections, to mitigate collision risks, and to increase intersection throughput significantly. Specifically, we design a vehicular network protocol that integrates with mobile wireless radio communication standards such as Dedicated Short Range Communications (DSRC) and Wireless Access in a Vehicular Environment (WAVE). This protocol relies primarily on using V2V communications, GPS and other automotive sensors to safely navigate intersections and also to enable autonomous vehicle control. Vehicles use DSRC/WAVE wireless media to periodically broadcast their position information along with the driving intentions as they approach intersections. We used the hybrid simulator called GrooveNet [1, 2] in order to study different driving scenarios at intersections using simulated vehicles interacting with each other. Our simulation results indicate that very reasonable improvements in safe throughput are possible across many practical traffic scenarios.

INTRODUCTION

Current human driver-based intersections which are managed by stop signs and traffic lights are not entirely safe, based on Federal Highway Administration (FHWA) statistics [12]. Our goal is to design new methods to manage intersections, which lead to fewer collisions and less travel delay for vehicles crossing at intersections. Various driverless vehicles have been developed and tested at intersections, such as in the DARPA Urban Challenge [3] and General Motor's EN-V, which has been recently unveiled in Shanghai, China [4]. Our focus is to use vehicle-to-vehicle (V2V) communication as a part of co-operative driving in the context of autonomous vehicles to manage intersection traffic efficiently and safely.

Past work in this domain includes the use of Vehicle to Infrastructure (V2I) communication by having a centralized system in which all vehicles approaching an intersection communicate with the intersection manager. The intersection manager is the computational infrastructure installed at intersections and to make reservations for each approaching vehicle and manages all vehicles crossing the intersection [5,6,7,14,15,18].Installing centralized infrastructure at every intersection is not very practical due to prohibitively high total system costs. In this work, we advocate the use of Vehicle-to-Vehicle (V2V) communications and a distributed intersection algorithm that runs in each vehicle. Our focus in this paper is on (a) designing new protocols for V2V based-intersection management, (b) extending an advanced mobility simulator for vehicles, and (c) comparing our protocols to the operational efficiency of conventional intersections with stop signs and traffic lights.

The rest of this paper is organized as follows. Section 2 introduces the collision-detection algorithm used in our proposed intersection protocols. Section 3 contains intersection protocols used to manage various intersection scenarios. This section consists of a stop-sign model, a traffic light model and three V2V-based protocols: V2V Stop-Sign Protocol (SSP), Throughput-Enhancement Protocol (TEP) and Throughput-Enhancement Protocol with Agreement (TEPA). In Section 4, we describe the implementation of our protocols using the GrooveNet hybrid simulator, with new mobility and trip models. Section 5 contains the evaluation of our intersection protocols. Section 6 presents our concluding remarks.

COLLISION DETECTION AT INTERSECTIONS

We currently define an intersection as a perfect square box which has predefined entry and exit points for each lane connected to it. The trajectory of the vehicle crossing the intersection, is supposed to be the path taken by the vehicle from the entry to the exit point. We assume that each vehicle has access to a map database that provides routing, lane and road information, in which each segment of the road has a unique identifier (ID). Intersections are also identified by unique IDs in this map database.

Suppose *Arrival-Time* is the time at which a vehicle arrives at an entrance of the intersection and *Exit-Time* is the time at which the vehicle exits the intersection area. We refer to the part of the road that a vehicle is currently on as its **current road segment (CRS)**, and the part of the road that the vehicle will be moving to after the current road segment as the **next road segment (NRS)**. In the context of an intersection, CRS corresponds to the road segment that a vehicle is on before the intersection, and NRS represents the road segment that the vehicle will be on after crossing the intersection.

Each vehicle broadcasts CRS, NRS, current lane number, as well as the Arrival-Time and the Exit-Time, to all the other vehicles in its communication range. Vehicles are also assumed to have access to a global positioning system (GPS) with locally generated Radio Technical Commission for Maritime (RTCM-104) corrections to achieve Real-time Kinematic (RTK) solution.

Vehicles use this information to determine the other vehicles' turn types. Figure 1 shows an example of this, wherein a vehicle intends on entering the intersection from the east and exiting to the south. Based on the CRS, NRS and lane number, we can figure out that the vehicle is going to make a right turn. We assume in this paper that vehicles can make different turns regardless of their current lane number but they should stay in the same lane after passing any intersection and do not switch lanes. It is relatively easy to restrict this behavior, assumed for convenience here.

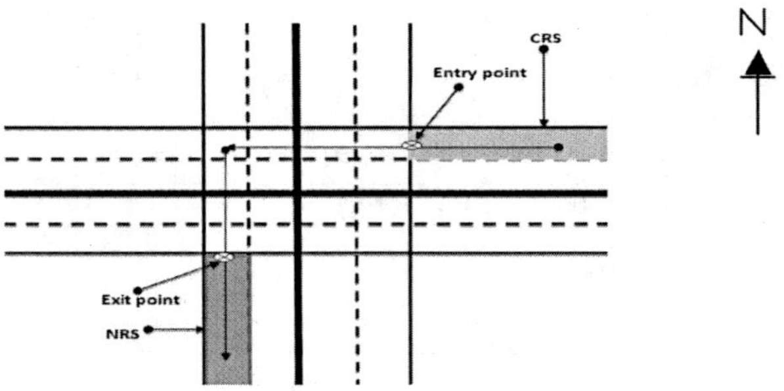

Figure 1.

We first identify the conditions required for two or more vehicles to collide at an intersection.

If a vehicle enters an intersection while another vehicle is in the intersection area, their (*Arrival-Time*, *Exit-Time*) intervals must overlap. Two vehicles being inside the same intersection at the same time is a necessary, but not sufficient condition for a collision. In Figure 2 (a), two vehicles are within the intersection at the same time but not occupying the same space. Figure 2(b) shows a scenario in which a vehicle is coming from the south and turning right while the other vehicle is coming from the north and also turning to its right. In this case, both vehicles can cross the intersection at the same time without a collision.

A collision occurs if the following conditions are all true:

1. Same Intersection: vehicles are at the same intersection.

2. Time Conflict: vehicles have overlapping (Arrival-Time, Exit-Time) intervals.

3. Space Conflict: vehicles occupy the same space while crossing the intersection.

If any of the above three conditions is false, then there will be no collision and vehicles can safely continue along their trajectory.

Our **Collision Detection Algorithm for Intersections** (CDAI) will be run on each vehicle that crosses a transaction, with information exchanged among vehicles approaching, crossing and leaving the intersection. The algorithm uses path prediction to determine any space conflicts with other vehicles trying to cross the intersection. Each lane on the road is considered to be a polygon, which starts from the previous intersection and ends at the next approaching intersection. Then, CDAI predicts the space (or region) which will be occupied by the vehicle during its trajectory. Utilizing the CRS (current road segment), current lane, and NRS (next road segment) information for each vehicle, CDAI predicts the path taken by the vehicle to cross the intersection and generates two polygons: the first polygon is related to the vehicle's CRS and current lane, and the second polygon is

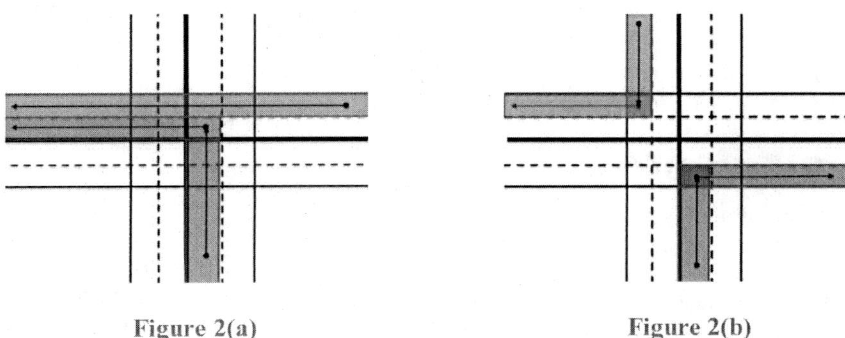

Example scenarios in which no space conflict occurs at the intersection

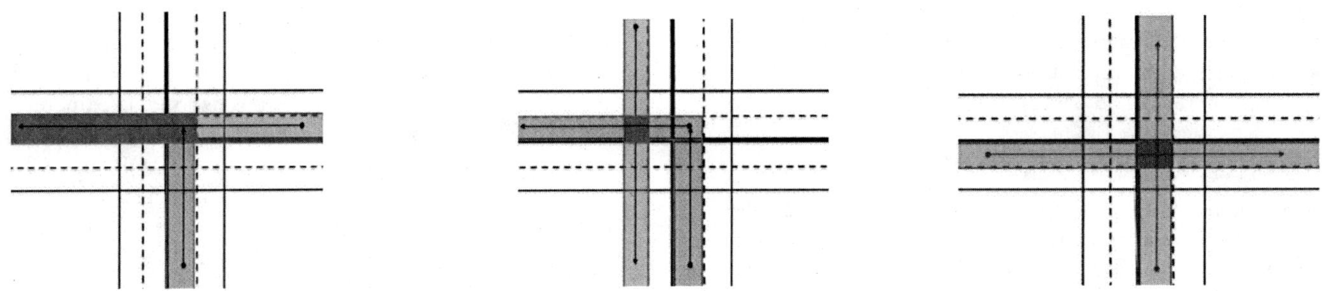

Figure 3. Three example scenarios of space conflict

related to the vehicle's NRS. Each polygon's height is the length of the road between two consecutive intersections and the polygon's width is the lane width. So, for each vehicle, these two polygons together form the complete spatial region related to its path, which we refer to as its **Trajectory Box (TB)**. As illustrated in Figure 3, a collision can potentially occur if two vehicles approaching the same intersection have intersecting "TBs".

To find out if the TBs of two vehicles intersect, we use the *Separating Axis Theorem* [8, 9]. The Separating Axis Theorem states that, for a pair of convex polygons that are not in a state of collision, there exists an axis perpendicular to an edge of one of the polygons that has no overlap between the projected vertices of the two polygons. This theorem can be simplified for our purposes since we are only dealing with two-dimensional rotated rectangles. Therefore, each polygon is tested against the four axes of the other polygon and if all projections overlap, a collision is detected. An optimization on this theorem exists for two-dimensional rotated rectangles, wherein, the polygon-under-test is rotated and centered on the intersection of the *x*-axis and the *y-axis*, and hence projections need to occur for only 2 axes [13].This solution works for any collision possibility, even for *cross-collisions* where a collision occurs between two polygons perpendicular to each other.

If a potential collision is detected by CDAI, it uses a **priority-based policy** to assign priorities to vehicles so that an unambiguous and repeatable precedence order in which vehicles cross the intersection can be established. For prioritizing the movement of vehicles at the intersection, the "first come, first served" (FCFS) algorithm is used. Based on FCFS, the first car arriving at the intersection is the first one crossing and leaving the intersection. Even though FCFS is an efficient algorithm, it can lead to a deadlock situation in particular scenarios such as when several vehicles get to an intersection at the same time (or very close to each other). To avoid any deadlocks, three **tie-breaking policies** are applied for vehicles with the same arrival time in the following sequence:

1. Roads are categorized as *primary roads* and *secondary roads* based on the roadmap database information, and higher priority is assigned to vehicles arriving at an intersection using a primary road than vehicles arriving using a secondary road.

2. If vehicles arrive at an intersection at the same time and using the same type of (primary or secondary) road, priorities are assigned based on their trajectories and whether turns are required. Specifically, vehicles going straight have higher priority than vehicles turning right, with vehicles turning left getting the lowest priority.

3. If all the previous conditions still result in a tie among two or more vehicles, the *Vehicle ID* (VID) which is unique for each vehicle is used to break ties - the vehicle holding a higher VID is given higher priority to cross the intersection.

The CDAI decision is made within each vehicle based on the information communicated using V2V.The algorithm will alert a vehicle if it can cross the intersection safely or, if any collisions are predicted, the vehicle must stop.

INTERSECTION PROTOCOLS

In this section, we describe three protocols[1]: the Stop-Sign Protocol (SSP), the Throughput-Enhancement Protocol (TEP) and Throughput-Enhancement Protocol with Agreement (TEAP). We will specify their functionality under various scenarios. The contents of messages communicated among the vehicles will be detailed in the next section. In the following three protocols, we assume that all vehicles have the same shape and physical dimensions. They do not use any controller model, which means that there is no consideration about a vehicle's movement during acceleration and deceleration. The communication medium has been assumed to be perfect; therefore, no packet loss occurs.

STOP-SIGN PROTOCOL (SSP)

In this protocol, we assume that stop-signs are not physically present at the intersection but vehicles obey the stop-sign rules when they approach an intersection. Vehicles only use V2V communications. Let t_s be the minimum amount of time in seconds that a car must wait at an intersection before proceeding. When vehicles approach an intersection, they must obey the rules of a stop sign which is to waitt_s seconds even if there is no other vehicle around. The FCFS priority policy mentioned in the previous section is obeyed by each vehicle. Vehicles also use STOP and CLEAR safety messages at the intersection in order to inform other vehicles in range about their current situation and movement parameters. The following rules are applicable.

• **Sending STOP:** As a vehicle approaches an intersection, it transmits a STOP safety message. Any vehicles within range will receive that message. When the vehicle arrives at the intersection, it also comes to a complete stop for t_s seconds.

• **On Receiving STOP:** On receipt of a STOP message, a vehicle uses the CDAI scheme described earlier, except for the *Space Conflict* rule. If more than one car arrives at the same intersection and will be inside the intersection area for an overlapping interval of time, priorities will get assigned to them and the vehicle with the highest priority will cross the intersection aftert_s seconds pass. Lower-priority vehicles will remain stopped even after t_s seconds, waiting to receive a CLEAR message.

• **Sending CLEAR:** When a vehicle crosses the intersection secondary and travels a distance defined by a threshold parameter D_{TC}, it broadcastsCLEAR messages indicating that the intersection is now safe to pass.

• **On Receiving CLEAR:** On receiving this message, the vehicle checks if it has stopped for at leastt_s seconds and, if true, it then checks if the sender of the CLEAR message is the same as the sender of the STOP message. The FCFS, priority and tie-breaking rules are again applied. Ift_s seconds have not passed as yet, the vehicle remains stopped while processing received messages to make a decision when the t_s seconds ends.If several vehicles are stopped at the intersection, by re-applying the priority policy, each vehicle decides if it should remain stopped or it can cross the intersection next as it has the highest priority among all stopped vehicles at the intersection.

THROUGHPUT ENHANCEMENT PROTOCOL (TEP)

This protocol is designed to manage intersection crossings by pure V2V communication without using any infrastructure such as stop-signs, traffic lights, sensors and cameras. The goal is to enhance the throughput at intersections without causing collisions. Vehicles again use STOP and CLEAR safety messages to interact with other vehicles. We define the throughput of an intersection based on the delay of all vehicles trying to cross the intersection. The following rules are applicable to each vehicle.

• **Sending STOP:** Every vehicle has access to its own GPS coordinates, speed and also to the map database; using these values, it computes the distance to the approaching intersection. The accuracy of this distance prediction is directly related to GPS accuracy. If the current distance of the host vehicle from the other vehicle is not greater than a threshold parameter D_{STOP}, then it starts sending periodic STOP messages (with the goal of informing other vehicles within range that it is getting close to the intersection). The STOP message will be sent with frequencyf_{STOP}.

• **Sending CLEAR:** When the vehicle exits the intersection, it sends periodic CLEAR messages with frequencyf_{CLEAR} until it travels further than a threshold value D_{CLEAR} from the exit point of the intersection. This behavior lets other vehicles know that the intersection is no longer in use by this vehicle.

• **On Receiving STOP:** On receiving a STOP message, the vehicle checks if all three collision conditions are satisfied. If even one of the conditions is not satisfied, then it means that the vehicle can cross the intersection without a collision with the sender of the STOP message. Otherwise, the vehicle acts based on the priority assigned to it using the priority policy. If it has lower priority than the sender of the STOP message, it comes to a complete stop at the intersection. Else, it has higher priority and ignores this message. In the latter case, the vehicle will have precedence at the intersection. Note that a

[1]Our protocols are inspired at least in part by Kurt Dresner's work [11]. Our focus is exclusively on V2V-based protocols, and our contributions include support for intersection management protocols in GrooveNet [1, 2], detailed evaluations and ongoing implementations in real vehicles.

vehicle which first started sending STOP messages may be superseded by a later vehicle due to priority considerations.

- **On Receiving CLEAR:** Each vehicle stores the information within received STOP messages which made it stop at the intersection. On receiving a CLEAR message, the vehicle checks if this message is sent from the sender of the last STOP message that has higher priority and because of which the vehicle is waiting at the intersection. This check is possible by just looking at the unique VID embedded in the message. If the VID of the CLEAR message is the same as the VID of the last processed STOP message, then the space that the vehicle needs to occupy for crossing the intersection is now clear.

Using TEP, vehicles stop at the intersection only if the collision detection algorithm predicts a collision and assigns a lower priority to them based on the messages it receives from all vehicles at the intersection. If no collision potential is detected or the highest priority is determined among contending vehicles, a vehicle can ignore other STOP messages, broadcast its own STOP messages to notify other vehicles, and cross the intersection safely. Multiple vehicles can be inside the intersection area at the same time if no space conflict occurs based on the collision detection policy's results. These rules increase the throughput of the intersection by decreasing the average delay time relative to the situation that vehicles should stop at the intersection. (We are currently studying enhancements to this protocol which will enable vehicles to slow down instead of coming to a complete stop when there are vehicles with higher priority entering the intersection. Evaluations of this scheme will be reported in the near future).

A reader might note that TEP implicitly assumes that V2V messages are not lost. While TEP will indeed work better with a very reliable wireless medium, the periodic transmission of STOP and CLEAR messages is targeted at a lossy communications medium and the protocol can tolerate *some* lost messages.

THROUGHPUT ENHANCEMENT PROTOCOL WITH AGREEMENT (TEPA)

This protocol is built on TEP and is explicitly designed to handle lost V2V messages. Additional CONFIRM and DENY messages are used to perform explicit handshaking between vehicles approaching the same intersection. Each vehicle makes its own local decision as in the previous protocols, but each vehicle announces its decision to cross the intersection by sending a CONFIRM or DENY message to either adhere to or override a decision made by another vehicle. On receiving a STOP message from another vehicle, the receiver will also send a message to acknowledge the reception of the message. The following rules are used by each vehicle *in addition to* the rules used by TEP:

- **Sending CONFIRM:** if no collision with the sender of a STOP message is predicted by CDAI, this message is sent first. It is also sent if a collision is predicted and a lower priority is assigned to the receiver of the STOP message. In this case, the receiver of the STOP message comes to a complete stop and waits for a CLEAR message.

- **Sending DENY:** If a collision is predicted and the receiver of the STOP message has a higher priority than its sender, the vehicle will send a DENY message to inform the sender of the STOP message that the latter's decision has been overridden and that this vehicle will *not* stop at the intersection.

- **On Receiving CONFIRM:** if the vehicle had sent a STOP message earlier, it has higher priority than the sender of the CONFIRM message and continues to proceed with its current decision.

- **On Receiving DENY:** if the vehicle had sent a STOP message later, it now has lower priority than the sender of the DENY message and must wait for a CLEAR message when it must re-evaluate the situation.

The collision detection scheme used in our intersection protocols ensures that two vehicles will not occupy the same space at the same time while crossing the intersection. Essentially, if there is any trajectory conflict, then one of the cars will be assigned a higher priority based on the priority policy, and the other one will wait for a CLEAR message without entering the intersection area. This prevents any collision between vehicles crossing the intersection.

DISTANCE KEEPING

In order to ensure a safe distance between cars, a distance-keeping protocol known as the *Car-Following Model* is used. This model is designed to control the mobility of vehicles while moving towards and exiting the intersection. A message of type *Generic* is sent at a regular interval and contains information about a vehicle's position, current lane, as well as current and projected map DB locations. On receiving this message, each vehicle checks if it is on the same road segment and the same lane as the sender. If this is the case, then by comparing its current GPS position with the sender's position, the vehicle determines if the sender is in front or behind it. In case of being behind the sender's vehicle, the vehicle adjusts its current velocity to the speed of the vehicle in front to prevent any collision. The vehicle does not need to have the same speed as the leader vehicle unless the distance between them is less than a threshold D_{follow}. Otherwise, it can maintain its current velocity which is related to the road's speed limit.

INTERSECTION SAFETY MESSAGE TYPES

We now describe in detail the content of transmitted messages.

The STOP message contains 9 parameters:

• **Vehicle ID:** Each vehicle has a unique identification number.

• **Current Road Segment:** Identifies the current road that the vehicle is using to get to the intersection.

• **Current Lane:** Identifies the lane being used.

• **Next Road Segment:** The next road taken by the vehicle after crossing the intersection.

• **Next Vertex:** The next intersection that the vehicle is getting close to.

• **Arrival-Time:** The time at which the vehicle gets to the intersection.

• **Exit-Time:** The time at which the vehicle will exit the intersection.

• **Message Sequence Number:** A unique number for each message from a vehicle. This count gets incremented for each new message generated by the same vehicle. This helps a receiver since it only needs to process the last message received from a particular sender.

• **Message Type:** The type of the message which is STOP in this case.

The CLEAR message contains 3 parameters: **Vehicle ID, Message Sequence Number**, and **Message Type:** CLEAR.

The CONFIRM message contains 3 parameters: **Vehicle ID, Message Sequence Number**, and **Message Type:** CONFIRM.

The DENY message contains 3 parameters: **Vehicle ID, Message Sequence Number**, and **Message Type:** DENY.

IMPLEMENTATION

In this section, we describe the implementation of the V2V protocols and the messages described in the previous two sections. To implement and analyze intersection protocols, traffic at intersections needs to be simulated. For this purpose, we use a tool called GrooveNet [1, 2] built at Carnegie Mellon University. We first give a brief introduction to GrooveNet and describe the extensions made.

GROOVENET[2]

GrooveNet [1, 2] is a sophisticated hybrid vehicular network simulator that enables communication among simulated vehicles, real vehicles and among real and simulated vehicles. By modeling inter-vehicular communication within a real street map-based topography, GrooveNet facilitates protocol design and also in-vehicle deployment. GrooveNet's modular architecture incorporates multiple mobility models, trip models and message broadcast models over a variety of links and physical layer communication models. It is easy to run simulations of thousands of vehicles in any US city and to add new models for networking, security, applications and vehicle interaction. GrooveNet supports multiple network interfaces, GPS and events triggered from the vehicle's onboard computer. Through simulation, message latencies and coverage under various traffic conditions can be studied.

New models can easily be added to GrooveNet without concern of conflicts with existing models as dependencies are resolved automatically. Three types of simulated nodes are supported: (i) vehicles which are capable of multi-hopping data over one or more DSRC channels, (ii) fixed infrastructure nodes and (iii) mobile gateways capable of vehicle-to-vehicle and vehicle-to-infrastructure communication. GrooveNet's map database is based on the US Census Bureau's TIGER/Line 2000+ database format [10]. Multiple message types such as GPS messages, which are broadcast periodically to inform neighbors of a vehicle's current position, are supported. On-road tests over 400 miles within GrooveNet have lent insight to market penetration required to make V2V practical in the real world [1].

Mobility Models

One major extension to GrooveNet that we made is the inclusion of lane information for roads. The TIGER map database has no information concerning the number of lanes along each road. We used the heuristic of adding lane information based on road-type information present in the database. GrooveNet has several mobility models, such as the Street Speed, Uniform Speed and Car-Following models. We have modified these models for our current purposes. In addition, we have also created new mobility models that support the presence of multiple lanes with vehicles now also having the ability to switch lanes. Cars can switch lanes either at randomly chosen times or using predefined starting lanes. Specifically, the new mobility models that were implemented are as follows:

1. Stop-Sign Model: When a simulated vehicle approaches an intersection managed by stop-signs at each entrance, it comes to a complete stop regardless of the situation of any other vehicle at the intersection. In other words, the velocity of the vehicle becomes zero even if there is no other car trying to cross the intersection. In discussions, police

[2]GrooveNet is an acronym that stands for "Geographical Routing of Vehicular Networks".

recommend 3 seconds of complete stopping even at an empty intersection. This stop delay will increase in proportion to the number of cars that arrived earlier at the intersection.

2. Traffic-Light Model: The traffic-light model follows the same basic logic as the stop sign model except that stop signs are now replaced by traffic lights. The *Green-Light Time* of the traffic light has a default value that can be changed by the user. Both the Stop-Sign and Traffic-Light models have been designed to simulate the behavior of vehicles at intersections equipped with stop-signs or traffic-lights. In these two models, vehicles do not communicate with each other.

3. V2V Stop-Sign Model: This model represents the implementation of the Stop-Sign Protocol (SSP) described earlier. Each intersection in the map has a unique number which is called its *Vertex Number*. Based on the vertex number, each vehicle determines the next intersection it is approaching and also all the roads connecting at this intersection. The vehicle sends out a periodic safety message as described earlier. These messages are processed by other vehicles receiving them to know if multiple vehicles are approaching the same intersection. A priority-assignment policy decides which vehicle gets to cross the intersection first. In case of distinct arrival times, a first-come-first-served policy is used. In case of ties, tie-breaking rules are applied. Any vehicle with a lower priority comes to a stop at the intersection. The vehicle then checks if other vehicles have exited the intersection. Based on the V2V stop-sign protocol, if the vehicle should remain stopped, the velocity stays zero until its next update cycle, after which the tests are executed again. This continues until the vehicle gets the permission to cross the intersection and sets its velocity to the street speed limit.

4. Throughput-Enhancement Model: The Throughput-Enhancement Protocol (TEP) is implemented by this model. This model uses the complete collision detection algorithm (CDAI) including *Space Conflict*. Vehicles obey the car-following rules on the road before getting to the intersection such that their speed gets adjusted to the vehicle in front based on the information received within periodic Generic safety messages. As the vehicle arrives at the intersection, it follows the V2V-based intersection rules and uses CDAI to determine if it is safe to cross the intersection. All safety messages including STOP, CLEAR and GENERIC are sent with a frequency of 10Hz. All safety messages utilize the same 10Hz V2V Basic Safety Message (BSM) formats defined by SAE J2735 Dedicated Short Range Communications (DSRC) Message Set Dictionary. Data elements in Part II of BSM are used to specify the type of the safety message and also encapsulate related elements defined in the previous section. The safe distance maintained between two contiguous vehicles is selected to be 10 m.

5. Throughput Enhancement with Agreement Model: This model is designed to use all five types of safety messages: STOP, CLEAR, CONFIRM, DENY and GENERIC. Each vehicle moves based on the car-following protocol before approaching an intersection as well as after exiting the intersection area. Vehicles follow the Throughput-Enhancement with Agreement Protocol (TEPA) rules to get to an agreement on the sequence that the vehicles at the intersection should cross and also inform each other about their decision.

EVALUATION

In this section, we present a detailed evaluation of the proposed protocols using the models added to GrooveNet. The evaluation is carried out under different types of traffic scenarios and using different kinds of intersections. We compare the different mobility models: the Stop-Sign Model, the Traffic-Light model and V2V-interaction models. Two instances of the Traffic Light model are used, one with green light duration of 10 seconds and another with duration of 30 seconds.

In this paper, we do not consider any lost messages due to a lossy communication medium and we have assumed a GPS system with high accuracy. Under these assumptions, the TEP and TEP-A will behave in exactly the same manner. This also holds true for the V2V stop sign model as compared to the normal stop-sign model. Therefore, as part of our evaluation, we only consider TEP and the stop-sign model.

METRIC

We calculate the trip time for each simulated car under each model and compare that against the trip time taken by the car assuming that it stays at a constant street speed and does not stop at the intersection. The difference between these two trip times is considered to be the **trip delay** due to the intersection. We take the *average trip delays* across all cars in a simulation sequence as our metric of comparison.

The trip route for each car is calculated using the DjikstraTripModel in GrooveNet which calculates the shortest route between two points using Djikstra's algorithm. The route is chosen with a waypoint at the intersection forcing the route to pass through the intersection. The logging mechanism in GrooveNet was modified to enable logging of start and end times of cars to measure their trip times.

SCENARIOS

Since there is a large variation in intersection types, we restrict our attention to the following three categories of intersections:

- **Four-way Perfect-Cross Intersections:** The intersection legs are at perfect right angles to the neighboring leg.

- **T-junction:** Two roads are perpendicular to each other, and one of the roads ends at the intersection.

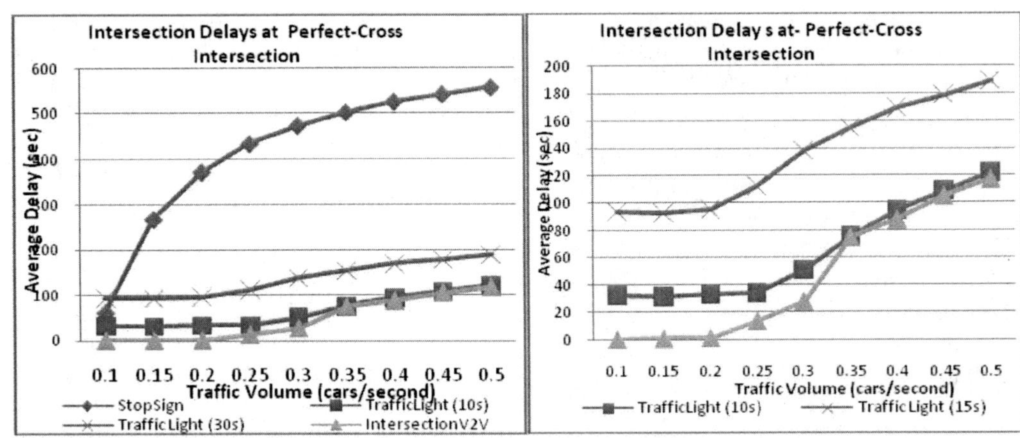

Figure 3. Delays for Perfect-Cross Intersection. Figure (a) shows all protocols. Figure (b) shows more detail w/o the Stop-Sign Protocol.

• **Four-way Angled Intersections:** These intersections are four-way intersections where they do *not* intersect at a right angle.

We run all our simulations on 4-lane roads, with 2 lanes in each direction. The intersection type, vehicle-birthing sequence, vehicle routes and turn-types are generated offline. Each vehicle is removed from simulation when it reaches its destination. This file is then fed into GrooveNet to simulate the intersection protocols. Traffic volume is specified on a per intersection-leg basis, allowing intersection legs to have different traffic levels. Each simulations uses 250 vehicles, and each run is terminated when the last vehicle reaches its destination. The simulation model in GrooveNet was modified to prevent a vehicle from becoming active if vehicles with earlier start times are already present within 10 meters of its starting position in its lane. This feature prevents cars from starting if the lane is already completely backed up.

EXPERIMENTAL RESULTS

In our first experiment, we compare different protocols for a perfect-cross intersection with an equal amount of traffic volume in every lane and an equal amount of turn ratios (that is, a vehicle has equal odds of going straight or making a turn at an intersection). The results are presented in Figure 3-(a). As can be expected, the Stop-Sign model results in higher average delays than the other protocols. As the traffic volume increases above 0.1 vehicles per second, the performance of the Stop-Sign model drops dramatically and significant traffic backlog results. In contrast, both Traffic-Light models behave at a near-constant level until the traffic volume reaches 0.25 cars per second for the Traffic-Light model with a green-light time of 10 seconds and 0.3 cars per second for the Traffic-Light model with a green-light time of 30 seconds. After that, the average delay jumps until it settles down at a higher near-constant level at about 0.35 cars per second. Beyond this traffic volume, the Traffic-Light models behave the same regardless of the green-light duration as all the lanes are completely saturated and traffic is backed up significantly. The V2V Intersection model performs the best, doing very well at low traffic volumes up to 0.2 vehicles per second resulting in very negligible delay. As traffic volume increases, the average delay increases and beyond 0.3 cars per second, it performs very similar to the Traffic-Light model with a green-light time of 10 seconds. However, the overall performance improvement is about 26% as compared to the latter Traffic-Light model. Figure 3-(b) zooms into the plot of Figure 3-(a) to show a detailed comparison between the Traffic-Light models and the Intersection model, by not showing the poorly performing Stop-Sign model.

According to classical queueing theory, the average delay will asymptotically become *very* high when the arrival rate (i.e. traffic intensity) exceeds the service rate (throughput) at the intersection. This delay, however, occurs under steady-state conditions only after a considerable amount of time. Due to practical considerations, our simulations are run for finite durations, and hence capture only transient delay behaviors after overload conditions have been reached. Nevertheless, our results clearly indicate that before overload conditions are reached, the service rate (i.e. throughput) with the V2V-Intersection protocol is noticeably better than the Traffic-Light models.

We then repeated the above experiment for a T-junction and the corresponding results are shown in Figure 4. For the T-junction, the V2V-Intersection protocol has an 83% overall performance improvement over the Traffic-Light model with a 10-second green-light time, and a 94% overall performance improvement over the Traffic-Light model with a 30-second green-light time. The T-junction has fewer conflicts to deal with than at a perfect-cross intersection, resulting in less stopping at the intersection for the V2V-Intersection model leading to its much better performance than before.

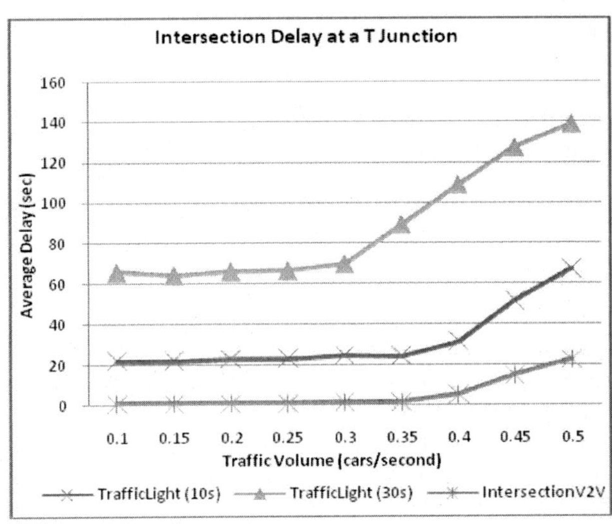

Figure 4. Delays at a T-Junction

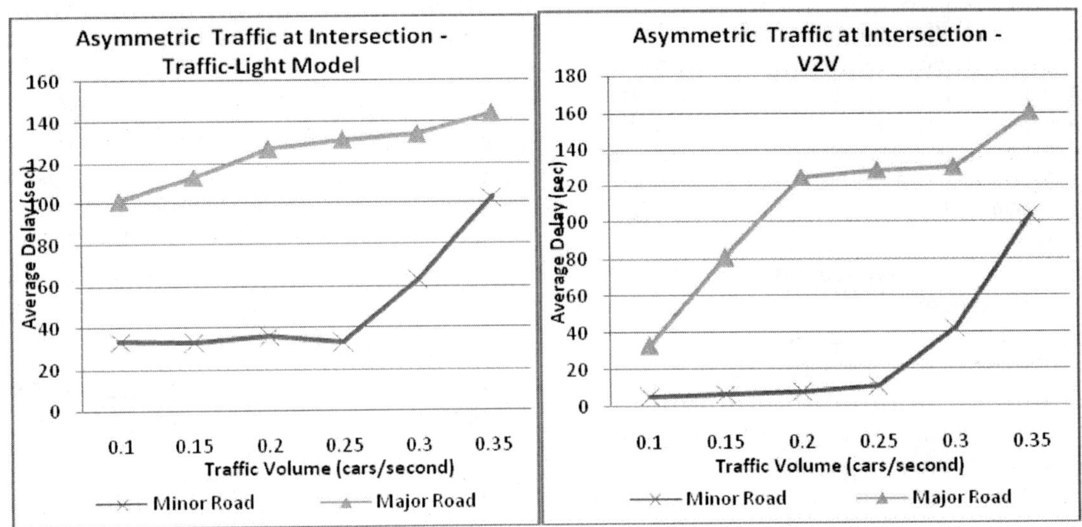

Figure 5. Delays with Asymmetric Traffic Traffic-Light Model (b) V2V Intersection Model

Next, we studied the scenario where traffic varies along different intersecting roads. That is, when two roads intersect, one road has more traffic than the other. However, we still assume that both roads have the same type and hence one does not have priority over the other. The corresponding results are given in Figure 5-(a) for the Traffic-Light Model and Figure 5-(b) for the V2V-Intersection Model. Again, the V2V-Intersection Model performs better than the Traffic-Light Model.

SUMMARY/CONCLUSIONS

A substantial fraction of automotive collisions occur at intersections. Furthermore, intersections are often traffic bottlenecks contributing to significant trip delays. In this paper, our goal was to design intersection management protocols using only vehicle-to-vehicle communications to address these two core issues of safety and throughput. We believe that intersection collisions can be reduced and throughput improved significantly using only V2V protocols. Since installing wireless infrastructure at every intersection to support vehicle to intersection protocols can be prohibitively costly, a V2V-based approach seems more practical for deployment. We have described and evaluated four V2V-based protocols namely Stop-Sign, Traffic-Light, Throughput-Enhancement and Throughput-Enhancement with Agreement protocols. We have also compared these protocols to conventional stop-signs and traffic lights, and have evaluated the average delays encountered at an intersection. We extended GrooveNet [1], a sophisticated hybrid vehicular network simulator, to support these protocols. Our results indicate the potential of these new V2V-based protocols to manage intersections with minimal dependency on infrastructure. Although our protocols are designed for autonomous vehicles that use V2V communication for co-operative driving, they can be adapted to a driver-alert system for manual vehicles at traffic intersections.

LIMITATIONS

The protocols we evaluate do not take into account any controller model for the cars. Since they assume a simplistic movement model based on current speed and current heading, certain assumptions will be violated when applied to real cars, especially when considering the throughput enhancement model where we look at polygon intersections. The ability to integrate several different controller models needs to exist and their effect on the protocols needs to be studied. Currently, we also do not deal with position inaccuracies and packet losses with wireless communication. Position accuracy will affect the protocols since each vehicle depends on its position and the known position of the other vehicles to make safety-critical decisions. Wireless packet loss results in dropped messages between vehicles and this can lead to vehicles not being able to sense other vehicles around them. We also make assumptions at a global level such as the constant speed of all cars (unless they are using the car-following model), and these assumptions are made at a global level. Hence, protocol changes will also need to be made for adapting to scenarios involving different types of cars traveling at different speeds.

Future Work

We intend to overcome the limitations described above and extend the V2V protocols in the context of real cars. We are working on hybrid simulations with real and simulated vehicles to take advantage of GrooveNet's hybrid environment. We are also working on extending our protocols to support enhancements, which will allow a vehicle to slow down, and not come to a complete stop, at an intersection to allow another vehicle to cross. We have indeed already implemented a version of this protocol on real-world Segway robots but it is not captured in this paper. There is also ongoing work to look at integration of Vehicle-to-Infrastructure (V2I) technologies within these protocols to take advantage of statically known entities at intersections. We intend to design new protocols which use the integration of V2I and V2V for managing intersections, where autonomous and human-driven vehicles are both present.

REFERENCES

1. Mangharam, R., Weller, D. S., Rajkumar, R., Mudalige, P. and Bai, Fan, "GrooveNet: A Hybrid Simulator for Vehicle-to-Vehicle Networks", Second International Workshop on Vehicle-to-Vehicle Communications (V2VCOM), San Jose, USA. July 2006

2. Mangharam, R., Meyers, J., Rajkumar, R., Stancil, D. et al., "A Multi-hop Mobile Networking Test-bed for Telematics," SAE Technical Paper 2005-01-1484, 2005, doi: 10.4271/2005-01-1484.

3. DARPA. The DARPA urban challenge, 2007.http://www.darpa.mil/grandchallenge.

4. GM, EN-V concept vehiclesat expo 2010, http://media.gm.com/content/media/us/en/news/news_detail.globalnews.html/content/Pages/news/global/en/2010/0817_env_capabilities

5. Dresner, Kurt & Stone, Peter (2008), Replacing the Stop Sign: Unmanaged Intersection Control, The Fifth Workshop on Agents in Traffic and Transportation Multiagent Systems. pp. 94-101, Estoril, Portugal.

6. Dresner, Kurt & Stone, Peter (2008), A Multiagent Approach to Autonomous Intersection Management. Journal of Artificial Intelligence Research(JAIR)

7. Drabkin, V., Friedman, R., Kliot, G., and Segal, M.. Rapid: Reliable probabilistic dissemination in wireless ad-hoc networks. In The 26th IEEE International Symposium on Reliable Distributed Systems, Beijing, China, October 2007.

8. Golshtein, E. G.; Tretyakov, N.V.; translated by Tretyakov, N.V. (1996). Modified Lagrangians and monotone maps in optimization. New York: Wiley, p. 6.

9. Shimizu, Kiyotaka; Ishizuka, Yo; Bard, Jonathan F. (1997). Nondifferentiable and two-level mathematical programming. Boston: Kluwer Academic Publishers, p. 19.

10. U.S Census Bureau. http://www.census.gov/geo/www/tiger/

11. Dresner, Kurt, Ph.D. Dissertation, "Autonomous Intersection Management", University of Texas at Austin, May 2010.

12. US Department of Transportation-Federal Highway Administration Publication, National Agenda for Intersection Safety http://safety.fhwa.dot.gov/intersection/resources/intersafagenda/

13. Meythaler, Eric, 2D Rotated Rectangle Collision http://www.gamedev.net/reference/programming/features/2dRotatedRectCollision/

14. Maile, M., Ahmed-Zaid, F., Basnyake, C., Caminiti, L., Kass, S., Losh, M., Lundberg, J., Masselink, D., McGlohon, E., Mudalige, P., Pall, C., Peredo, M., Popovic, Z, Stinnett, J., and VanSickle, S. Cooperative Intersection Collision Avoidance System Limited to Stop Sign and Trafic Signal Violations (CICAS-V) Task 10 Final Report: Integration of Subsystems, Building of Prototype Vehicles and Outfitting Intersections. Washington, DC: National Highway Traffic Safety Administration.

15. Vehicle Safety Communications 2 consortium. Cooperative Intersection Collision Avoidance System Limited to Stop Sign and Trafic Signal Violations (CICAS-V) Final Report. Washington, DC: National Highway Traffic Safety Administration.

16. Ahmed-Zaid, F., Bai, F., Bai, S., Basnayake, C., Bellur, B., Brovold, S., Brown, G., Caminiti, L., Cunningham, D., Elzein, H., Ivan, J., Jiang, D., Kenny, J., Krishnan, H., Lovell, J., Maile, M., Masselink, D., McGlohon, E., Mudalige, P., Rai, V., Stinnett, J., Tellis, L., Tirey, K., VanSickle, S. *Vehicle Safety Communications - Applications (VSC-A)* First Annual Report. Washington, DC: National Highway Traffic Safety Administration.

17. Vehicle Safety Communications 2 consortium. *Vehicle Safety Communications - Applications (VSC-A)* Final Report. Washington, DC: National Highway Traffic Safety Administration

18. Bouraoui, Laurent, Petti, Stéphane, Laouiti, Anis, Fraichard, Thierry and Parent, Michel, INRIA Rocquencourt, "Cybercar cooperation for safe intersections", Proceedings of the IEEE ITSC, 2006 IEEE Intelligent Transportation Systems Conference, Toronto, Canada, September 17-20, 2006

CONTACT INFORMATION

Reza Azimi
sazimi@andrew.cmu.edu

Gaurav Bhatia
gnb@ece.cmu.edu

Ragunathan (Raj) Rajkumar
rai@ece.cmu.edu

Priyantha Mudalige
priyantha.mudalige@gm.com

2009-01-0959

Nomadic Device Connectivity Using the AMI-C HMI Architecture

Frank Szczublewski, Laci Jalics and Mark Krage
Delphi Electronics & Safety

ABSTRACT

Nomadic mobile consumer electronic (CE) devices are growing in functionality and popularity. Some of these devices, such as navigation systems, are being used in vehicles as a lower cost alternative to integrated vehicle options. Other devices, such as MP-3 players, are becoming the preferred source of music on the go. Wireless nomadic devices are now capable of accessing E-mail and other Internet-based functions. Automakers are beginning to recognize the importance of integrating support for such devices to facilitate their use in vehicles. A key element of this integration is the ability of the vehicle HMI to support both the operation of nomadic devices as well as the display of content from such devices. This paper presents an example of how a nomadic device can be properly integrated with the vehicle HMI using the AMI-C HMI architecture. In particular, a commercial nomadic device was used to stream MP3 content to a vehicle radio using an 802.11 wireless connection. The goal of this effort was to produce a "good user experience" similar to that of commercial, integrated product. The result is a compelling implementation of MP3 streaming audio playback that users perceive as being built into the vehicle

INTRODUCTION

The term "nomadic devices" refers to portable consumer electronic devices that provide one or more computer (e.g., personal digital assistants), communication (e.g., cellular phones, GPS), or entertainment (e.g., MP-3 players) functions. The popularity of such devices has grown along with the ability to provide more and more functionality in highly portable form factors with nearly constant or decreasing prices. As the users of nomadic devices become comfortable with and, in some cases, dependent on these devices, their use has grown to new environments. People now carry on phone conversations while shopping in stores or driving to work; people now listen to recorded music while jogging or working in the yard; and people surf the Internet while having coffee at a café. Nomadic devices are becoming increasingly pervasive and are now competing with some integrated automotive electronic entertainment and communication devices. This is especially true in the areas of playback of recorded music and navigation assistance.

Automakers are beginning to recognize that integrating nomadic devices with vehicle HMIs may provide a way to get around the differences in life cycles of personal electronic communication and entertainment devices and related integrated product options in vehicles. Typically, it takes automotive OEMs two to or more years to validate the design of a product or option, while consumer electronic devices often have lifecyles of six months. As a consequence, consumer electronic devices developed by OEMs for vehicles can be out of date when they are initially offered for sale. A way around this dilemma is to develop an interface in the vehicle that accommodates current consumer electronic products. A real-world example of this type of accommodation is the recent introduction of the *Sync*,[1] a product that connects to mobile electronic devices to the vehicle. Sync's HMI is well blended into that of the vehicle's and is used to provide access to e-mail.

In a presentation by Johan Engstrom [2], in 2006, the AIDE (Adaptive Integrated Driver-vehicle InterfacE) solution for "Nomadic device integration" was presented. This approach advocates the use of a nomadic device gateway in the vehicle, which enables connectivity between nomadic devices and the vehicle HMI (Human Machine Interface). The nomadic device gateway concept is presented in more detail in reference [3]. The basic idea in this integration is to use on-board HMI for both the control of the nomadic device and for the display of information from the nomadic device. Since most nomadic devices are not designed for use when driving a vehicle, the integration with the vehicle HMI is an influential element for their safe use in vehicles. Gardner[3] lists a number of possible benefits from the nomadic gateway: minimization of driver workload; increased consumer convenience; more enjoyable user experiences and potential new applications.

As an example application, the Engstrom presentation mentions streaming media content from a nomadic device to an in-vehicle output device. The authors of this paper, in fact, have developed a working demonstration of the integration of cellular phone with the vehicle HMI for the purpose of streaming audio to the HMI of a vehicle. This integration was based on the AMI-C HMI architecture, which is discussed in the following section.

THE AMI-C HMI ARCHITECTURE

AMI-C was a collaboration of OEM's and suppliers, led by the suppliers, for developing mobile-multimedia standards. This collaboration was initiated in the late 1990's and disbanded in the early 2000's. The focus of AMI-C HMI specification [4] was a "content-based" approach to help multi-media devices work in harmony with vehicle HMI's and in different vehicle brands. In the content-based approach, the multi-media device sends its content (the basic information to be displayed) to an OEM specified HMI manager. This manager, in turn, formats the content into a form appropriate for the particular vehicle's HMI. With this approach, the same content can be sent to multiple, different HMI managers, and result in multiple, different rendered displays of the same information. A key element of this specification is an XML (eXtensible Mark-up Language) based VUIML (Vehicle User Interface Mark-up Language) that is used to for communication between the AMI-C compliant device and the HMI Manager.

The software "drivers" for the applications of interest can either reside in the nomadic device or can be downloaded to the software environment of the vehicle HMI. If the driver resides in the nomadic device, a (wireless) standard, such as Bluetooth, which comprehends the application, must be used to communicate with the HMI manager. On the other hand, if the driver has been downloaded to the software environment of the vehicle HMI, only the content needs to be transmitted to the HMI manager. The latter approach was used for the application of this paper.

By mapping the existing HMI controls to the controls of the nomadic device in the HMI Manager, it is possible use the vehicle HMI controls for both their normal vehicle functions and nomadic device control. However, some vehicle display functions may have a need to override the nomadic device display due to, for example, a malfunction or an urgent message, e.g., "low fuel." Similarly, the timing of a navigation instruction may need to override the display of the normal vehicle display. For this reason, the HMI manager also needs to manage the priorities of the displayed content on a message-by-message basis.

THE USER EXPERIENCE OF THE NOMADIC DEVICE CONNECTED WITH THE HMI OF A VEHICLE

Ideally, the user experience associated with using an integrated nomadic device should appear seamless to the user, i.e., it should be as easy to use as an original equipment feature using the HMI controls of the vehicle. Our initial demonstration used the cellular phone as the source information for a Java based MP3 player, which was resident on a personal computer (PC). Psinaptic supplied a variant of Java's Jini, *Jmatos*, for this demonstration. The complete song list and the controls for play, play next, play previous and stop were handled from the PC's player. This integration was so seamless that the people viewing this demonstration often had to be reminded that the MP3 audio streaming from the phone and was not resident in the PC.

In the next phase, the demonstration moved to the radio in the vehicle, where the bonding between the radio and the nomadic device took place without user intervention. When the bonding was complete, the list of songs and the same set of music playback controls were available. The phone provided the MP3 service to the vehicle radio and upon discovery of the MP3 service, the song list was provided to the client and the control interface to the MP3 client was enabled. The player plays the song list from a start point until the last song in the list has been played.

To show the flexibility of this concept, the radio head was a standard unit with the basic tuner and volume knobs and only two lines of text display. The upper line of text was used to show the music sort type (Genre, Artist, playlist, album) and the lower line was used to show the song to select. The tuner knob rotation was used to scroll the song list and pressing this knob selected the song. The presets were used to choose the music sort type.

This control scheme fit well with the normal radio functions and provided the user with a familiar, user-friendly experience. From a safety perspective, it is believed that using the vehicle HMI controls in place of the CE device controls should be safer.

A DESCRIPTION OF THE INTEGRATION OF A NOMADIC DEVICE WITH THE HMI OF A VEHICLE

In order to evaluate the integration of nomadic device with a vehicle HMI, we decided to develop a concept vehicle to demonstrate a cellular phone that streams audio to the audio system of a vehicle. A Chevrolet Malibu was selected as the vehicle and its radio was modified to include a production Hitachi SH4 daughter board (from another vehicle program), which is capable of WiFi and MP3 playback. An XV6700 Verizon phone was used as the nomadic device. The hardware for this demonstration system is shown in Figure 1.

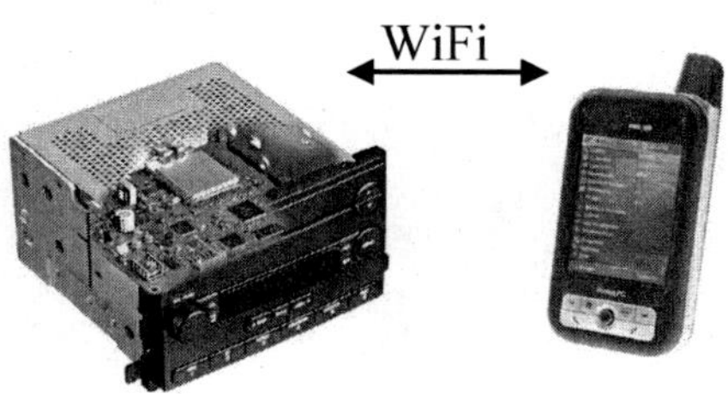

Figure 1: Streaming audio demonstration hardware

A block diagram of the software architecture for the demonstration is shown in Figure 2. A content based HMI Manager was created that enables applications to present their content without knowledge of how it is presented. A Jini [5] service discovery server was put into the radio so that wireless Jini1 clients could register their services; in this case a MP3 WiFi phone brings with it an MP3 application. The phone uses Jini to register with the discovery server and then transfers its MP3 application to the radio. The MP3 application that is now in the radio interrogates the phone to access its playlist.

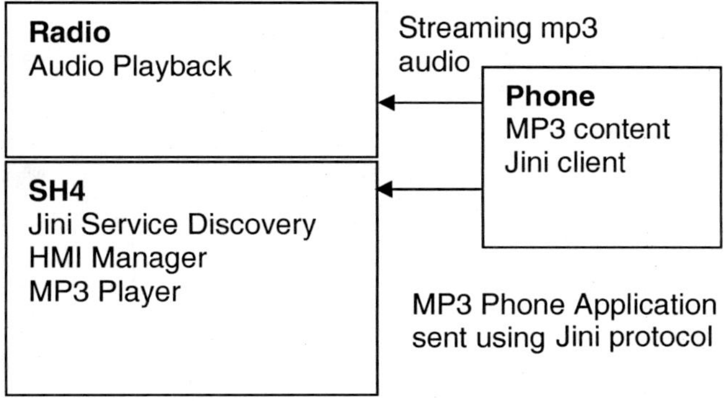

Figure 2. Streaming audio demonstration hardware

The radio ran a java based OSGi [6] framework. One of the services on the OSGi Framework was a download service, which would look for an mp3 application using the Jini lookup service. Once it connected to the mp3 application, which was uploaded by the phone, it retrieved the playlist using the mp3 application API, which it passed to the player running on the radio. Using HTTP, the radio opened a stream connection to the MP3 file on the phone and played it through the audio system of the radio. The HMI manager on the radio allowed the user to select songs from the playlist by rotating the knob and pushing the select button.

The reaction of evaluators of the nomadic device concept vehicle has been very positive. Bringing a nomadic device to the vehicle and having it wirelessly communicate with the entertainment system is compelling. In addition, the user interface with the nomadic device was made to look and feel as if it were designed for that vehicle. Combining service discovery (Java, OSGi, Jini) with a content-based HMI Manager, and a wireless connection provides a powerful user experience.

POST AMI-C HMI SPECIFICATION DEVELOPMENTS

AMI-C was a collaboration of OEMs that was formed to develop some needed standards related to future multi-media applications in vehicles. While AMI-C did produce an HMI content-based specification, not all OEM's having an initial interest in this development completed their tenure with AMI-C. BMW, for example, initially was an AMI-C participant but decided to withdraw and developed a separate "content-based" approach, which is described in reference [7]. It remains to be seen if the AMI-C HMI specification will gain traction with other OEMs or if other OEMs will collaborate with BMW or develop yet other separate approaches to accomplish the same objectives. In any case, it is clear that OEMs are giving the "content-based" solutions serious consideration for the integration of CE devices with vehicle HMIs..

CONCLUSIONS

Due to the significantly different life cycles of vehicles and consumer electronics technologies, in-vehicle Internet-based services, a new paradigm is needed to accommodate the seamless use of current consumer electronic devices in vehicles.

Changes are needed to the vehicle user interface to enable the display of content from nomadic devices. The implications of such changes potentially involve both hardware and software upgrades to the vehicle during its useful life.

The AMI-C architecture used in our concept vehicle proved to be very effective in the development of a seamless integration.

REFERENCES

1. http://fordvehicles.com/sync , Ford/Microsoft joint effort
2. Johan Engstrom (Volvo Technology Corporation), "Toward the automotive HMI of the future: Mid-term results from the AIDE project," 2006 Advanced Microsystems for Automotive Applications - Berlin, April 25-27, 2006.
3. Mike Gardner, Motorola, May 2006 "position" paper, "Nomadic Devices" (www.morotola.com/mot/doc/6/6459_MotDoc.pdf)
4. Laci Jalics and Frank Szcziblewski, "AMI-C Content-Based Human Machine Interface (HMI)," SAE 2004-01-0272
5. http://www.sun.com/software/jini/
6. http://www.osgi.org
7. Reinhard Stolle, et. al., "Integrating CE-based Applications into the Automotive HMI, SAE 2007-01-0446.

"Verify-on-Demand" - A Practical and Scalable Approach for Broadcast Authentication in Vehicle-to-Vehicle Communication

2011-01-0584
Published
04/12/2011

Hariharan Krishnan
General Motors Company

Andre Weimerskirch
escrypt Inc.

doi:10.4271/2011-01-0584

ABSTRACT

In general for Vehicle-to-Vehicle (V2V) communication, message authentication is performed on every received wireless message by conducting verification for a valid signature, and only messages that have been successfully verified are processed further. In V2V safety communication, there are a large number of vehicles and each vehicle transmits safety messages frequently; therefore the number of received messages per second would be large. Thus authentication of each and every received message, for example based on the IEEE 1609.2 standard, is computationally very expensive and can only be carried out with expensive dedicated cryptographic hardware. An interesting observation is that most of these routine safety messages do not result in driver warnings or control actions since we expect that the safety system would be designed to provide warnings or control actions only when the threat of collision is high. If the V2V system is designed to provide too frequent warnings or control actions, then the system would be a nuisance to the driver. Therefore it is reasonable to define an approach where messages are first processed and then authenticated using verification on-demand. In this paper we describe such an approach and discuss its implementation for V2V safety system. It is shown that Verify-on-Demand (VoD) is a practical and scalable approach for broadcast authentication in V2V safety communication while conforming to the IEEE 1609.2 standard.

INTRODUCTION

In Vehicle-to-Vehicle communication (V2V), vehicles equipped with a short range wireless transceiver and a Global Positioning System (GPS) receiver regularly exchange safety-related information including time, location, and further vehicle status data amongst neighboring vehicles [1]. The communication, in general, is done as a single-hop, periodic broadcast although multi-hop routing may also be used to extend the geographical range and region of message reception [2]. It is expected that periodic vehicle broadcast of safety information would be around 10 messages per second with an average message size about 200 bytes [3]. The required transmission range of safety messages is approximately 300 meters for V2V safety communication applications. It is expected that V2V would employ the wireless communication protocol based on IEEE 802.11p Dedicated Short Range Communications (DSRC) in the 5.9 GHz band [4], although other short range wireless protocols may also be used.

Security is a core issue for V2V safety communication [5]. In particular, vehicles need to be able to authenticate that a received message originated from a properly certified vehicle and that the message was not manipulated on its way between the sender and receiver vehicles. It is assumed that there is a Public Key Infrastructure (PKI) deployed and the messages are authenticated using digital signatures in accordance with the IEEE 1609.2 standard specification [6]. IEEE 1609.2 describes a message format of secured safety messages in a V2V network. IEEE 1609.2 suggests an API and message format for using security features based on Elliptic Curve

Digital Signature Algorithms (ECDSA) [8] and certificates, namely attaching a digital signature and a certificate, a certificate digest or a certificate chain (if a hierarchical PKI is used) with each message. While this solution is robust there are concerns regarding Over-The-Air (OTA) bandwidth overhead and run-time performance in a V2V safety application setting. Network simulations [10] suggest that a certificate once or twice per second and a digest otherwise is sufficient in order to reduce the OTA bandwidth due to certificate size.

V2V safety applications require that vehicles are able to verify a large number of messages at short delay. As the penetration of V2V vehicles increases, the number of received messages per second could become very large. Estimation for the number of messages to verify is potentially beyond 1,000 per second, whereas a delay of 10-20 ms due to security overhead is acceptable. Attaching a digital signature and a certificate to each message impose a considerable amount of OTA bandwidth overhead as well as high demands in the computing device's resources. In particular, a customized application-specific elliptic curve cryptographic processor is required to handle the computational load. Such an additional custom-specific co-processor might be commercially infeasible and hinder V2V deployment. The main requirements for a proper security protocol are efficiency, in particular low computational and OTA bandwidth overhead, as well as small latency due to security overhead and scalability. The security protocol is expected to run on embedded computer that can be found in vehicles today.

For V2V safety applications that require verification of a large number of messages per second, we look at further solutions. In general, security authentication is performed for every received wireless message by conducting verification for a valid digital signature, and only messages that have been successfully verified are processed further. However, as stated earlier, verifying digital signatures consumes a significant amount of the share of the automotive processor [7]. Thus verification of each and every received message, for example based on the IEEE 1609.2 standard, is computationally very expensive and cannot in general be carried out even with specialized hardware. An interesting observation is that, most of these periodic safety messages will not result in driver warnings since we expect that the vehicle safety system would be used to provide warnings only when the threat of collision determined by vehicle safety applications is high. Therefore, we define an approach where messages are first processed and then verified only on-demand. The solution is more efficient regarding running-time and CPU overhead and is especially suited for V2V safety applications and requires no additional security-specific computing processor.

In this paper, we first introduce the V2V safety communication system. Next, we describe the conventional Verify-and-Then-Process approach normally used for broadcast authentication in V2V safety communication. Then we describe a novel approach called Verify -on-Demand (VoD) which provides practical and scalable broadcast authentication for V2V safety communication. The details of the security implementation on a 400 MHz processor, analysis of its pros and cons will be discussed. System implementation and supporting data are used to conclude that, for V2V safety applications, 1609.2 ECDSA with VoD (i.e., verification of prioritized, application-filtered threats) achieves the desired performance.

V2V SAFETY COMMUNICATION AND MESSAGE AUTHENTICATION

NOMINAL V2V SAFETY COMMUNICATION SYSTEM

Figure 1 shows a simple nominal architecture of a V2V safety communication system. The Sensor Data Handler (SDH) processes Host Vehicle (HV) GPS data such as vehicle location, time, etc. and also the vehicle-bus data such as speed, acceleration, etc. The DSRC radio periodically (for e.g. 10 times per second) transmits and also receives safety broadcast data required for vehicle safety communication. Messages received from Remote Vehicles (RVs) by the DSRC Radio are then processed by the Wireless Message Handler (WMH). Safety applications and algorithms within the Threat Processing & Threat Arbitration module evaluate the collision or other safety threat level of the HV with other communicating RVs in its vicinity. If a certain vehicle safety threat threshold is exceeded, determined by the Threat Level being above a calibrated threshold, then this module issues a threat notification via the Driver Notification module, and the driver of the HV is made aware of the safety threat via appropriate driver vehicle interfaces inside the vehicle (e.g. haptic, visual, auditory warnings).

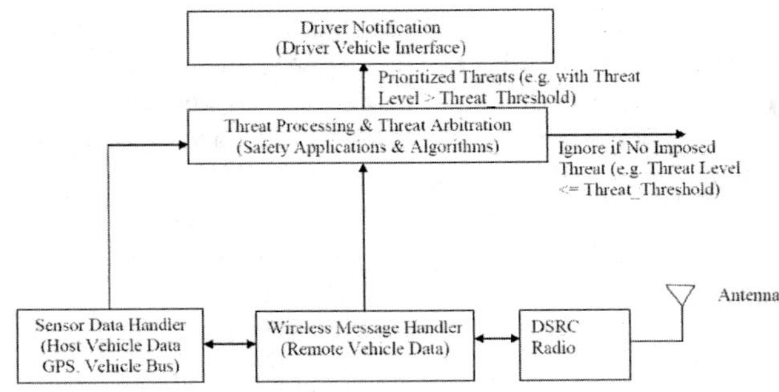

Figure 1. Nominal Architecture of V2V Communication System.

MESSAGE BROADCAST AUTHENTICATION WITH DIGITAL SIGNATURES

In V2V, received safety messages have to be authenticated. The straightforward method of providing message broadcast authentication is to implement digital signatures. The sender signs the safety message and broadcasts the signature along with the message. Receivers can then verify the message. Before message verification, the receivers need to be able to get a hold of the sender's certificate [6, 7]. A brief description of the protocol parameters and expected performance is provided next.

Protocol Parameters and Structure

- H: H(m) describes the hash of message m computed using the Secure Hash Algorithm (SHA). |H| is the hash length of H, in case of SHA-256 we have |H| = 32 bytes [9].
- Sig: Sig(m, A_{SK}) describes the signature of a message m with secret key A_{SK}. In ECDS A-256 the signature length of Sig is |Sig| = 64 bytes [8].
- TS: describes the 6-byte time-stamp to avoid replay attacks.
- Safety messages may include the certificate digest or certificate as part of the data packets [6]. Certificates digest in an eight byte hash of the certificate, so it is more bandwidth efficient. It is a short reference to the certificate but is not of use unless a receiver has already received and cached the certificate. Thus, one model is to transmit certificates every second and use certificate digest for messages in-between [10].
- When certificates are sent in a piggy-back fashion to form data-certificate packets, we have the following data structure:

Cert. (117 bytes)	m	$Sig_A(H(m)\|\|TS)$ (64 bytes)	TS (6 bytes)

- Data packets structure for safety messages that include a certificate digest is as follows:

Cert. Digest (8 bytes)	m	$Sig_A(H(m)\|\|TS)$ (64 bytes)	TS (6 bytes)

- Ver: Ver(m, s, A_{PK}) describes the verification process of a signature s against message m and public key A_{PK}. The result is either 'success' or 'failure'.

Expected Performance

ECDSA-256 and SHA-256 are used to compute digital signatures [6]. The computational overhead due to hashing is negligible for the considered message sizes. The time delay is computed as the sum of computation time at the sender and receiver side. OTA overhead per message consists of the digital signature but no additional network layer overhead since signatures are sent together with the message. As stated earlier, note that additional overhead is introduced by the certificate distribution compared to certificate digest distribution. The security overhead and expected performance measures for a 400 MHz computing platform are shown in Table 1.

Table 1. Message Authentication with Digital Signatures

Over-the-air overhead	70 bytes per message + 8 bytes per message (certificate digest) or 117 bytes per message (certificates)
Computations for sender per message	H(m) + Sig ≈ 6 ms
Computations for all receivers per message	H(m) + Ver ≈ 23 ms
Maximum Signature Generations per second	≈ 166
Maximum Signature Verifications per second	≈ 43

For V2V safety communications, it is clear from Table 1 that the required transmissions (e.g. 10 messages per second) can all be signed before being broadcasted without much computational complexity. However, only a small fraction of received messages may be authenticated in a 400 MHz computing platform (i.e. only from 4 RVs at the rate of 10 messages per second per vehicle). Thus message authentication is a significant and overwhelming challenge in V2V communication, which is being addressed in this paper.

VERIFY-AND-THEN-PROCESS

As stated earlier, broadcast message authentication is of primary importance for vehicle safety communication. In particular, vehicles need to be able to authenticate that a message originated from a properly certified vehicle and that the message was not manipulated on its way between the sender and receiver vehicles. In order to accomplish the above, the message signature verification functionality may be performed at the Security Module, as shown in Figure 2 with the primary aim of performing broadcast authentication and filtering bogus messages (i.e. those messages with the correct format but invalid signature or authentication tag). The verify-and-then-process approach first verifies the signatures of all received safety messages for trustworthiness. If the signature verification is successful, then the message is processed further. Otherwise the message is a bogus message and hence discarded. There is a delay in initiating message threat processing due to time taken for verification.

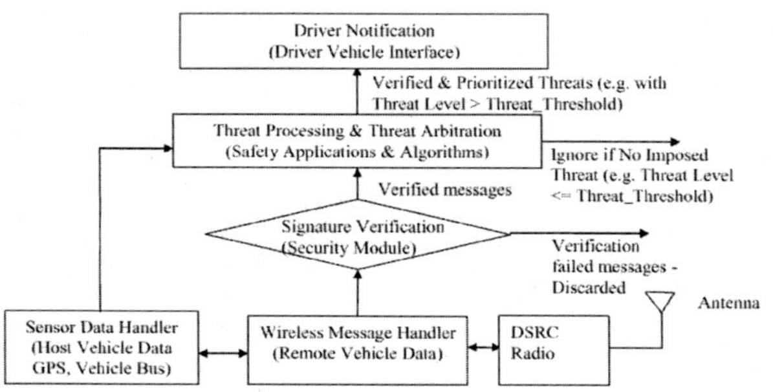

Figure 2. Verify-And-Then-Process Flow.

Thus only verified messages are processed further. If the Threat Processing & Threat Arbitration module determines that an RV message causes the safety threat level to be larger than a calibrated threshold (representing a potential threat), Driver Notification provides the needed information to the driver of the HV in the form of safety warning / notification in the most appropriate and intuitive manner. The Threat Processing & Threat Arbitration module typically works on a message-by-message basis when evaluating the safety threat level caused by an RV's V2V safety message. Driver Notification only passes a warning to the vehicle driver after evaluating a potential safety threat level. Also, refinements may be used in Driver Notification so that the driver is not repeatedly annoyed by the safety warnings or notifications, e.g. Driver Notification might decide to suppress warnings to the driver even in the case of a continuing potential threat level if an earlier warning was just provided to the driver.

From Table 1, it is quite clear that verifying digital signatures is computationally very expensive. Typical requirement of vehicle safety communication is that each vehicle broadcasts safety messages (periodically) about 10 times per second, and up to a transmission range of about 300 m. It should therefore be clear that, as the penetration of V2V vehicles increases, the number of received messages per second could be very large and would exceed 1000 messages per second. Thus verification of each and every received message, for example based on the IEEE 1609.2 standard, would consume all of the share of the automotive processor and cannot in general be carried out, even with specialized hardware, at low cost.

We therefore conclude that the verify-and-then-process approach for broadcast message authentication based on the IEEE 1609.2 standard does not provide the scalability needed for practical automotive implementations. Novel methods for message authentication in V2V are a necessity to enable deployment in the near-future.

VERIFY-ON-DEMAND

The verify-and-then-process approach, described in the previous section, for broadcast authentication is based on the underlying assumption that all received safety messages need to be verified before they are processed by the application layer. An interesting and powerful observation is that, most of these periodic safety messages will not result in driver warnings or control actions since we expect that the vehicle safety system would be designed to provide warnings or control actions only when the threat of a collision determined by vehicle safety applications is high. Therefore it is reasonable to define an approach where messages are first processed and then verified only on-demand. VoD is a novel approach that provides practical and scalable broadcast authentication for vehicle safety communication.

Assuming that only messages that evaluate to a safety threat level larger than a calibrated threat threshold (representing a potential threat) have an actual impact to a vehicle's safety level, it is reasonable to only verify those received safety messages that result in a safety threat level above that threshold value. Note that this approach does not affect the signature generation. All messages are still signed before being broadcasted.

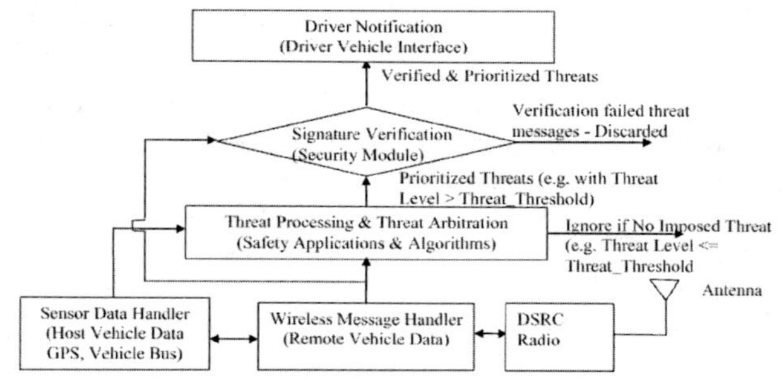

Figure 3. Verify-On-Demand Flow.

As shown in Figure 3, VoD can be implemented by introducing the signature verification functionality at the Security Module in-between the Threat Processing & Threat Arbitration and the Driver Notification modules. The Threat Processing & Threat Arbitration module evaluates the safety threat level caused by each wireless safety message received from RVs. Only for messages that evaluate to a safety threat level larger than a calibrated threshold, the Security Module initiates on-demand signature verification. Thus on-demand signature verification is required only on safety messages that result in a safety threat level that demands warnings or control actions. It waits for a verification to be completed and, when successful, forwards prioritized threats to Driver Notification.

For the vehicle safety communication system, Threat Processing & Threat Arbitration is based on the current safety message received from remote vehicles. Current processing does not use past wireless data. Therefore for periodic messages, only the recent wireless safety message from each remote vehicle may need to be buffered to enable VoD. The life-time of the buffer can be set to the maximum processing delay expected. If there are a-periodic event-driven safety messages, then those wireless safety messages may need to be buffered as well.

Note that the security verification functionality is put in-between two application processing blocks such that separation of concerns is removed and cross-layer architecture is introduced. Therefore the implementation of this approach tends to be different than in the verify-and-then-process approach. VoD should be seen as a practical approach to broadcast authentication for vehicle safety communication since digital signature verification is computationally very expensive and there are a large number of messages in the system but only very few of those raise safety warnings or control actions during typical driving conditions. This basic principle can be used with existing security protocols and standards, such as the IEEE 1609.2, right away while the research continues into the design of other efficient authentication protocols for vehicle safety communication. The approach also allows implementation of V2V systems today on existing automotive grade (i.e. 400 MHz processor) hardware platforms, and then over time one may chose to verify more and more messages as the computational hardware platform becomes faster.

V2V safety systems should surely not be designed to raise a large number of safety notifications in a short span of time to the driver. Otherwise the driver will be annoyed by the system. If the safety system is designed with the assumption that at most 10 new messages in a given second will raise safety warnings of importance to the driver, even that will stretch the requirements for VoD. Therefore it is reasonable to expect that VoD would need to conduct at most 10 digital signature verifications per second. In comparison to the overall number of received messages, which could exceed 1,000 messages per second, this is a significant reduction of the signature verification load compared to the standard verify-and-the-process approach. Again this approach does not affect the signature generation. All safety messages are still signed before being broadcasted.

Finally, we consider the security implications of such an approach. Let us assume that an attacker has complete knowledge of all involved decision algorithms in this approach and has full control over a DSRC radio including the secret key data. The attacker's goal is to generate malicious safety messages and sign them with valid digital signature such that there is no evidence of misbehaving. In such a case, the receiver will definitely choose the safety messages that impose a safety threat for verification and only accept the safety messages that pass verification. Such attacks need non-cryptographic methods of detection in every security approach that is employed for V2V safety communications.

We also consider denial-of-service (DoS) attacks. In general, V2V communication system can easily be overwhelmed by broadcasts of forged messages that impose a security threat and are therefore scheduled for signature verification. Thus, DoS can always easily be mounted regardless of the deployed broadcast authentication approach employed for vehicle safety communication. However, DoS attacks are easily detected by the system since verification of such messages will fail authentication. In such a situation, the system can inform the driver of a potential DoS attack and make the driver aware of such a situation.

Table 2 summarizes the pros and cons of both approaches.

Table 2. Pros and Cons of Verify-and-then-Process and VoD

	Verify-and-then-Process	***Verify-on-Demand (VoD)***
Pros	• No special security assumptions need to be made for the implementation. • Clear separation of security and application layer.	• Relieves the security module from its heavy load of verification. • Allows flexible balancing of verification load. • Stays easily compatible with future generation implementations and allows quick deployment.
Cons	• High processing burden.	• The approach introduces a cross-layer security design assumption on the application layer.

IMPLEMENTATION IN V2V TEST-BED

The Crash Avoidance Metrics Partnership-Vehicle Safety Communications 2 (CAMP-VSC2) Consortium initiated, in December 2006, a 3 -year collaborative effort in the area of wireless-based safety applications under the Vehicle Safety Communications - Applications (VSC-A) Project [10]. The VSC-A project was completed in December, 2009. Under the project, a vehicle test bed (this will now be referred to as the test bed in the remaining text of this document) was developed to serve as a prototype platform for the V2V system. The test bed was used to validate system specifications and performance tests that were developed as part of the VSC-A Project. The test bed also served as a flexible platform for testing various positioning, communication, and security solutions in a real-world setting and in safe and staged crash-scenario configurations to ensure the effectiveness of the applications.

Among other things, this project also focused on security for V2V safety messages with a main focus on efficient broadcast authentication of safety messages. Security protocols were implemented to run on the On-Board Equipment (OBE), which housed a 400 MHz processor. It was concluded that, for the VSC-A safety applications, 1609.2 ECDSA with VoD (i.e., verification of prioritized, application-filtered threats) achieved the desired performance. Therefore, this is the protocol that was used for the system objective testing in the project. Objective testing confirmed that ECDSA with VoD functioned properly under all test conditions for the VSC-A safety applications.

V2V TEST-BED

This section summarizes the test bed design and implementation. For a more detailed description, please refer to [10]. Figure 4 shows the block diagram developed for the V2V system test-bed. The following V2V safety applications were developed and implemented as part of the test-bed:

Emergency Electronic Brake Lights (EEBL)

The EEBL application enables an HV to broadcast a self-generated emergency brake event to surrounding RVs. Upon receiving such event information, the RV determines the relevance of the event and provides a warning to the driver, if appropriate. This application is particularly useful when the driver's line of sight is obstructed by other vehicles or bad weather conditions (e.g., fog, heavy rain).

Forward Collision Warning (FCW)

The FCW application is intended to warn the driver of the HV in case of an impending rear-end collision with an RV ahead in traffic in the same lane and direction of travel. FCW is intended to help drivers in avoiding or mitigating rear-end vehicle collisions in the forward path of travel.

Blind Spot Warning+Lane Change Warning (BSW+LCW)

The BSW+LCW application is intended to warn the driver of the HV during a lane change attempt if the blind-spot zone into which the HV intends to switch is, or will soon be, occupied by another vehicle traveling in the same direction. Moreover, the application provides advisory information that is intended to inform the driver of the HV that a vehicle in an adjacent lane is positioned in a blind-spot zone of the HV when a lane change is not being attempted.

Do Not Pass Warning (DNPW)

The DNPW application is intended to warn the driver of the HV during a passing maneuver attempt when a slower moving vehicle, ahead and in the same lane, cannot be safely passed using a passing zone which is occupied by vehicles with the opposite direction of travel. In addition, the application provides advisory information that is intended to inform the driver of the HV that the passing zone is occupied when a vehicle is ahead and in the same lane and a passing maneuver is not being attempted.

Intersection Movement Assist (IMA)

The IMA application is intended to warn the driver of an HV when it is not safe to enter an intersection due to high collision probability with other RVs. Initially, IMA is intended to help drivers avoid or mitigate vehicle collisions at stop-sign controlled and uncontrolled intersections.

Control Loss Warning (CLW)

The CLW application enables an HV to broadcast a self-generated, control, loss event to surrounding RVs. Upon receiving such event information, the RV determines the relevance of the event and provides a warning to the driver, if appropriate.

The test-bed modules are composed of support and application functions. The support functions interface to external equipment and calculate data to support the V2V application modules and engineering Driver-Vehicle Interfaces (DVIs). Since VoD is used mainly to determine the authenticity of received OTA messages, here we focus our discussion on the Wireless Message Handler (WMH) and Security Module (SM). WMH constructs and sends HV OTA messages and processes received RV OTA messages. V2V safety messages are defined in the Society of Automotive Engineers (SAE) J2735 Basic Safety Message (BSM) formats [11]. If security is enabled, WMH interfaces to the SM to generate signatures for transmitted messages and verify signatures for received messages.

Figure 4 shows the V2V test-bed with six safety applications. The application modules evaluate potential categorized safety threats based on the data and inputs from the support modules. The warning algorithm categorizes the safety output threat level of each application to be in one of the following threat states: NONE, DETECTED, INFORM or WARN. If the threat state is NONE, it implies no safety threat and so no driver notification is necessary. If the threat state is DETECTED, it implies Target Classification (TC) has determined that an RV is in a certain region of interest that resulted in the corresponding application threat processing, however, application threat processing evaluated to no safety threat and so no driver notification is necessary. If the threat state is INFORM, it implies that the application has determined a safety threat that may warrant a driver notification to caution as necessary. If the threat state is WARN, it implies that the application has determined an urgent safety threat that would warrant a driver warning and/or control action. Threat Arbitration (TA) prioritizes the concurrent threats produced by the applications. It uses the threat state of the applications as well as metrics that define crash severity for prioritization. Threat Priority (0) is the highest priority; Threat Priority (1) is the next one, and so on.

VOD IMPLEMENTATION

In VoD security implementation, received messages are first evaluated by applications and prioritized by TA (see Figure 4). If the output threat state of an application exceeds a predefined safety threshold (i.e. INFORM level), the signature of the received message that caused this safety alert level is verified. If the threat state does not exceed this predefined threat threshold, the message is discarded. This is shown in Figure 4 with a Security Verification Request.

Implementation details of VoD in the test-bed are provided below. Upon receiving an OTA message, WMH will:

- Decode the message using the decoding library, if the SAE J2735 message format is being used
- Parse the message (SAE J2735 format), perform validity checks, unpack, and scale the data
- Update an existing RV record or create a new record with the received data along with calculated Packet Error Rate (PER), latency, and security statistics
- Notify or signal the corresponding application process to trigger its execution
- Provide SM with the WMH assigned sequence number for the message for use in VoD security processing

When VoD security is being used, WMH will receive a signal or notification from the TA when messages must be verified (see Figure 4, Security Verification Request). Upon receiving such a request, WMH will:

- Read TA's shared memory to determine the WMH sequence number(s) of the message(s) to be verified and call SM to verify each message
- After all the messages have been verified, write the verification results to shared memory and signal or notify TA

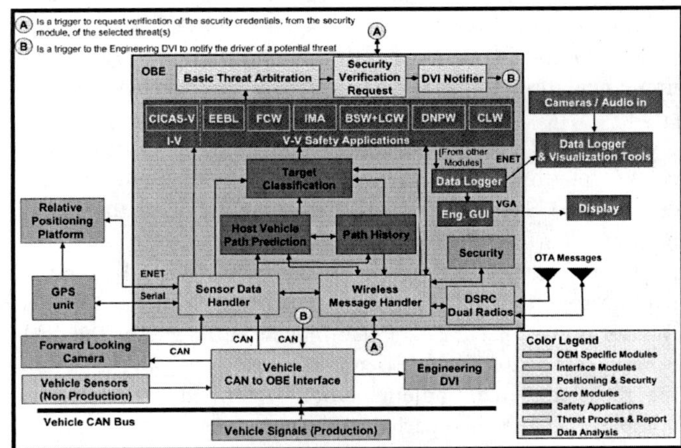

Figure 4. V2V System Test-Bed Block Diagram.

Finally, TA will only forward prioritized threats to DVI notifier (for driver warnings or control actions) only if the corresponding RV messages have successfully passed security verification.

We now discuss certificate queuing in conjunction with VoD implementation. If a signed OTA message is received and security is enabled, WMH will buffer the message and check the security approach. If the security approach is VoD, WMH sends a "certificate verify" request to SM if the security contents of the message contain a certificate, and sends the unverified OTA message data for further application processing. WMH provides a sequence number to be used for subsequent verify requests, and SM stores the number with the message. SM may choose one of the following approach to verify certificates: (1) Verify certificates as and when received, or (2) Store all received certificates without verifying these certificates. Only if a message is triggered for verification, then the previously stored certificate is verified. Storing received certificates is only necessary if certificates are not attached to each message. TA prioritizes the threat warnings from application modules. When VoD security authentication is being used, TA requests verification of signatures for messages that result in driver warning. Upon receiving a subsequent VoD verify request, WMH finds the message in its buffer based on the sequence number input and sends a verify request to SM. Upon receiving a response from SM, WMH provides the results to TA. TA only forwards prioritized threats that have passed security verification to DVI Notifier (for driver warnings or control actions).

IMPLEMENTATION PERFORMANCE

IEEE 1609.2 ECDSA security protocol with verify-and-then-process and VoD were implemented on a car-PC (a standard PC running at 2.4 GHz) and on-board the WAVE Safety Unit (WSU) (a 400 MHz industry computing platform [12]). The implementation for the WSU consists of the same source code with platform-specific assembly optimized cryptographic operations. Therefore, it is possible to use the car-PC platform with its variety of development tools to develop the SM and then to cross-compile it to the WSU platform. Performance measurements of the SM running on the WSU clearly show that the IEEE 1609.2 ECDSA protocol is too resource-demanding to run in software. This also holds for the powerful car-PC. Performance numbers for the SM running on-board the WSU are presented in Table 3. Note that these are actual implementation performance measures as compared to expected performance measures shown earlier in Table 1.

Table 3. Security Protocol Performance IEEE 1609.2 ECDSA

	IEEE 1609.2 ECDSA
Authentication generation (crypto only on idle system)	4.9 ms (ECC-224) / 6.6 ms (ECC-256)
Authentication generation*	6.6 ms (ECC-256)
Authentication verification (crypto only on idle system)	17.8ms (ECC-224) / 26.5ms (ECC-256)
Authentication verification*	28.5 ms (ECC-256)
CPU Load for 2 WSUs at 10 messages per second: Signing / Signing + Verifying*	8% / 34%
Latency: Avg. (no channel switching,)*	36 ms
Average OTA packet size (send certificate with each 3rd message)	115 bytes
*CPU load and latency was measured on a system that runs safety applications	

With VoD applied to ECDSA, the implementation proved that a security protocol can be efficiently implemented in software on-board of the WSU. The performance numbers per signature generation equal those of IEEE 1609.2 ECDSA. However, the CPU load of a receiving WSU is significantly lower due to the fact that only safety messages that result in a high threat level are verified. ECDSA VoD performed well with all VSC-A safety applications and was selected for the objective test procedures. Consider the verification error rate defined as the fraction of successfully verified packets over received packets that require verification. ECDSA VoD with certificates attached to each message is designed to have a zero verification error rate.

Overall the implementation strongly indicates that a security protocol can be efficiently implemented in software on board an automotive grade platform such as the WSU, if certain conditions such as advanced queuing techniques and VoD filtering are implemented.

VOD IMPLEMENTATION PROPERTIES FOR V2V SAFETY

- Very few of the V2V safety messages have actual safety impact that will result in driver warnings or control actions since the safety system would be designed to provide warnings only when the threat of collision is high. Therefore it is reasonable to define an approach where messages are first processed and then verified only on-demand.
- VoD only verifies received messages that result in potential impact on driver safety. It is secure since each safety message that results in driver warning or control action will certainly be verified.
- Question: In a V2V system, how many concurrent driver warnings or control actions are expected to be provided in practice? The answer to this question is, likely, one, i.e. driver workload studies and considerations would suggest that, at any moment, we present an appropriate driver warning or action corresponding to the highest priority threat produced by the V2V system.
- Question: Will this answer change if we were to add many more safety applications than the one prototyped in the test-bed and shown in Figure 4? The answer to this question is, likely, No, i.e. even if we had a much larger number of safety applications in a V2V system, the system should still select the highest priority threat for driver warning at any given moment.
- Question: In a V2V system, what is the lower bound update interval when we expect that the driver warnings to change? The answer to this question is, likely, 100 ms, i.e. based on the periodic update interval of V2V safety communication, and the system process cycle time, the highest priority threat produced by the system would not change faster than 100 ms.

• Question: In a V2V system, what is the upper bound on message verification requirements for VoD? The answer to this question follows from the previous answer. With driver warnings unlikely to change faster than 100 ms, a generous upper-bound would be 10 verifications per second. With concurrent certificate verifications, we will have at most 20 ECDSA digital signature verifications per second. With prior certificate verifications, we can reduce this to at most 10 ECDSA digital signature verifications per second.

• Question: In a V2V system, what is the upper bound on message verification time delay for VoD? Answer: VoD's message verification time delay has an upper-bound of 57 ms based on the current V2V implementation on WSU (i.e. 400 Mhz processor). This includes certificate and message digital signature verification times each having an authentication verification time of 28.5 ms (see Table 3). With prior certificate verification, this delay can be reduced to 28.5 ms (see Table 3). Optimization of the algorithms and additional processor resources would significantly reduce this delay.

• The implementation is practical since safety evaluation is based only on the current safety message received. The current safety message provides all the needed remote vehicle data for safety evaluations.

• ECDSA-VoD works as long as the application has well-defined decision logic for computing threat assessment states. The implementation of VoD requires an understanding of the application's decision logic as per design, identifying the remote vehicle(s) & message(s) that were used in the decision logic, and verifying those messages that result in threat alert on-demand (as per decision logic).

• ECDSA-VoD allows customized implementations on low-cost devices. Each automotive Original Equipment Manufacturer (OEM) can optimize their ECDSA-VoD implementation to accommodate application specific demands via proper processor capability, latency reduction, and heuristics for verifications, improved security algorithm execution, etc.

• ECDSA-VoD conforms with 1609.2 standards.

SUMMARY/CONCLUSIONS

The VoD processing method should not be seen as an alternative to efficient authentication protocols but as an orthogonal and practical approach. The basic principle can be used with existing security protocols right away (such as ECDSA) while research continues into the design of efficient authentication protocols for V2V safety communication. The design principle of verifying messages may be summarized as follows:

• If verification of all incoming messages can be done by designing an efficient authentication protocol, then we will be able to verify all incoming messages in a timely fashion.

• If verification of all incoming messages cannot be done in a timely fashion or is computationally expensive, then we can use the VoD approach to verify only the messages that result in potential safety threats to the host vehicle and its driver.

The VoD approach results in cross layer security design and introduces security assumptions in the application layer. However, the VoD approach allows balancing of the verification load at run-time in congested situations without any further compromise on the security properties of the V2V system. The approach also allows secure implementation of V2V applications today even on a computationally weak hardware platform, and, then over time, we may chose to verify more and more messages as the computational hardware platform becomes faster. Therefore, the VoD approach is inherently compatible to future versions and current standards and allows quick deployment today.

REFERENCES

1. Sengupta, R., Rezaei, S., Shladover, S. E., Cody, D., Dickey, S., and Krishnan, H., "Cooperative Collision Warning Systems: Concept Definition and Experimental Implementation," Journal of Intelligent Transportation Systems, 11(3): 143-155, 2007.

2. Bai, F., Krishnan, H., Sadekar, V., Holland, G., and Elbatt, T., "Towards Characterizing and Classifying Communication-based Automotive Applications from a Wireless Networking Perspective," IEEE AutoNet 2006 - The 1st IEEE Workshop on Automotive Networking and Applications, Co-located with the 49th Annual IEEE GLOBECOM Technical Conference, San Francisco, California, December, 2006.

3. Vehicle Safety Communications Project-Final Report, USDOT HS 810 591, http://www-nrd.nhtsa.dot.gov/departments/nrd-12/pubs_rev.html, April, 2006.

4. Wireless Access in Vehicular Environment (WAVE) in Standard 802.11 Information Technology Telecommunications and Information Exchange Between Systems, Local and Metropolitan Area Networks, Specific Requirements, Part 11: Wireless LAN Medium Access Control (MAC) and Physical Layer (PHY) Specifications}, IEEE 802.11p/D 1.0, Feb. 2006.

5. Raya, Maxim, Papadimitratos, Panos, and Hubaux, Jean-Pierre, "Securing Vehicular Communication", Infocom '06, April, 2006.

6. IEEE Trial-use Standard 1609.2TM-2006, WAVE - Security Services for Applications and Management Messages, 2006.

7. Raya, Maxim and Hubaux, Jean-Pierre, "The Security of Vehicular Ad Hoc Networks", Proceedings of the 3rd ACM workshop on Security of ad hoc and sensor networks (SASN'05), pp. 11-21, 2005.

8. ANSI, "Public Key Cryptography For The Financial Services Industry: The Elliptic Curve Digital Signature Algorithm (ECDSA)", ANSI X9.62, 1998.

9. NIST, Secure Hash Standard FIPS 180-2, August, 2002.

10. Vehicle Safety Communications - Applications (VSC-A) Project Final Report, Submitted to the Intelligent Transportation Systems (ITS) Joint Program Office (JPO) of the Research and Innovative Technology Administration (RITA) and the National Highway Traffic Safety Administration (NHTSA), May, 2010.

11. SAE International Surface Vehicle Standard, "Dedicated Short Range Communications (DSRC) Message Set Dictionary," SAE Standard J2735, Rev. Nov. 2009.

12. Wireless Safety Unit (WSU) Short Range Communications Module Data Sheet, Denso, June 2010.

CONTACT INFORMATION

Dr. Hariharan Krishnan
Staff Researcher
Electrical & Controls Integration Laboratory GM R & D Center
Mail Code: 480-106-390
30500 Mound Road
Warren, MI 48090-9055
Phone: (586) 986-6966
Fax: (586) 986 3003
hariharan.krishnan@gm.com

DEFINITIONS/ABBREVIATIONS

BSM
Basic Safety Message

BSW
Blind Spot Warning

CAMP
Crash Avoidance Metrics Partnership

CLW
Control Loss Warning

DNPW
Do-Not-Pass Warning

DoS
Denial-of-Service

DSRC
Dedicated Short Range Communication

DVIN
Drive Vehicle Interface Notifier

ECDSA
Elliptic Curve Digital Signature Algorithm

EEBL
Emergency Electronic Brake Lights

FCW
Forward Collision Warning

GPS
Global Positioning System

HV
Host Vehicle

IMA
Intersection Movement Assist

LCW
Lane Change Warning

OBE
On-Board Equipment

OEM
Original Equipment Manufacturer

OTA
Over-the-Air

PER
Packet Error Rate

PKI
Public Key Infrastructure

RV
Remote Vehicle

SDH
Sensor Data Handler

SHA
Secure Hash Algorithm

SM
Security Module

TA
Threat Arbitration

TC
Target Classification

USDOT
United States Department of Transportation

V2V
Vehicle-to-Vehicle Communication

VoD
Verify-on-Demand

VSC
Vehicle Safety Communications

WAVE
Wireless Access in Vehicular Environment

WMH
Wireless Message Handler

Communications - PHEV/EV Requirements

Communication Requirements for Plug-In Electric Vehicles

2011-01-0866
Published
04/12/2011

Richard A. Scholer
Ford Motor Co.

Dan Mepham
General Motors Company

Sam Girimonte
Chrysler Group LLC

Doug Oliver
Ford Motor Company

Krishnan Gowri and Nathan Tenney
Pacific Northwest National Laboratory

James Lawlis
Ford Motor Co.

Eloi Taha
Nissan Technical Center NA Inc.

John Halliwell
EPRI

doi:10.4271/2011-01-0866

ABSTRACT

This paper is the second in the series of documents designed to record the progress of a series of SAE documents - SAE J2836™, J2847, J2931, & J2953 - within the Plug-In Electric Vehicle (PEV) Communication Task Force. This follows the initial paper number 2010-01-0837, and continues with the test and modeling of the various PLC types for utility programs described in J2836/1™ & J2847/1. This also extends the communication to an off-board charger, described in J2836/2™ & J2847/2 and includes reverse energy flow described in J2836/3™ and J2847/3.

The initial versions of J2836/1™ and J2847/1 were published early 2010. J2847/1 has now been re-opened to include updates from comments from the National Institute of Standards Technology (NIST) Smart Grid Interoperability Panel (SGIP). Smart Grid Architectural Committee (SGAC) and Cyber Security Working Group committee (SCWG). These documents have been added to the NIST SGIP Catalogue of Standards[1] and it is expected the others to be added upon publishing. Additional efforts have continued with the Smart Energy Alliance (SEP2) as we coordinate the Application Specification with this SAE document for PEV utility messages.

J2836/2™ and J2847/2 are intended to be published early 2011 and include the requirements for DC energy transfer to the PEV where the PEV communicates with the Electric

Vehicle Supply Equipment (EVSE) that includes an off-board charger.

J2836/3™ and J2847/3 include the architecture and messages for reverse energy flow including the following four types that include specific architecture and communication requirements. Vehicle to Grid (V2G), Vehicle to Home (V2H), Vehicle to Load (V2L) and Vehicle to Vehicle (V2V). These also include options for either on-board or off-board energy conversion.

Two new set of J documents have been added to the task force effort. J2931 includes the communication requirements and protocol variations, while J2953 captures the communication interoperability requirements. J2931 has four parts. J2931/1, J2931/2, J2931/3 & J2931/4 have been added to the suite of documents to capture the communication protocol, test criteria and other items that tie to the J2847 series of messages. J2953 has started to identify the interoperability requirements and approach for the multitude of PEV and EVSE manufacturers to insure communication interoperability.

The objective of these documents is to publish initial versions that allow the task force to move into the implementation phase that will continue with simulation, test and evaluation of the systems and then re-publish with updates. This provides a two step approach to providing initial information and making it more complete as our progress continues. Initial publication also allows other groups and organizations to provide comments that will be addressed in the updated version as shown with J2847/1 noted above.

INTRODUCTION

The primary focus of the J2836/J2847 task force is to establish the protocols and requirements for communication between the electric vehicle (EV) and the electric vehicle supply equipment (EVSE) and the utility. The EVSE is, at least initially, the bridge (or proxy) to either the energy services interface (ESI) that includes the smart meter and potentially the Home Area Network (HAN). Power line communication (PLC) is identified as one of the preferred methods of communication. Several PLC types are continuing to undergo testing this year to identify robustness for use in the vehicle and EVSE with potential expansion to the HAN and smart meter.

Figure 1 shows the family of SAE documents addressing the various aspects of standardizing communication messages and system requirements. As Vehicle Manufacturers (VM) start, or continue the production of Plug-In Electric Vehicles (PEV) that is either a Plug-In Hybrid Electric (PHEV) or Battery Electric Vehicle (BEV), some consumers may desire faster charging. This can be accomplished with a larger off-board charger as an option for the customer, rather than including a larger on-board charger that burdens the PEV with additional weight and cost. The customers could also benefit since off-board chargers are not traded in with every vehicle change, and it could be moved with them, in the case they change homes. There are smaller chargers that could be included with the Electric Vehicle Supply Equipment (EVSE) at homes and larger ones at public locations where more power may be available. Regardless of the off-board charger size variations, the communication would be the same from the PEV. Communications is required to control off-board chargers and is explained in the J2836/2™ and J2847/2 documents. Additional safety features are required with DC charging and described in J1772™. There are also several approaches to energy flow and depending on the need and the cost, the PEV may also be a source for this. As a source of energy, communication is also required and is identified in J2836/3™ and J2847/3.

The task force has also spent significant portions of last year understanding how to include Power Line Carrier (PLC) for the communication medium using either the J1772™ power circuits (mains) or the control pilot (called Inband Signaling) to transfer the messages. J2931/1, J2931/2, J2931/3 & J2931/4 are the first in the series of documents that will identify the overall approach, requirements and include the protocol for PLC. Generally PLC has not been used in automotive applications. EMC testing has been performed by SAE and the ISO/IEC Joint working group Project Team 4, to determine the type that best fits the automotive market.

The objective of the SAE PEV Communication Task Force is to select an auto qualified production solution that can also be implemented in Europe and other market areas. In January, 2010 the ISO/IEC Joint Working Group (JWG) joined the SAE effort and started reviewing the use cases, messages and other items for the PEV to communicate with the utility. The general approach for Europe is to use single phase power and perhaps three phases for faster charge rates. The US, however only has single phase in homes and the approach is to use an off-board charger, transferring DC energy to the PEV for faster charge rates. Europe also has this off-board option of connecting the off-board charger to three phases and transferring DC to the PEV. SAE had also been looking at PLC types while the Smart Energy 2.0 communication stack is being completed and this is estimated to be a year or two out. We initially focused on the mains as the PLC path, and then expanded our attention to include using the J1772™ control pilot for the communication path, since it may have a shorter implementation time and would meet our communication needs with the off-board charger. The ISO/IEC Project Team 4 has joined SAE to evaluate and perform additional EMC and functional test.

SAE introduced Inband Signaling to the ISO/IEC JWG during the May, 2010 meeting and also provided the opportunity to review the utility messages including the

Summation of SAE Communication Standards (15 + new documents)

J2836™ – General info (use cases)

Dash 1 – Utility programs *

Dash 2 – Off-board charger communications

Dash 3 – Reverse Energy Flow

Dash 4 – Diagnostics

Dash 5 – Customer and HAN

J2847– Detailed info (messages)

Dash 1 – Utility programs *

*Dash 2 – Off-board charger communications ***

Dash 3 – Reverse Energy Flow

Dash 4 – Diagnostics

Dash 5 – Customer and HAN

J2931– Protocol (Requirements)

*Dash 1 – General Requirements ***

*Dash 2 – InBand Signaling (control Pilot) ***

Dash 3 – NB OFDM over control pilot or mains **

Dash 4 – BB OFDM over control pilot or mains **

J2953– Interoperability

Dash 1 – General Requirements

Dash 2 – Testing and Cert

Dash 3 –

* Two have initial versions published

** Six are expected to ballot 4Q 2010

Figure 1. SAE Documents Summary

association with the Energy Service Interface (ESI). The ESI may include a sub-meter (Energy Use Measuring Device - EUMD) if the PEV receives special tariffs and other items.

TESTING THE PLC SOLUTIONS

The task force started the PLC evaluation with a PLC competition in September, 2009 where 16 suppliers presented their products in a closed SAE meeting with vehicle manufacturers. This led to a second meeting in January, 2010, where five types were explored further and each of the five suppliers presented more general information to the SAE Task Force team. Product testing was planned next and requests were made to these five types to provide the hardware and evaluation boards so that these could be subjected to EMC compliance.

As the task force started looking at PLC over the mains (power circuits), it was realized that several issues would have to be resolved namely crosstalk, coexistence and association to the ESI. A common architecture is required, with both AC and DC power and this also needs a common interface with various utilities smart meters. Recent focus has been on the control pilot (referred to as Inband Signaling) instead of the mains as a communications path, since it provides a point-to-point circuit between the PEV and EVSE and most utility meters use ZigBee wireless to interface with the home. This would mean that the EVSE would be the bridge between the PEV and the smart meter or any external network, such as a HAN. Using the control pilot low voltage circuit may be less costly and more effective than using the mains with fewer issues to resolve, since the EVSE would be the bridge to the meter, in either case.

Initial EMC testing was started in May, 2010 at the Ford EMC labs, where four suppliers provided evaluation boards to be subjected to automotive tests. Intent was to obtain an initial evaluation for testing the lower layers of the communication stack identified as the (Physical) PHY and Link (MAC) layers. Two paths for the communication are being evaluated and are (1) Inband signaling is the approach defined to using the J1772™ control pilot and (2) the mains. Additional products were tested in July, 2010 with two technologies, Continuous Phase Frequency Shift Keying (CP FSK) and Narrow Band Orthogonal Frequency-division Multiplexing (NB OFDM) with favorable results. Functional test are planned with both AC and DC power applied to the system for both Inband signaling and mains.

In support of the SAE Hybrid J2836™&J2847&J2931 Committee, Electric Power Research Institute (EPRI) has undertaken evaluation of a set of power line carrier (PLC) technologies. EPRI Report TR 1019931, "Evaluation of Power Line Carrier Technologies for Automotive Smart Charging Appliances," documents Phase I activity (2010), where vendor hardware evaluation kits were operated and tested in the EPRI lab. The primary goals of this initial testing were to obtain evaluation hardware from each candidate communications method vendor, set the hardware up in the EPRI lab, and validate the function of the software provided by or recommended for testing by the vendor. Simple noise injection tests were run to load the links up to saturation point, allowing examination of the performance reporting software for each technology. This initial activity lays the groundwork for in-depth performance testing of the PLC types to occur in the first quarter of 2011. This will be done as several suppliers converge on one or possibly two solutions and lead to confirmation of the best approach. This will include inter-operability criteria that will reduce variability as multiple suppliers generate a common product for the PEV and EVSE/HAN.

SIMULATION AND MODELING OF THE SMART ENERGY 2.0 COMMUNICATION STACK

Though the initial EMC requirements[2] (http://www.fordemc.com) and testing, mentioned above, are available to test the PLC technologies PHY and MAC layers of the communication stack, there are no standard methods to test the application layer level communication of J2847/1 messages. EPRI is leading the effort to conduct a requirement survey and test plan to develop the criteria that can evaluate the performance of the PHY and MAC layer testing. The J2847/1 messages do not have an application protocol defined by SAE, and SAE is closely working with SEP 2.0 team to expand and support J2847/1 in the future. The SEP 2.0 protocol is still under development and requires a RESTful approach with IPv6 addressing. For testing communication using J2847 messages, a simple ASCII message protocol was implemented. Figure 2 shows the layers of the communication stack and potential separation of the DC messages from the utility messages may occur while using common upper layers for security and routing.

Several paths are planned as customers interface with their PEVs. The J2931 document series, that is further described later, focuses on the two most direct and understood paths, specifically from the PEV and the EVSE, and from the PEV to the ESI. An EVSE in a premise may or may not be a fixed device. Further, the physical communications technology used by the ESI on the premise may not always be the same and may change per customer. For the aforementioned reasons, a PEV may or may not be able to communicate directly with the ESI without requiring physical layer bridging of communications.

When the following conditions are satisfied, a PEV may communicate directly with the premise ESI:

- The PEV implements J2931/3 or J2931/4 communications on the AC mains
- The ESI implements the same MAC/PHY layer.

In case the above conditions are satisfied, the PEV may transmit onto the AC mains, and the ESI is capable of receiving these transmissions directly, without need for bridging or routing of packets.

In case either of the above criteria is not satisfied, then the PEV will not be able to communicate directly with the ESI without bridging, because the PEV and ESI implement a different physical communications layer. In this case, the EVSE, or other HAN device may implement a bridging function, which can transport packets from the physical layer supported by the PEV to that supported by the ESI and vice versa. If PLC over the power mains is used, then this bridging device may reside anywhere on the common ac circuit.

Still other communications paths are possible, but are outside the scope at this time and will be addressed later within the task force,

An example architecture of the PEV and EVSE is shown in Figure 3. In the event that the physical layer communications implemented by the ESI is different of that from the PEV, the approach is to use the EVSE as a bridging device to both the utility and the off-board charger, if included. The EVSE may

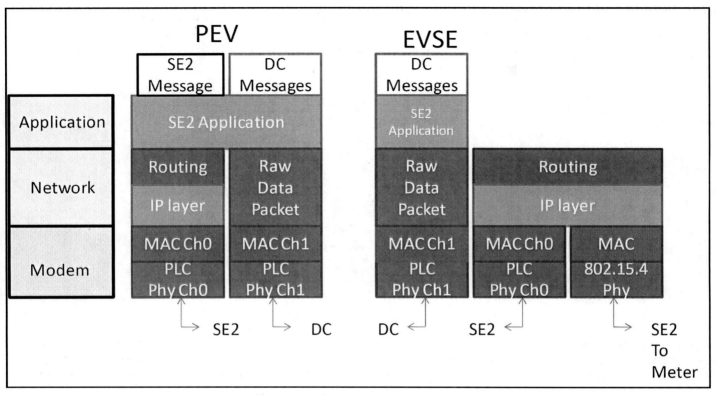

Figure 2. The Communication Stack

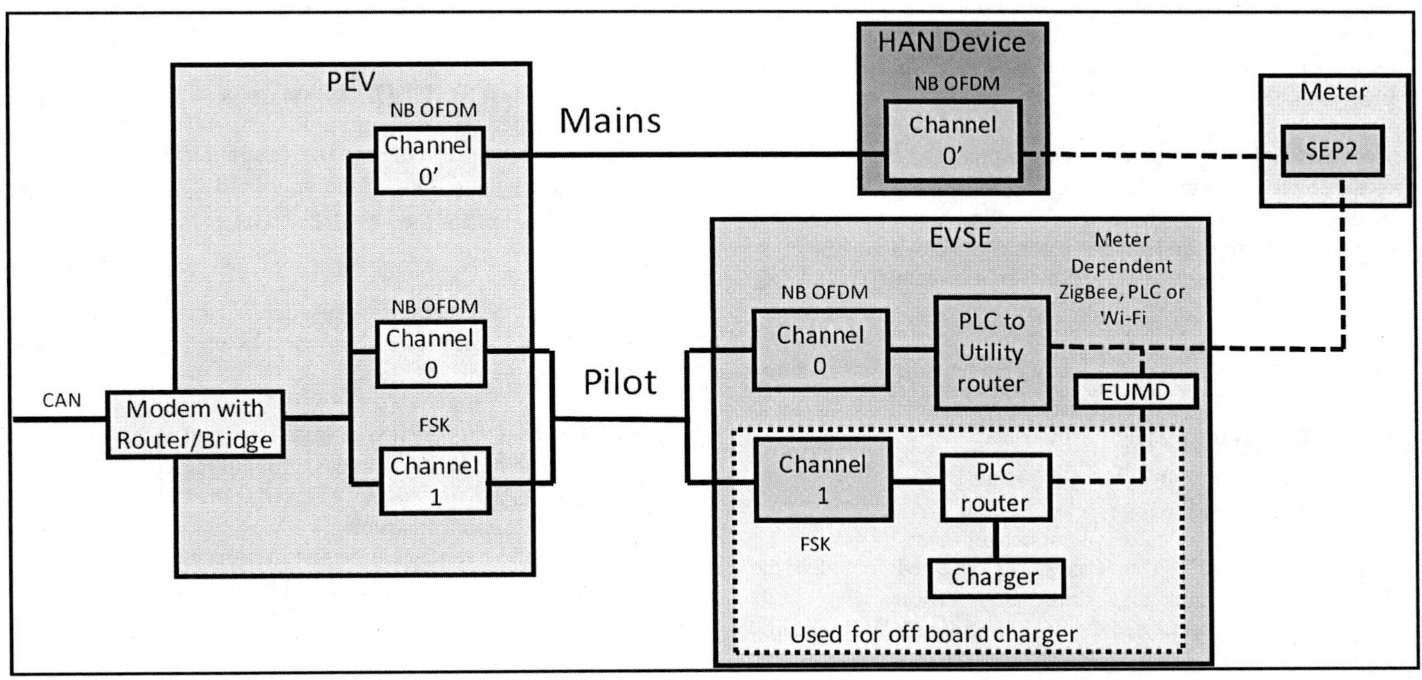

Figure 3. The Communication Path and PLC Options

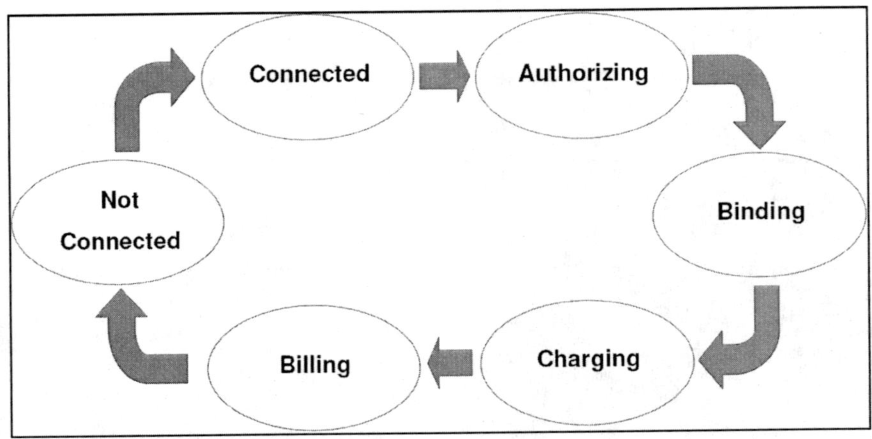

Figure 4. J2836/1™ Use case implementation for testing

be a HAN device and in that case would bridge to any other devices within the home.

Channel 1 is shown as the channel for the DC messages and association. Channel 0 is the channel for the utility messages and could also include association. Channel 0′ is the channel if the PEV communicates with a HAN device other than the EVSE. It is expected for some homes that the EVSE would also function as the HAN device. These three channels will be subjected to the simulation and testing described herein and further evaluation is expected to determine the advantages/disadvantages of each. Note that NB OFDM has only been used as an example in Figure 3, as the final form of PLC has not yet been determined within SAE.

The SAE J2836/1™ use case testing process begins with the vehicle connecting to the Charging Station, obtaining Authorization for charging, Binding communications, Charging, Billing, and disconnecting as shown in Figure 4. The J2847/1 messages for each action are identified in Table 1.

The communication module testing was carried out in three stages:

1. Develop a Human Machine Interface
2. Develop a communication module prototype
3. Install the modules and test the messages

The SAE J2836/1™ Use Cases implemented included the General Registration and Enrollment Process (E); TOU Program (U1), Direct Load Control Program (U2), Real Time Pricing (RTP) Program (U3), Critical Peak Pricing (CPP) Program (U4), Optimized Charging Program (U5), Level I (S1 - 120VAC), and Level II (S2 - 240VAC). The Enrollment Process and Program Selections are controlled by the HMI shown in Figure 5.

The panel on the left shows the price information obtained through the SAE J2836™ Authorization and Binding processes and shows the total charging cost. The icons in the middle panel indicate progress through the SAE J2836™ process including Connection, Authorization, Binding, and Charging. The only required user interaction is to select a charging program on the right panel of Figure 5. Selection of a different charging programs initiates communications to obtain pricing for the selected program and informs the user of the price associated for that program. The time by which the vehicle is expected to be charged is set as a default value and globally available to all pricing options.

In order to test the J2847/1 messages, a laboratory test bench was fabricated with a charging station, battery, charger, and a computer set up to connect the communication modules, as shown in Figure 6. One of the sample PLC evaluation kit was used to implement the J2847/1 messages identified in Table 1 for initial testing.

The initial test plan is focused on data transmission rate, latency and error rates for sample messages by repeated testing as described below:

A serial host was created to both send and receive serial data through the application processor interface boards, developed using an ARM7 processor. The serial host utilized was based around a National Instruments PXI system, which off-loaded the testing to a dedicated device to reduce external influences on the timing. The dedicated serial host also enabled precision timing of the message latency to be observed.

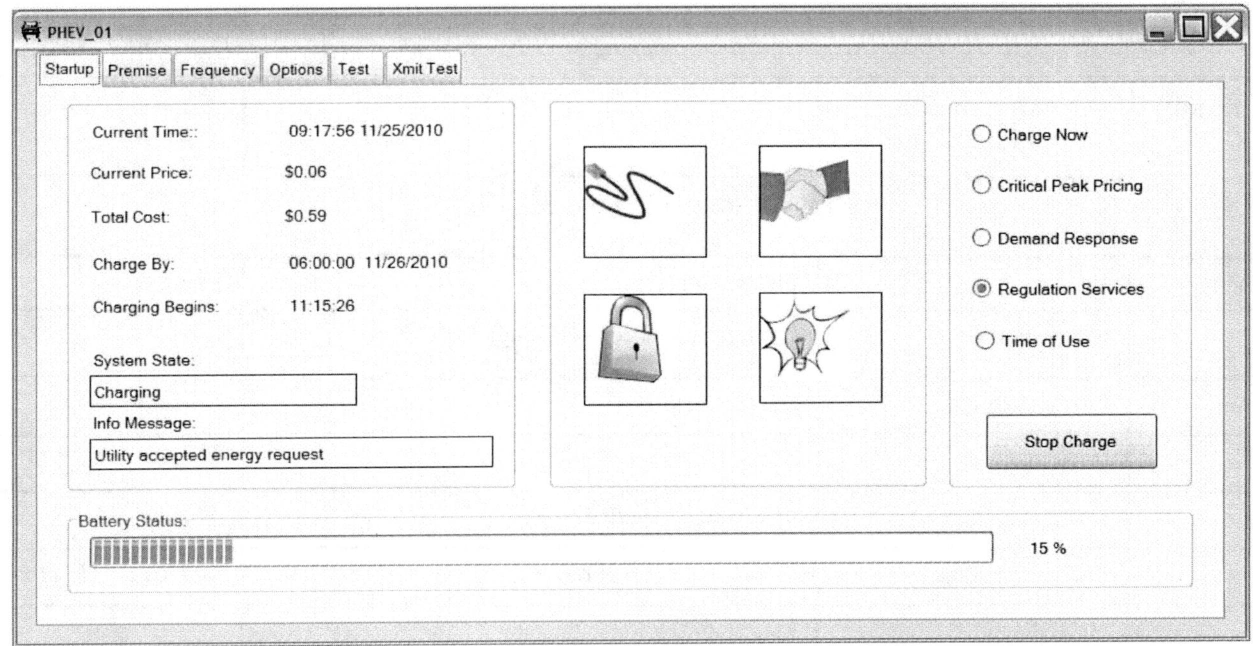

Figure 5. HMI prototype for laboratory testing

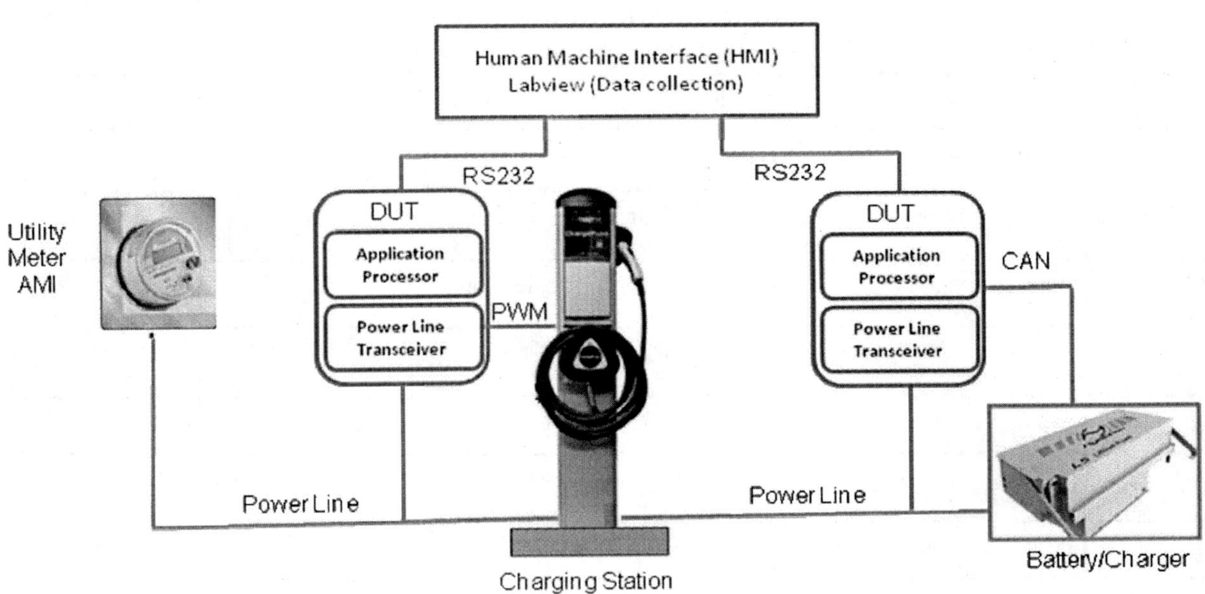

Figure 6. Laboratory test setup for powerline communication modules

Table 1. SAE J2847/1 messages and their mapping to SEP 2.0 variables implemented for testing

Initiator	Information (J2847/1)	SEP2.0	Notes
Connecting Messages			
PHEV	VIN	ElectricVehicle/ID	VIN & PIN
PHEV	Billing Request	BillingData/CustomerBillingInfo/ID	
PHEV	Energy Request	Charger/batChgPwr	Amount
PHEV	Program	BillingData/TariffProfile	TOU/DL/RTP/CPP/Active
Utility	Energy Available	PrepaymentData/RealEnergy	Amount & Rate
Utility	Energy Schedule	PricingData/TariffProfile	TOU/DL/RTP/CPP/Active
EVSE	Meter Reading	MeteringData/MeterReading	
Charging Messages			
PHEV	Battery Status	SOC: PEVData/BatteryStatus /stateOfCharge Voltage: PEVData/Battery/BatVnom Current: PEVData/CurrentFlow Capacity: PEVData/Battery/ahrRtg Power: PEVData/ActivePower	
PHEV	Charge Control	ChargerStatus / batChaSt	
PHEV	Date / Time	Base/Time	
PHEV	Premises Limits	Mains/Voltage	
Utility	Energy Available	DRLCEvent/EndDeviceControl	
Charge Complete Messages			
PHEV	Charge Complete	PEVData/ElectricVehicle/BatteryStatus/batSt	
PHEV	Power Used	BillingData/RealEnergy	
EVSE	Meter Reading	MeteringData/MeterReading	
Utility	Invoice Amount	BillingData/ErpInvoice/amount	
PHEV	Odometer Time	PEVData\ElectricVehicle\OdometerReadDateTime	
PHEV	Odometer Reading	PEVData\ElectricVehicle\OdometerReading	

- AC L1: 120V AC single phase
  - Configuration current 12, 16 amp
  - Configuration power 1.44, 1.92kw
- AC L2: 240V AC single phase
  - Rated Current ≤ 80 amp
  - Rated Power ≤ 19.2kw
- Δ AC L3: TBD
  - AC single or 3ϕ ?

- Δ DC L1: 200 – 450V DC
  - Rated Current ≤ 80 amp
  - Rated Power ≤ 19.2kw
- Δ DC L2: 200 – 450V DC
  - Rated Current ≤ 200 amp
  - Rated Power ≤ 90kw
- Δ DC L3: TBD
  - 200 – 600V DC ?
  - Rated Current ≤ 400 amp?
  - Rated Power ≤ 240kw?

Δ = Under Development

Figure 7. SAE Charging Configurations and Ratings Terminology

The PXI system sent a command to one PLC device as if it were a selection a GUI, or other input panel. The system then waited to read a corresponding result from the receiving end, which could typically go into another GUI or input panel as a display item. For the initial studies performed, two different messages (and sizes) were utilized. The first of these was the transmission of a 17-character Vehicle Identification Number (VIN) from the vehicle to the electric vehicle supply equipment (EVSE). The second message was a two byte battery state of charge data. Both messages were tested using a similar methodology, outlined below.

Testing began by sending the desired message to the sending device. As soon as the message successfully left the host serial buffer, a millisecond timer process is started. This millisecond timer continues to increment until one of two things occurs. Either a message is received up to a carriage return character, or a pre-determined time out expires. If a time out condition occurs, the entire data pack is just flagged as an error. The system then clears the buffers and waits for the next message transmission. During this process, the average latency is continually updated, along with a rough estimate of the bit-error-rate (as estimated by mismatches in the VIN sent and the VIN received).

The test set up has been validated with verifying the communication paths and initial tests are being carried out with a sample PLC transceiver. The data transmission rate varies from 1.33 bits per second to 213 bps depending on the data packet size. Further testing is underway to determine the error rates and compare the data rates for other packet sizes. Further work is in progress to develop the communication modules for all the selected PLC technologies and provide the latency and error rate data at the application layer so that SAE can define the performance requirements for automobile application.

OFF-BOARD CHARGING

J1772™ describes the architecture and power levels for the PEV and EVSE hardware for both AC (on-board) and DC (off-board) charging and summarized in Figure 7. The expected customer charging locations are shown in Figure 8 for the home, workplace and public, along with the expected type of EVSE and charging levels expected at these locations. DC L1 may occur more at homes than workplace and public locations since the power levels match the AC L2 levels. DC L2 is expected to include a larger off-board charger and is the "public gas station" model for fast charges. DC L2 is expected along interstates. Commercial locations, gas stations, and truck & bus stops. Note that higher power levels suggested for DC L1 may require significant upgrades both in a home's electrical system and the service feeding the home.

PEV charge rates vary between a Battery Electric Vehicle (BEV) and Plug-In Hybrid Electric Vehicle (PHEV) since the BEV is expected to include at least three times the battery size since it is all electric where the PHEV also includes an engine for propulsion. Various options as to on-board charger size are also going to affect charge rates and whether the PEV is "equipped" for the DC capabilities to plug into and charge from any off-board chargers.

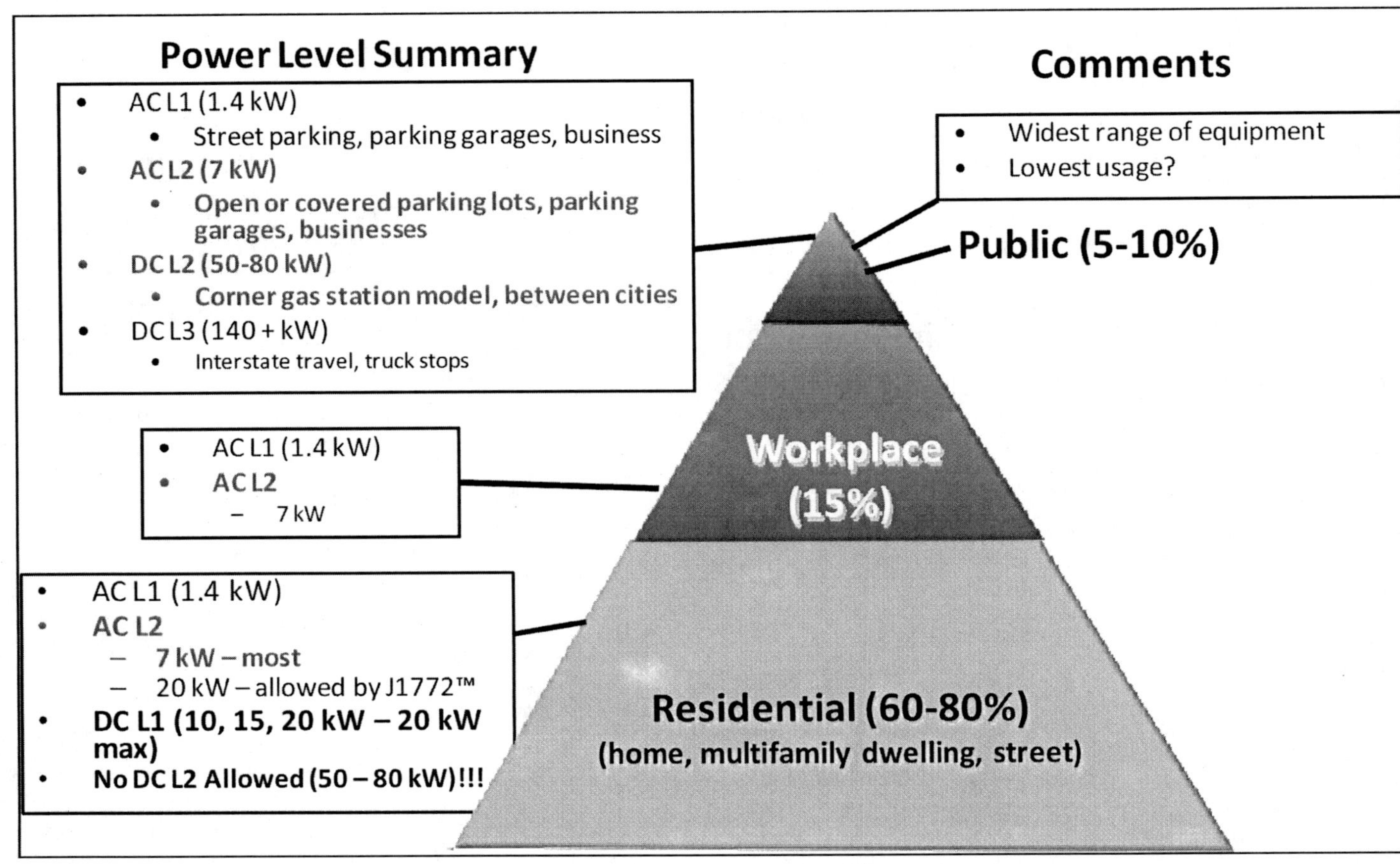

Figure 8. Expected Charging Locations and EVSE's Available

Some of the initial PEVs include a 3.3 kW on-board charger that may be air cooled. If VMs offer larger on-board chargers, they may require liquid cooling to meet a vehicle packaging criteria. Thus, increasing the power capacity of an on-board charger, increases the size, weight and cost of a PEV. The added weight also reduces the mileage and may require added battery capacity to offset this loss. An off-board charger does not have the restrictive packaging requirements of a vehicle and may still be air cooled since there is more space in a garage or other outdoor facilities. Larger fans and cooling systems could be used whereas, thiswould not be practical onboard a PEV.

Thus SAE Recommended Practice J2847/2 establishes requirements and specifications for communication between plug-in electric vehicles and the DC off-board charger. Where relevant, this document notes, but does formally specify, interactions between the vehicle and vehicle operator. The specification supports DC energy transfer via Forward Power Flow (FPF) from grid-to-vehicle and provides the communication to achieve battery pack charging control irrespective of battery pack variations. While J2847/1 supports AC or DC energy transfer, J2847/2 supports the additional messages for DC energy transfer. Even though DC charging communication was included in J2293, this new approach utilizes new connectors and PLC for digital communication that still requires more evaluation and effort in production programs.

The signals within J2847/2 document apply to the use case for off -board DC charging. The expectation is that the system behaves similarly as the onboard charger. The document includes messages used to control the off-board charger level. The intelligence for charge control is provided by the PEV and the chargeer responds accordingly. This document does not attempt to describe Home Area Network (HAN) signals.

A charging session is defined as the time span from a when vehicle is plugged in to the time it is unplugged. Each charging session has three phases:

1. Hand Shaking (Initialization): This includes the PEV and Charger negotiations for their operating limits, and confirms that conditions are correct to allow charging.

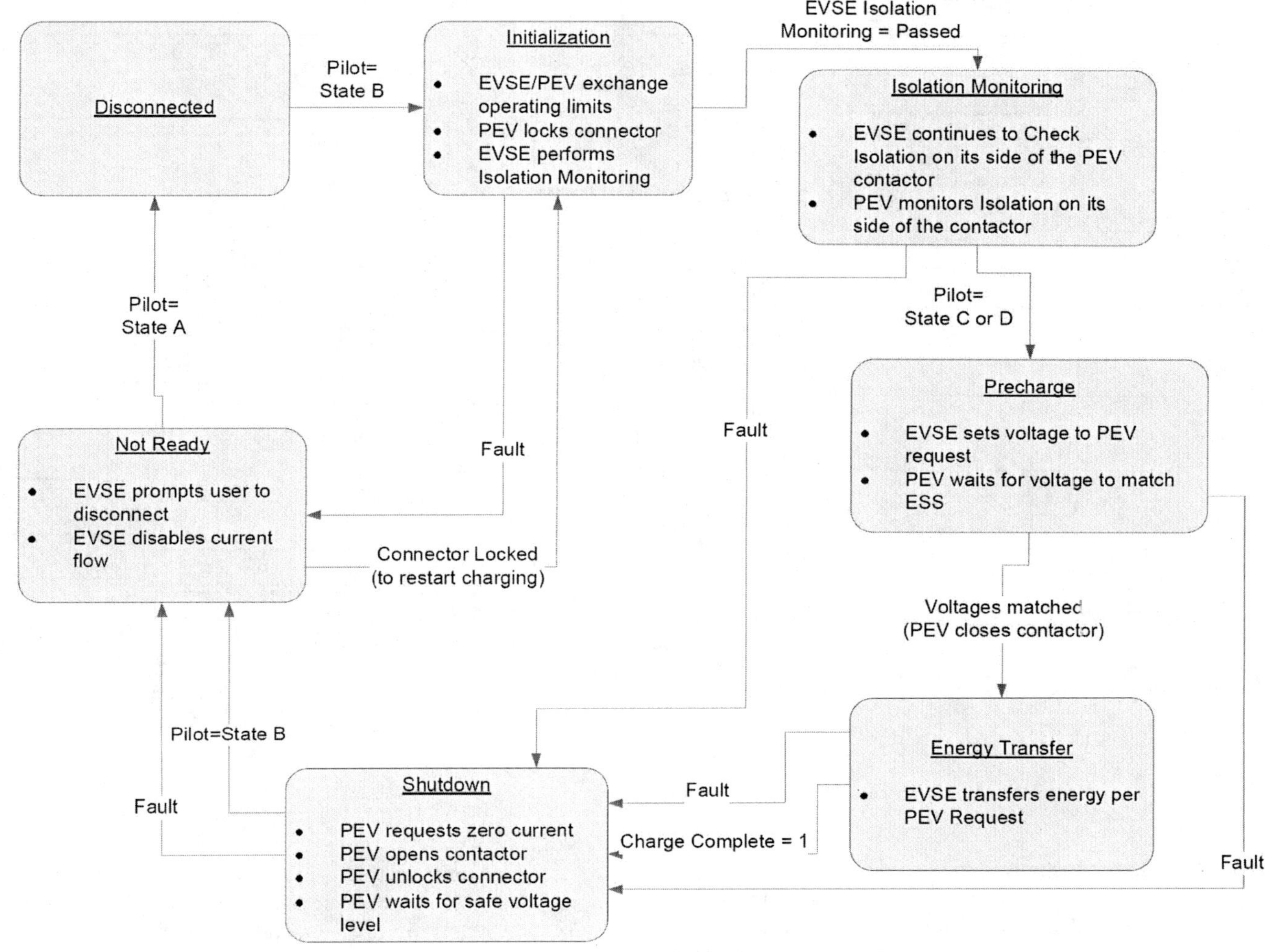

Figure 9. DC Charging System State Diagram

2. Energy Transfer: In this phase, the PEV will continuously request the charger output level from the charger. The charger maintains the output to within the negotiated limits.

3. Normal shutdown: When the vehicle determines that enough energy has been transferred, it will request the charging station to stop sending power. The HV systems will be placed in a safe disconnected state.

Figure 9 shows the DC Charging System State Diagram for the various stages of operation. Figure 10 provides the timing diagrams for DC charging.

The task force effort for DC charging will be moving into the implementation phase over the next year and updates are planned for both J2836/2™ and J2847/2 as the systems mature.

REVERSE ENERGY FLOW

J2836/3 and J2847/3 documents address the use cases and messages for reverse energy power flow. The initial concept was to provide load leveling for the energy needs of a home, the transformer that feeds multiple homes but also apply to public buildings. The load leveling would reduce the peaks while filling in the low levels and make the energy delivered more balanced for the utility. This can be accomplished by the PEV providing both forward and reverse energy flow as needed. The bi-directional conversion can either be on-board or off-board but is not expected in the initial PEVs. The task force turned our attention to delaying or merely reducing load requirements during peak times of the grid as a more effective approach to leveling the loads. As homes need more

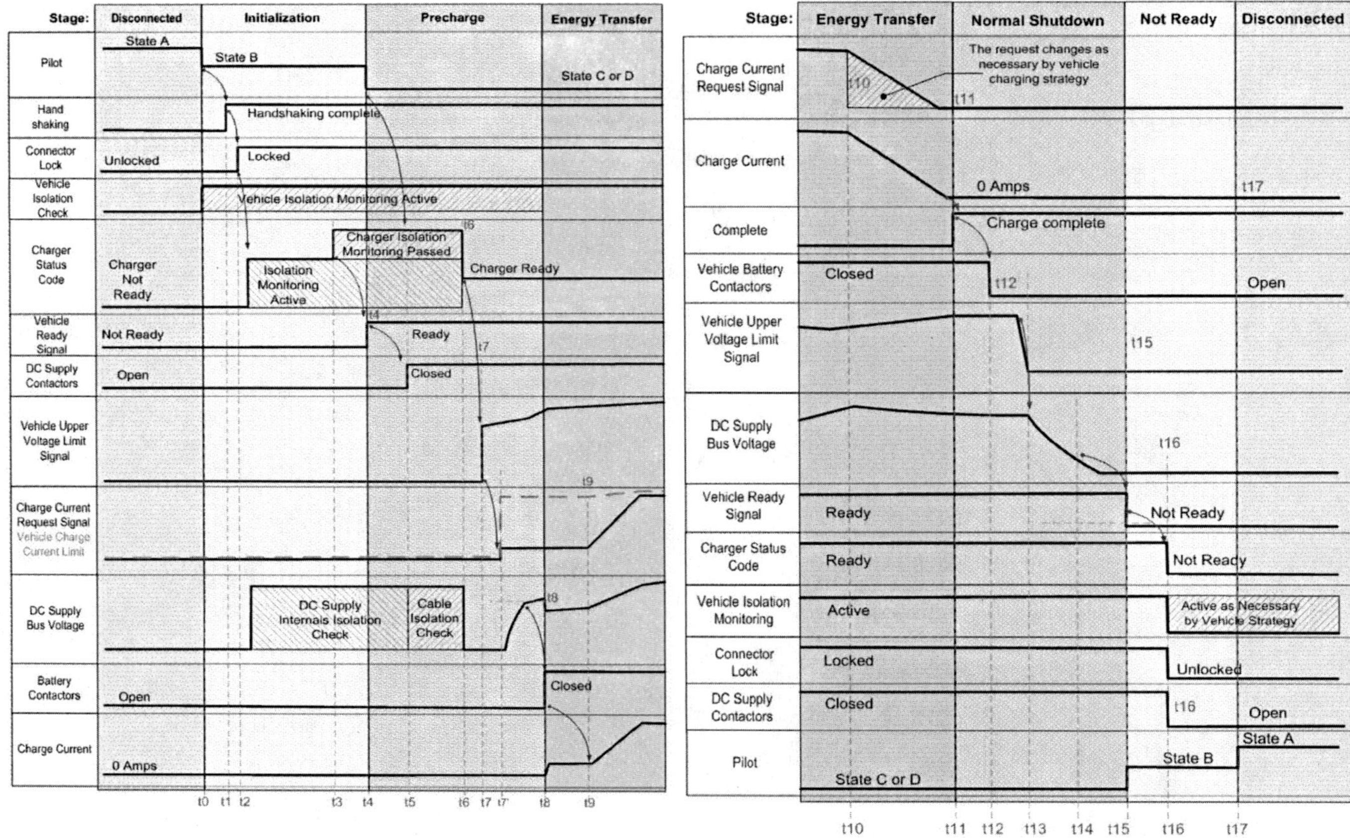

Figure 10. DC Charging Timing Diagram

energy, Demand Response (DR) devices are more commonplace to balance supply and demand and these devices could be used by the utility to delay or reduce the energy to the PEV as peak loads occur. Reverse energy flow can provide a partial amount of the energy to offset the home consumption during peaks or even supply energy to multiple homes fed by a common transformer within a neighborhood. The use cases are still being developed along with the architecture and this information will be fed into J2836/3™ that will include additional fundamental safety items.

There are still cases where reverse energy flow is desired and are labeled into four categories.

1. Vehicle to Grid (V2G) is where the PEV would supply synchronized power back onto the grid via J1772 port and a bi-directional EVSE.

2. Vehicle to Home (V2H) is where the PEV would be used as a home generator and synthesize AC grid. Additional communication to switching and controls would be used to isolate common loads either from grid or PEV to source their needs.

3. Vehicle to Load (V2L) is where synthesized power is supplied where grid power is not available. The EVSE would provide outlets that are sourced from the PEV

4. Vehicle to Vehicle (V2V) is where power could be supplied from one PEV to another. Primairly road side assistance or Mechanic Jump start model.

The architecture has been identified into four categories as shown below.

1. AC Level 2 - Onboard Conversion

- Offboard charger not included

2. AC Level 2 - Onboard Conversion DC Level 1 - Offboard Conversion

- Offboard charger included

3. AC Level 2 - Onboard Conversion DC Level 1 or 2 - Offboard Conversion

- Hybrid connector on PEV

4. DC Level 3 - Future development based on ongoing effort

- Hybrid plus 2nd connector on PEV

1) Reverse Energy Flow
J1772™ connector, on-board conversion.

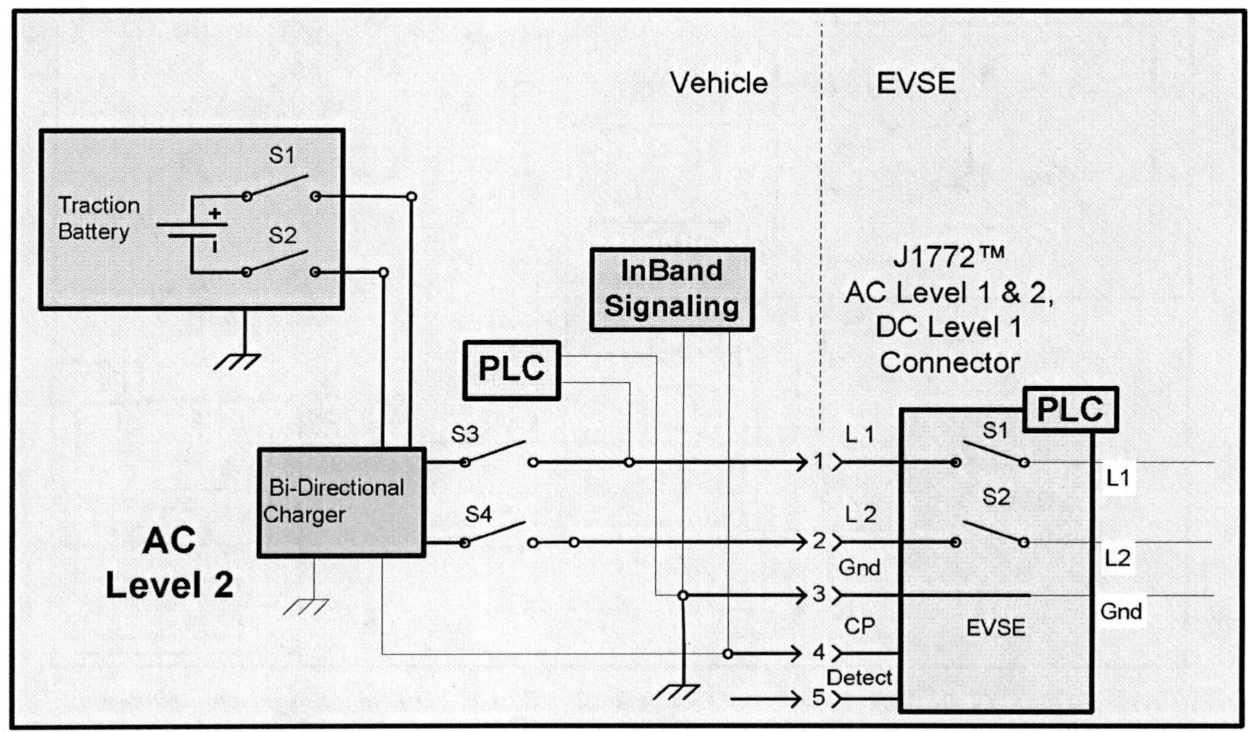

Figure 11. AC Level 2 - Onboard Conversion Architecture

Figure 11, 12 and 13 show architectures #1, #2 and #3. The architecture #4 is under development. The EVSE connector is noted as AC L1, L2 and DC L1, which is correct, however all types of reverse energy flow is not expected to offer AC L1 power levels. AC L1 may be delivered from a PEV with on-board conversion for V2L but not expected for V2H where higher power levels are planned.

The effort for reverse energy flow is being developed and being tracked with DOE funded programs that will assist in this development. Additional safety items are also being considered for future updates to J1772™ and the digital communication will be added to J2847/3 as the Use Case Document matures.

COMMUNICATION PROTOCOL AND REQUIREMENTS

As the task force developed the J2836™ (use cases and general info) and J2847 (detail for messages, sequence diagrams, etc.) series of documents, another series of documents was required to define lower levels of communication. This document is numbered J2931 J2931 is divided into four subsections. J2931/1 (titled Digital Communications for Plug-in Electric Vehicles) identifies general information and requirements for digital communications protocols. Examples of the requirements that will be captured in J2931/1 include association with a meter or Energy Service Interface (ESI), enrollment and registration, timing, security, and general interoperability requirements. Efforts are being made to harmonize these requirements, particularly regarding registration, enrollment, and security, with other prevailing industry standards such as Smart Energy Protocol 2.0.

J2931/2 (titled Inband Signaling Communication for Plug-in Electric Vehicles - Analog Modem) will encapsulate lower layers of the communications stack (i.e. the PHY and MAC layer) for communications using a modulated signal over the J1772™ Control Pilot wire, otherwise known as "inband" communications. As previously mentioned, several OEMs are seeking to implement vehicles that employ off-board DC charging in the very near future, and with the long lead time for practical implementation of PLC communications, it is anticipated that many OEMs may choose to implement off-board DC charging using Pilot wire communications as defined by this document.

2) Reverse Energy Flow
J1772™ connector, on-board or off-board conversion

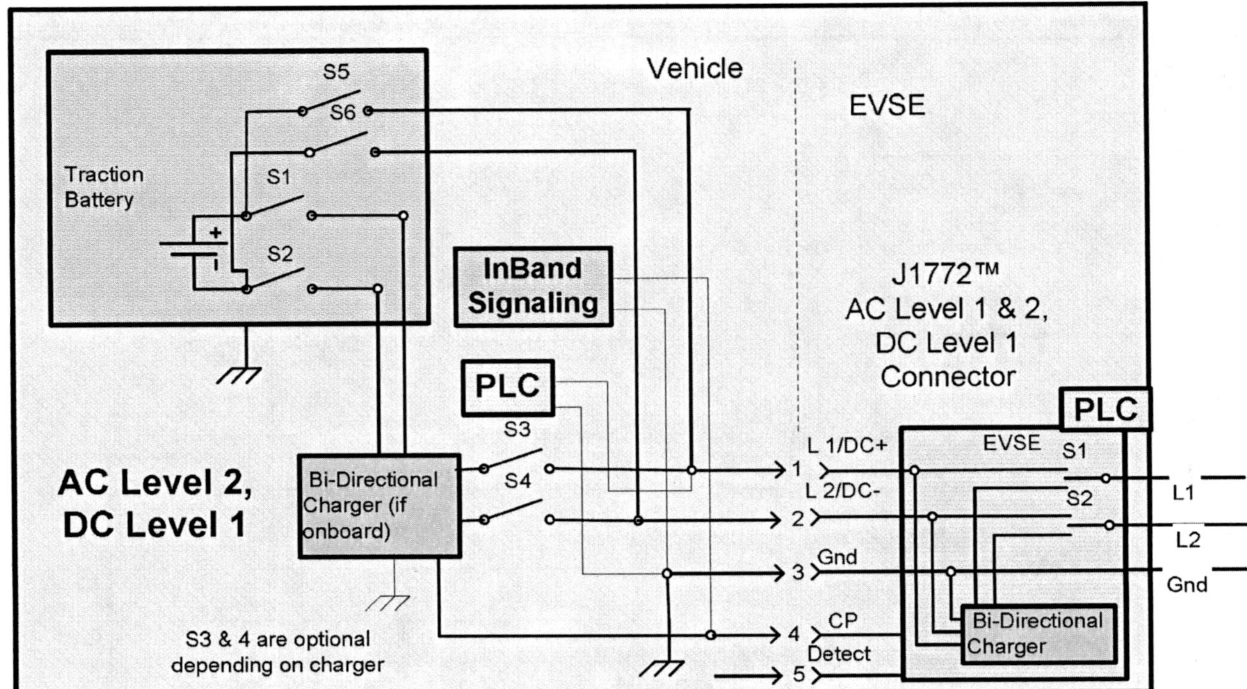

Figure 12. AC Level 2 - Onboard Conversion & DC Level 2 Offboard Conversion Architecture

3) DC Level 2 Charging (up to 80 kW)
J1772™ hybrid connector - on-board or off-board conversion.

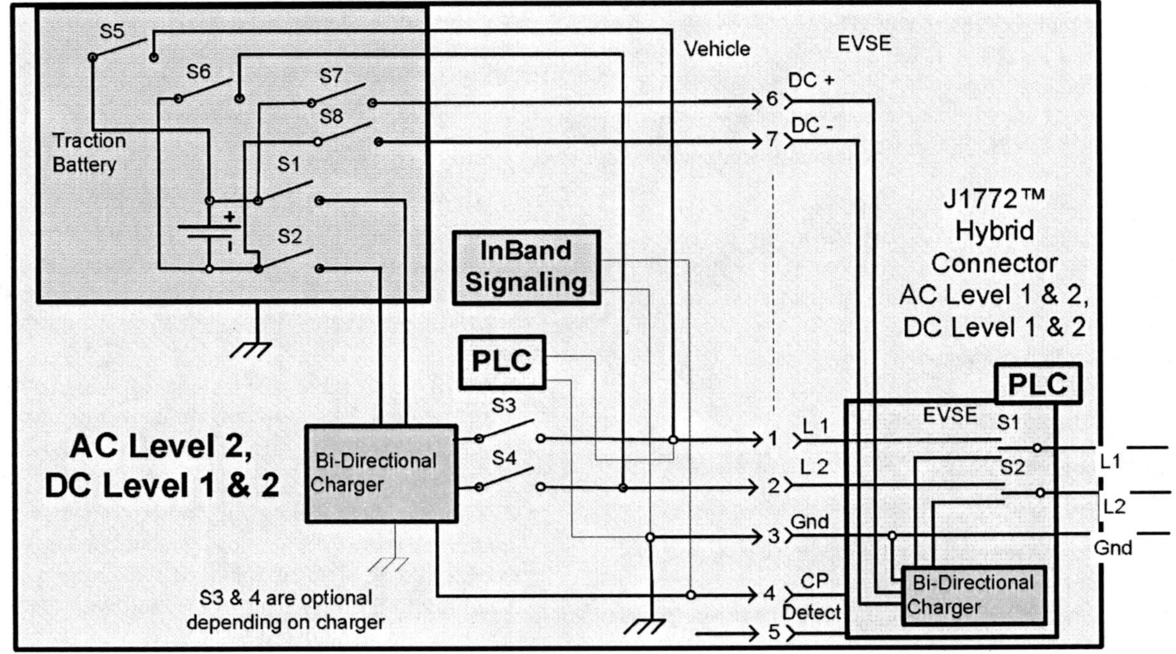

Figure 13. AC Level 2 - Onboard Conversion & DC Level 1 or 2 Offboard Conversion Architecture

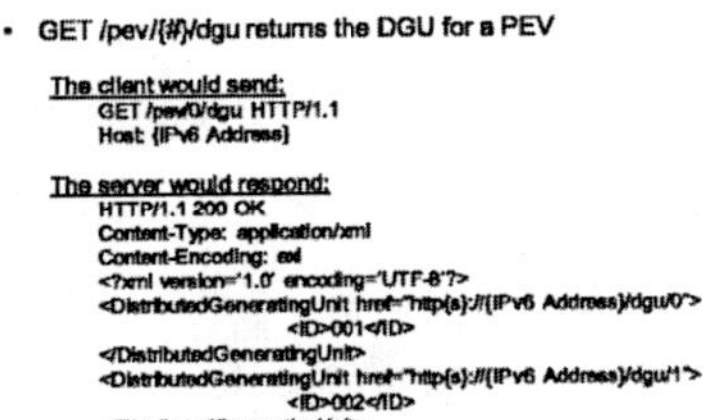
• GET /pev/{#}/dgu returns the DGU for a PEV
The client would send:
GET /pev/0/dgu HTTP/1.1
Host: {IPv6 Address}
The server would respond:
HTTP/1.1 200 OK
Content-Type: application/xml
Content-Encoding: exi
<?xml version='1.0' encoding='UTF-8'?>
<DistributedGeneratingUnit href="http{s}://{IPv6 Address}/dgu/0">
<ID>001</ID>
</DistributedGeneratingUnit>
<DistributedGeneratingUnit href="http{s}://{IPv6 Address}/dgu/1">
<ID>002</ID>
</DistributedGeneratingUnit>

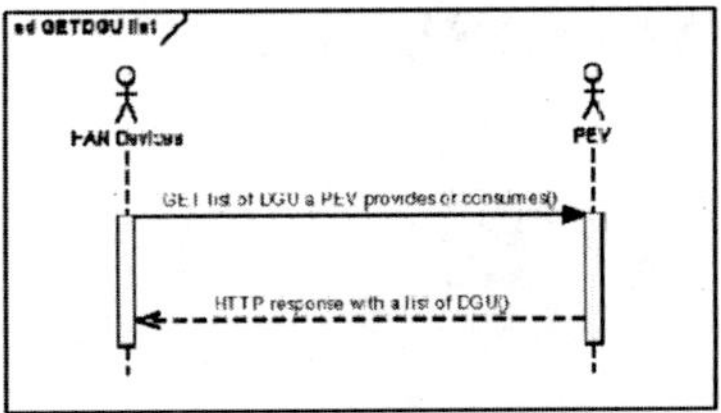
sd GETDGU list
PEV
GET list of DGU a PEV provides or consumes()
HTTP response with a list of DGU()

Figure 14. Sample of the SEP 2.0 Protocol

Similarly, J2931/3 (titled Narrow Band PLC Communication for Plug-in Electric Vehicles) will encapsulate the lower layer details for communications using the J1772™ Control Pilot or mains "power circuits" for NB OFDM PLC. For vehicle to grid communications during AC charging, for the purpose of participating in utility programs such as Time-of-Use rates, Demand Response, etc., it is anticipated that communication using PLC on the J1772™ Control Pilot or power circuits may be the approach for these messages.

J2931/4 (titled Broad Band PLC Communication for Plug-in Electric Vehicles) is the BB OFDM that could be used over the control pilot or mains.

All four sections of the document are presently under development by the task force with initial versions published mid 2011.

SUMMARY/CONCLUSIONS

This paper presented the current state of development of electric vehicle communication standards and progress in PLC testing. Though the task force has made significant progress in developing uses cases and messages for utility communication, off-board charging and reverse energy flow, several important issues need to be resolved for developing communication module performance requirements, field testing and verification. As the task force moves into the implementation phase by simulating and modeling as described above, then into initial production programs along with additional testing SAE, ISO/IEC and EPRI plan this year, we should be able to narrow our choices to one or two solutions.

The DC messages have significantly different requirements than with the utility messages such as latency and a private approach may be applied to this since the messages are strictly between the PEV and the off-board charger in the EVSE. Attempting to combine the DC message approach with the utility messages in SEP2 may or may not be an optimal approach and needs more time to determine the best approach for our standards.

REFERENCES

1. NIST SGIP Catalogue of Standards (http://collaborate.nist.gov/twiki-sggrid/bin/view/SmartGrid/SGIPCatalogOfStandards

2. EMC requirements (http://www.fordemc.com).

CONTACT INFORMATION

Rich Scholer
Ford Motor Company
EESE Product Design Engineer
Plug-In & Fuel Cell Vehicles
Phone: 313-323-0460
rscholer@ford.com

ACKNOWLEDGMENTS

The authors of this paper have been the leads in identifying the Use Cases and initial communication requirements. As subsequent papers are presented, the content and authors will change as we discuss PLC, testing and other tasks as the task force continues to move forward.

Dan Mepham led the J2847/1 effort and is currently leading the J2931 suite of documents. Eloi Taha is the lead for the J2847/2 document. Sam Girimonte is the lead for the J2836/3™ & J2847/3 documents. Doug Oliver provided significant technical support to J2847/1, J2847/2 and J2931. Jim Lawlis led the EMC automotive testing at Ford for qualifying and disqualifying the PLC types. John Halliwell is the lead for identifying the PLC requirements, testing & interoperability. Krishnan Gowri and Nathan Tenney are supporting the application layer level testing of utility communication messages.

DEFINITIONS/ABBREVIATIONS

BB OFDM
: Broad Band Orthogonal Frequency-division Multiplexing

BEV
: Battery Electric Vehicle

CP FSK
: Continuous Phase Frequency Shift Keying

EMC
: Electromagnetic Compatibility

EPRI
: Electric Power Research Institute

ESI
: Energy Service Interface

EUMD
: End Use Measuring Device

EVSE
: Electric Vehicle Supply Equipment

GUI
: Graphic User Interface

HAN
: Home Area Network

HMI
: Human Machine Interface

JWG
: Joint Working Group

MAC
: Medium Access (Link) Layer

NB OFDM
: Narrow Band Orthogonal Frequency-division Multiplexing

NIST
: National Institute of Standards Technology

PEV
: Plug-In Electric Vehicle

PHEV
: Plug-In Hybrid Electric Vehicle

PHY
: Physical Layer

PLC
: Power Line Carrier

SCWG
: Cyber Security Working Group committee

SEP2
: Smart Grid Alliance

SGAC
: Smart Grid Architectural Committee

SGIP
: Smart Grid Interoperability Panel

V2G
: Vehicle to Grid

V2H
: Vehicle to Home

V2L
: Vehicle to Load

V2V
: Vehicle to Vehicle

VM
: Vehicle Manufacturers

Communication between Plug-in Vehicles and the Utility Grid

2010-01-0837
Published
04/12/2010

Richard A. Scholer
Ford Motor Company

Arindam Maitra
EPRI

Efrain Ornelas
PG&E

Michael Bourton
Grid2Home

Jose Salazar
SCE

ABSTRACT

This paper is the first in a series of documents designed to record the progress of the SAE J2293 Task Force as it continues to develop and refine the communication requirements between Plug-In Electric Vehicles (PEV) and the Electric Utility Grid. In February, 2008 the SAE Task Force was formed and it started by reviewing the existing SAE J2293 standard, which was originally developed by the Electric Vehicle (EV) Charging Controls Task Force in the 1990s. This legacy standard identified the communication requirements between the Electric Vehicle (EV) and the EV Supply Equipment (EVSE), including off-board charging systems necessary to transfer DC energy to the vehicle.

It was apparent at the first Task Force meeting that the communications requirements between the PEV and utility grid being proposed by industry stakeholders were vastly different in the type of communications and messaging documented in the original standard. In order to understand and adequately capture the communication structure between plug-in electric vehicles and the electric power grid, the task force generated two new documents. The original J2293 was also re-issued to keep it intact for legacy equipment.

The two new SAE documents are J2836™ and J2847. SAE J2836™ is an Information Report that captures the communication requirements between plug-in electric vehicles and the electric power grid based on use cases. A use case is simply a "story" that includes various "actors", and the "path" they take to achieve a particular functional goal. By considering the actions of the actors working to achieve this functional goal, a completed use case results in the documentation of multiple scenarios, each containing a sequence of steps that trace an end-to-end path. These sequential steps describe the functions that the proposed systems and processes must provide, directly leading to the requirements for the given use case.

SAE J2847 is a Recommended Practice which builds upon the Use Cases defined in J2836™ and defines the detailed messages and specifications for vehicle to utility communication.

Both SAE J2836 & J2847 have a series of dash 1 through dash 5 in order to keep the task force focused as we start, beginning with the fundamental requirements and migrate to other more advanced options. Dash 1 identifies the Utility rate and incentive programs, dash 2 includes the detail for an EV Supply Equipment (EVSE) for off-board DC chargers (updated version of J2293), dash 3 is for reverse energy flow, dash 4 includes diagnostics of the charging system and PEV and dash 5 contains vehicle manufacturer specific options.

The primary purpose of SAE J2836™ and J2847 is to achieve grid-optimized energy transfer for plug-in electric vehicles - that is, ensuring that vehicle customers have sufficient energy for driving while minimizing system impact on the electric grid. This can be accomplished, for example, by voluntary participation in utility controlled-charging programs in return for financial incentives, and hence the specification therefore supports information flows that enable such mechanisms.

These specifications support energy transfer via both Forward Power Flow (FPF) (grid-to-vehicle), and Reverse Power Flow (RPF) from vehicle-to-grid. Forward Power Flow is used to charge the vehicle's rechargeable energy storage system (RESS); support for FPF is optional, though encouraged, for any plug-in vehicle implementation. Reverse Power Flow may be used to discharge the RESS, in order to provide support to the grid, or to power local loads during a grid outage; support for RPF is also optional and may be limited with various RESS applications.

Beyond its primary purpose of energy transfer, these documents enable other applications between vehicles and the grid, such as vehicle participation in various utility rates and incentive programs, utility-controlled charging plans (e.g. Demand response and direct load control), and participation in a Home-Area Network (HAN) of utility-managed electrical devices. The protocol established is designed to be extensible, so that as new applications emerge, additional messages can be added while maintaining support for the existing message set.

INTRODUCTION

In order to better understand the communications between the PEV and the utility, it was important to understand the existing standards and then later apply them to the entire system. This is the same approach the Charging Controls Task Force used to generate both J1772™ and J2293 SAE documents and furthermore link them with new UL documents and updates to the National Electric Code to insure system compatibility. We also had to understand customer requirements, especially as to how they plan to purchase and operate their PEVs. Furthermore, it is important to understand the EVSE installation at the home or how they plan to use them in public locations. Once this system is properly identified and understood, we can offer optimized designs to customers that meet their electric transportation needs and lessens the impact to the utility distribution network. Since the customer will charge a PEV more frequently than they fuel a gasoline vehicle, we also want to make the process simple, intuitive, familiar and quick so it requires as few steps from the customer as possible.

SAE STANDARDS BACKGROUND

The initial J1772™ document identified the fundamental interface to the EVSE including the connector. Since SAE J1772™ describes the vehicle requirements, UL2202, 2231 & 2251 were also generated to identify the requirements off-board the vehicle. SAE and UL both showed the interface at the vehicle, since they need to be the same to make the system function correctly. Article 625 was also added to the National Electric Code to identity the infrastructure requirements and other sections described the wiring sizes and other aspects of these new Electric Vehicles as they were connected to the Utility Grid.

The EVSE has three architectures that include: (1) Level 1 - a cordset with a mobile EVSE that connects the EV to a standard 120V/15A or 20A outlet, (2) Level 2 - requires a premise mounted unit at 240V - single phase up to 100A circuit and (3) DC Charging - a premise unit that also includes an off-board charger. This 3rd version requires additional communication between the Vehicle Charging Controller and the off-board charger. Since off-board chargers transfer DC energy to the vehicle instead of AC as in cases 1 & 2, then the communications described in J2293 are still applicable. On-board chargers are typically smaller (power output) than off-board units and hence DC charging is typically used for higher energy transfer resulting in shorter charge times.

When the Hybrid Electric Vehicle (HEV) Committee formed a few years ago the J1772™ Task Force was re-formed to update the standard with a new connector, increased power levels for AC, and other associated changes. The original connector was designed for battery electric vehicles (BEVs) and was generally placed in the vehicle location where the gasoline fuel door had been traditionally placed. With the introduction of Plug-In Hybrid Vehicles (PHEVs), the fuel door was still needed for gasoline, and a smaller "plug-in" connector in an alternate location was desired.

As the connector was redesigned the three original communication terminals were removed in order to make the unit smaller, especially since electric utilities were expected to use ZigBee wireless or Power Line Carrier (PLC) communication mediums. In addition, most vehicle manufacturers (VM) have also migrated from J1850 to CAN and are looking at other mediums for future programs and hence reinforced the idea for removing the communication terminals.

The J2293 Task Force wanted to complement J1772™ updates by generating new SAE documents for advanced grid communication. The team began by analyzing the fundamental control of energy to the PEV that utilizes the J1772™ control pilot by "waking up" the PEV and identifying the Available Line Current (ALC) available to the

PEV from the premise. By the customer or the utility including a device to control the EVSE control pilot, they can also control the PEV and meet their objectives. Furthermore, some existing EVSEs and PEVs that do not have communicate abilities nor are all future products expected to have communications. The customer and utility still need to be able to modify grid demand with these combinations. As higher power is transferred however, EVSEs and PEVs that include communication would offer more information for advanced planning of the grid usage and are expected to be more prevalent.

UTILITY BACKGROUND

Use cases were the first tool utilized to identify requirements for utility communications. The Electric Power Research Institute (EPRI) and several electric utilities had already begun to generate use cases for their systems but now needed to add the PEV and its connected to the EVSE.

We spend several months working with the Electric Transportation Infrastructure Working Council (IWC), Electric Power Research Institute, Southern California Edison (SCE) & Pacific Gas & Electric (PG&E) among other utilities on top level use case development. These top level use cases were selected based on high-level requirements which included:

- Supports secure two-way communication with the Energy Services Communication Interface (i.e., Utility)
- Supports time- or price-based charging preferences based on current electric rate/tier
- Supports vehicle charging at any voltage
- Support vehicle load correlation (end use metering of the PEV)
- Support Demand Side Management Integration
- Support vehicle charging regardless of utility metering and/or communication availability
- Supports vehicle roaming and unified billing infrastructure
- Supports Customer override/opt-outs
- PEV-to-Utility communications technology based on open standards

Detailed use cases were then developed from these top level cases. The intent of these detail use cases was to separate the functions and actions to make them more complete, exclusive, and simple to understand. Figure 1 shows the approach used for the detail use cases and starts with "E" for the general information (that applies to all subsequent ones), then progress to the five utility programs (U1-5), then the three connection architectures (S1-3) per J1772™, then the location for the energy transfer is next (L1-4) and finally the desired function (PR1-4). If a utility or customer wanted a complete use case set, from this selection, they would select any combination of these to match what is offered or desired.

<figure 1 here>

PEV roaming is another topic we are currently addressing, and to do this we wanted to demonstrate the variations of Plug-In Vehicles (PEV) to Hybrid Electric Vehicles (HEV), that do not plug into the grid for charging. Combining the variations in vehicle technologies with those of the EVSE (J1772™) can create a better understanding the challenges posed by roaming. Current PEV technologies can be separated into two categories as follows:

- Plug-In Hybrid Electric Vehicle (PHEV): has a high voltage battery roughly 5 to 6 times more capacity than a HEV
- Battery Electric Vehicle (BEV) which increases the battery capacity by roughly 3 times more than a PHEV.

Vehicle manufacturers and utilities have studied the adoption rate of EVs in the past, however future demands for new PHEVs is expected to be considerably different. One of the reasons for the difference is directly related to the battery size which will vary due to the electric driving range of these vehicles. A PHEV may use all of its capacity before recharging since it can rely on a small Internal Combustion (IC) engine, unlike a BEV which is completely dependent on electric energy and will be required to recharge after reaching its maximum range. This implies that even though the capacity varies, either technology may both take the relative same amount of energy for each charge session, however the BEV may desire a faster rate at the session. Both a PHEV and BEV may also have the same on-board charger size and to obtain faster charge times, customers may rely on DC energy from an off-board charger. Many times it is difficult and expensive to package a larger charger on-board the vehicle and may not be very cost effective for the customer to include this in their EVSE. This allows faster charge and would allow customers to purchase larger off-board chargers with the initial EVSE and accommodate future vehicle purchases instead of with every vehicle purchase. Figure 2 shows the variations between vehicles and EVSEs.

<figure 2 here>

Since the amount and recharge rate of energy to PEVs can be different for each charging session, use case U5 includes messages from the PEV "requesting" an amount and rate of energy from the premise and the premise responds with "available" energy or default settings as to not overload the grid. In the home, the EVSE is considered similar to other appliances but since the PEV energy needs vary each time, and any PEV can also connect to this EVSE, making this request allows more advanced planning for the charge cycle to satisfy both the customer and utility needs.

Figure 3 shows typical charge times with vehicle variations and on-board vs. off-board charger selections. J1772™ included DC energy transfer in past releases, but since the focus on this update was Level 1 & 2, DC was identified as "under development". Our task force and the J1772™ task force is continuing the effort to define the DC system and further defining "fast" charge as it also has a variety of meanings. Figure 2 shows "fast", "faster" and "fastest" to signify three levels of DC energy that will be discussed in future meetings of these groups.

<figure 3 here>

COMMUNICATION MEDIUMS

In order to offer consumers pricing incentives for charging their PEVs off-peak and/or allow some form of a charge control by a utility in exchange for monetary compensation, there is a need for PEV to Utility communications. Power Line Carrier (PLC) is considered the communication medium of choice amongst leading Utility members, EPRI, and SAE, between the PEV and the EVSE since many utility programs will offer these special rates for charging PEVs, and requires the PEV load to be measured separately for billing and roaming purposes. In these cases the utility system would require a sub-meter in the EVSE circuit to measure the energy to the PEV. This also requires an association from the PEV to assure it is not a conventional appliance connected to the EVSE for this lower energy rate.

All methods under consideration that would solve the association problem use the contactor in the EVSE to isolate or validate the PLC communication to an EVSE, which is associated with a particular sub-meter.

There are several PLC technologies with associated standards and our task force has had multiple meetings to help update voting members on these and to make a choice on which one to include. This is an on-going effort and will continue with a reduced selection of products in the next year to simulate, test and validate these to determine a final selection. Utility smart meters could include PLC or wireless and the EVSE or the HAN may bridge between these if different.

Figure 4 shows the simplistic approach to the EVSE and as it gains features, a modularization approach is planned as shown below. The simplistic communication paths are shown on the left of the diagram whereas the main actors are the Utility, Customer and PEV. The EVSE is centered and the communication path can take any direction between these. As we attempt to separate the information the utilities desire from the customer expectations, we show both the Utility network and Consumer network on the right with the plan to modularize these to offer selections as the customer wants added content. The customer could also have the communication bridge in the Energy Management System (EMS).

<figure 4 here>

OTHER ORGANIZATIONS

SMART ENERGY PROFILE FOR HOME AREA NETWORK COMMUNICATIONS SEP 2.0

ZigBee(TM) Smart Energy Profile (now know as SEP 1.) was originally adopted in 2007 by key members of the utility industry and meter manufacturers to enable communication within home area network (HAN) among the smart meter and other smart devices or loads within the premises.

In general, ZigBee focuses on a network to application-layer standard[1] that builds on the wireless MAC and Physical layer defined by the IEEE 802.15.4 standard. In addition, its primary focus is on communications related to energy efficiency, usage, price, and messaging. It is designed not to specify behavior, but to specify communications to support behavior. In that sense, it specifies more "the method" of communication (how to communicate) rather than the 'what' to communicate. It is also important to note that when the ZigBee committee began this work, it did not want to recreate existing standards for home automation. In addition, both the range of backhaul bandwidths and cost to implement were kept in mind during development. ZigBee radios also boast of very low power consumption while operating, making them amenable to battery-operated distributed appliances. There are about 30 million meters with ZigBee currently under contracts in the US today.

The ZigBee 'ecosystem' or stakeholders currently include utilities & retail energy providers, government and regulators, outreach to other standards bodies such as HomePlug Alliance. It also includes security experts, AMI and meter vendors, demand response and load control equipment and system vendors, and manufacturers of Thermostats, Displays and Smart Appliances. Also included in ZigBee standards-making are system integrators and semiconductor manufacturers.

Since its inception, the ZigBee Smart Energy Profile has evolved rapidly, with version 1.0 released in December 2007. Version 1.5 will have incremental enhancements to v1.0 such as Over The Air upgrade.

During 2008, with a realization, that total coverage could not achieved by a wireless standard, a liaison was created between ZigBee and HomePlug Alliance, (ZBHP) with the mandate of creating a hybrid of wired and wireless Technologies. Power Line carrier provides range and Wireless gives mobility and in the case of 802.15.4 low power. At the same time, consensus was reached with the key stakeholders that the Network layers should be IP based. This resulted in the creation of what is now known as Smart Energy Profile 2.0. (SEP 2.0)

In late 2009, SAE created a liaison with the SEP 2.0 workgroup and is now where most of the attention of automotive and utility communities are focused.

The goals for SEP 2.0 are to support an IP stack modeled with the Smart Energy Profile clearly described in UML. In addition, version 2.0 will have objects derived from IEC CIM (61968) and 61850, and is thought to be leaning towards web services based architecture. In addition, it will incorporate feedback and lessons learned from SDOs. The overarching objective is to create a well-defined end-to-end architecture.

Utilities are currently installing Smart Meters that include Smart Energy 1.0 and plan to migrate to SEP 2.0 as it is completed. The differences between Smart Energy 1.0 and SEP 2.0 are outlined in J2847/1 but Smart Energy 1.0 did not include any PEV effects that are now being considered and included.

In general, ZigBee wireless will be used for battery powered or non-mains connected devices such as Gas, Water Meters and thermostats and Wired (PLC) will be used to connect to devices that might be outside of the range or where a clear association is required such as Electric Vehicles.

In the EPRI IWC meetings, we started identifying the aspect of no communications vs. communications from either/or the PEV or EVSE. The objective is to manage the Distributed Energy Resources within a home that now includes a PEV.

Figure 5 shows the current state - Battery EV connection with no communication

One of the major objectives for PEV users is to obtain a reduced for charging their vehicle. Many utilities offer these rates, but usually require separate dedicated meters to differentiate between the home and charging load. This is accomplished using the dual meter as shown.

The dual or dedicated meter for the vehicle, may also enable a future feature known as "roaming", which would allow a vehicle to be charged at other locations besides the home premise, while the cost allocated to the user's electricity account.

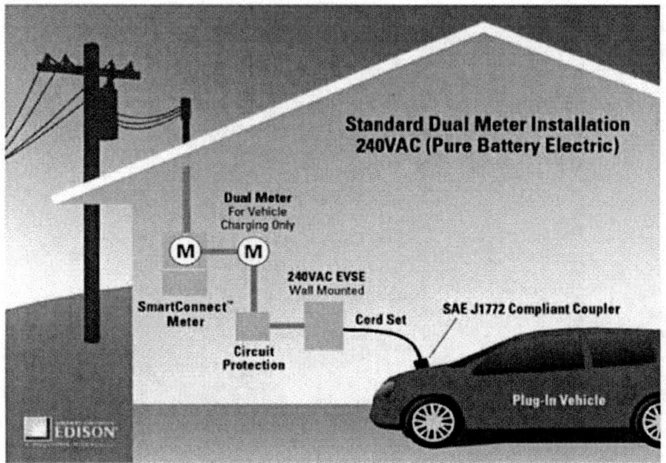

Figure 5. PEV connection with no communication

Figure 6 depicts a proposed future state with a customer using a level 1 cordset identified in the detailed use case S1. The dual meter has been replaced with a Smart Socket (that integrates a utility provided plug-in revenue grade meter) that will accomplish the sub-metering to the PEV and ultimately allow the electric utility to separate the PEV load from the rest of the home.

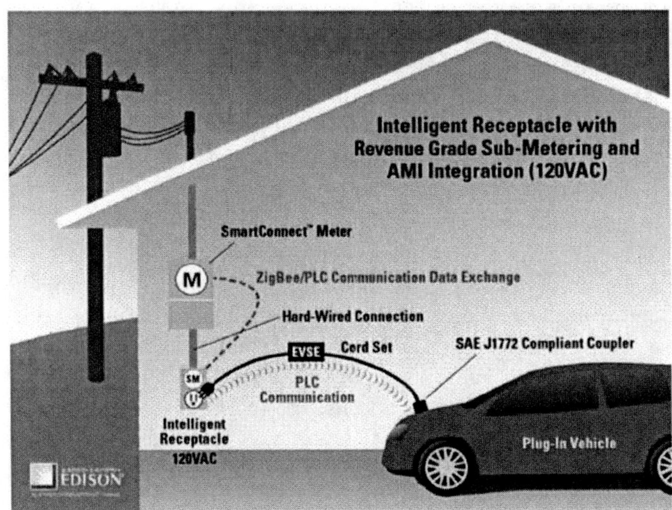

Figure 6. PEV using cordset EVSE

Figure 7 demonstrates a proposed level II EVSE with built-in sub-metering capabilities. This metering approach is also applied to the detail use case S2 or S3 whereas the EVSE is permanently mounted to the home. Like the previous case, this method allows the electric utility company to provide customers with a system for PEV specific rates.

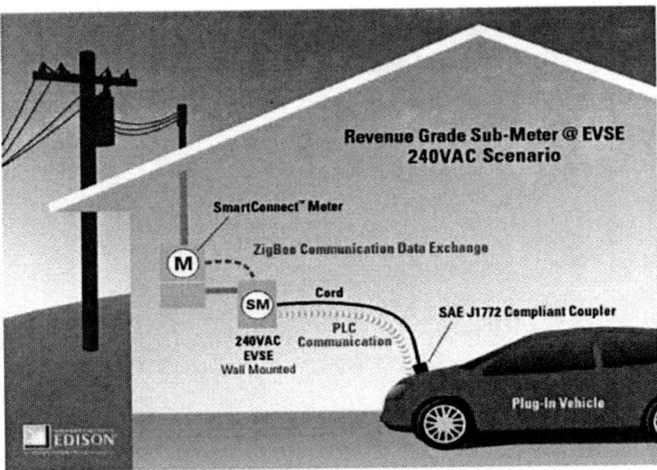

Figure 7. PEV using premise EVSE

NATIONAL INSTITUTE OF STANDARDS TECHNOLOGY (NIST)

Historically, a lack of common standards has been a barrier to rally electric utilities, smart appliance manufacturers and other smart grid vendors, to accelerate the adoption of smart grid technologies which will ultimately enable the entire US energy system (electricity and transportation) to function in a more optimized manner. To break this impasse, the US Department of Commerce and more specifically the National Institute of Standards and Technology (NIST) has been tasked by congress (under the 2007 Energy Security Act) with creating a roadmap for accelerating smart grid adoption. As a part of this activity, NIST's Smart Grid Interoperability Roadmap creation began with EPRI taking a leadership role in developing the Smart Grid Interoperability Framework[2]. This task primarily involved defining the scope of smart grid across utility power value chain spanning generation through end use, identifying existing standards, their overlap as well as gaps in capabilities and recommending a plan of action to create interoperable set of standards that the entire community can develop and use. A subsequent report based on peer review of the EPRI report was released by NIST in August[3].

NIST effort is significant in several ways - firstly, it recognizes PEVs as a key enabling technology requiring the need to be integrated with the smart grid to be a part of the national energy system. In addition, it recognizes and validates the existing standards-making efforts from SAE, ZigBee/HomePlug Smart Energy Alliance and helps reinforce the existing standards structure for PEV to grid communications by adding to it requirements such as cybersecurity and interoperability in the form of leveraging existing robust data structures and standards from IEC and others. So, while NIST's primary role is that of a coordinator, by both highlighting the importance of existing standards and identifying areas of strengthening them, it is serving the dual benefits to PEV- smart grid communications standards-making against the backdrop of a fragmented history of standards-making in this arena.

SAE is also supporting National Institute of Standards Technology (NIST) since they are assigned to collect information on Smart energy.

NIST has completed Phase 1 that included three workshops that were intended to develop the roadmap to release Smart Grid 1.

Phase 2 is to create Public-Private partnerships for longer term evolution.

Phase 3 is to establish testing and certification framework.

More information on NIST can be found on their site http://www.nist.gov/smartgrid/

ISO/IEC

ISO & IEC has a joint working group and is developing common standards with both groups. This joint group is identified as ISO/TC 22/SC3/JWG V2G CI and has also had a joint meeting with our J2293 task force. The goal is to standardize on PLC communications from the PEV that satisfy both SAE and ISO/IEC requirements.

ITU

The ITU has been developing a new standard for home networking that work over powerline, coax and telephone wiring in the home.

IEEE

There are several IEEE documents that relate to electric transportation such as P1809, 1547, P1901 & P2030. Our task force is starting to have joint meetings with IEEE and plan to understand the objectives of current and proposed effort so we can maintain consistency to these standards. SAE is also forming a liaison to ANSI to be better connected with these IEEE groups. IEEE 1547 has a series of documents issued but the others are under development.

P1809 is a Guide for Electric-Sourced Transportation Infrastructure

IEEE 1547 is for Interconnecting Distribution Resources with Electric Power Systems. This is connected to our task force dash 3 documents for reverse energy flow from the PEV.

P1901 is an IEEE draft standard for broadband over power line networks defining medium access control and physical layer specifications.

P2030 is a Guide for Smart Grid Interoperability of Energy Technology and Information Technology Operation with the Electric Power System (EPS) and End-Use Applications and Loads

ASSOCIATION FOR HOME APPLIANCE MANUFACTURERS (AHAM)

AHAM is developing standards for home appliances that include an Energy Management System (EMS) with customer interfaces. The EVSE is similar to other home appliances and requires a common communication to this EMS. The EMS is also the link to other mediums the customer may select for the home such as Universal Remote Control (URC), internet and other wireless mediums. This EMS also includes the link to the smart meter for the utility needs.

SUMMARY/CONCLUSIONS

The SAE J2293 Task Force has focused its effort on the utility programs but is also making progress on dash 2 for the DC messages and others. The intent is to publish our efforts to date and continue with the simulation/modeling, build, test and validation phase and then update the documents with the results. The following schedule is the current plan to ballot the initial versions of the series of documents. The dates indicated are the start of the balloting process and several steps are required before the documents are available.

J2836/1™ & J2847/1 - Utility programs - Initial ballot January, 2010

J2836/2™ & J2847/2 - DC charging - Initial ballot 1Q, 2010

J2836/3™ & J2847/3 - Reverse energy - Initial ballot mid, 2010

J2836/4™ & J2847/4 - Diagnostics - Initial ballot 4Q, 2010

J2836/5™ & J2847/5 - VM specific - Initial ballot 4Q, 2010

Upon the initial ballot of these documents, the testing and validation phase is expected to occur within a year so these can be updated with their results in a 2nd ballot.

CONTACT INFORMATION

Rich Scholer
Ford Motor Company
EESE Product Design Engineer
Plug-In & Fuel Cell Vehicles
Phone: 313-323-0460
rscholer@ford.com

ACKNOWLEDGMENTS

This paper has been generated as a result of the SAE J2293 Task Force effort over the last two years. Several individuals have contributed to the overall objective of a PEV communicating with the Utility. The Task Force membership is diverse and interacts with other organizations that has guided us to this point and will continue to direct this team as we continue towards our goals. We started with a few members and have grown to almost 200. We also interact with the J1772™ Task Force to insure compatibility with the system architecture. Our continued effort will be to include DC charging and three phase systems into the communication as the J1772™ Task Force also includes these items.

Our immediate focus is to establish the communication between the PEV and the EVSE. The EVSE is, at least initially, the bridge (or proxy) to either the HAN or Smart Meter. The PLC communication types we are testing this year will lead us to using this on the PEV and EVSE with potential expansion to the HAN, Smart Meter and other home appliances.

The authors of this paper have been the leads in identifying the Use Cases and initial communication requirements. As subsequent papers are presented the content and authors will change as we discuss PLC, testing and other tasks and the task force continues to more forward.

REFERENCE

1. Simpson, R., "ZigBee Smart Energy 101", 10/16/2009, ZigBee Alliance, http://www.zigbee.org/imwp/download.asp?ContentID=16600

2. Report to NIST on Smart Grid Interoperability Standards Roadmap, EPRI publication, June 17, 2009, http://nist.gov/smartgrid/InterimSmartGridRoadmapNISTRestructure.pdf

3. NIST Framework and Roadmap for Smart Grid Interoperability Standards, Release 1.0 (Draft), Office of the National Coordinator for Smart Grid Interoperability, NIST, September 2009 http://www.nist.gov/public_affairs/releases/smartgrid_interoperability.pdf

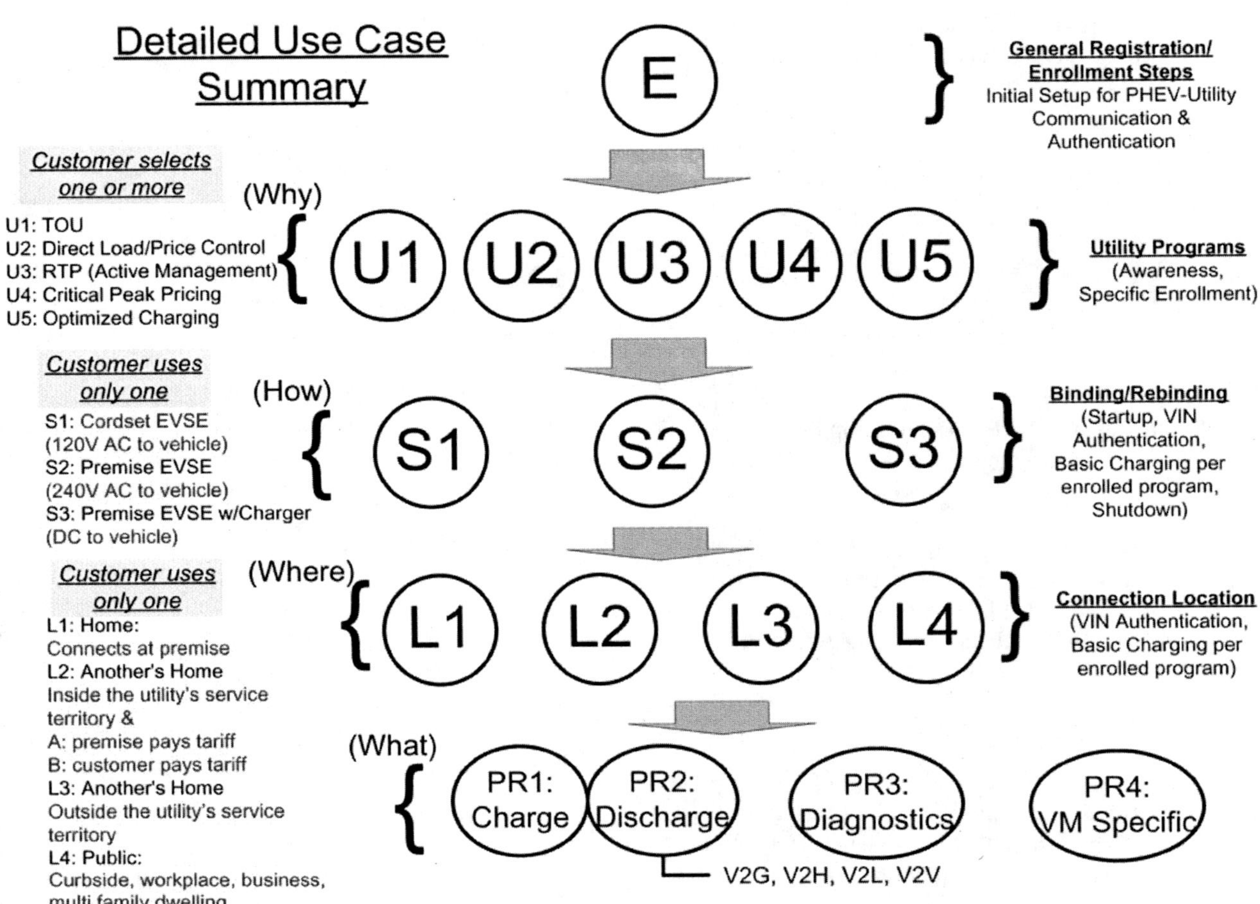

Figure 1. Detail Use Cases

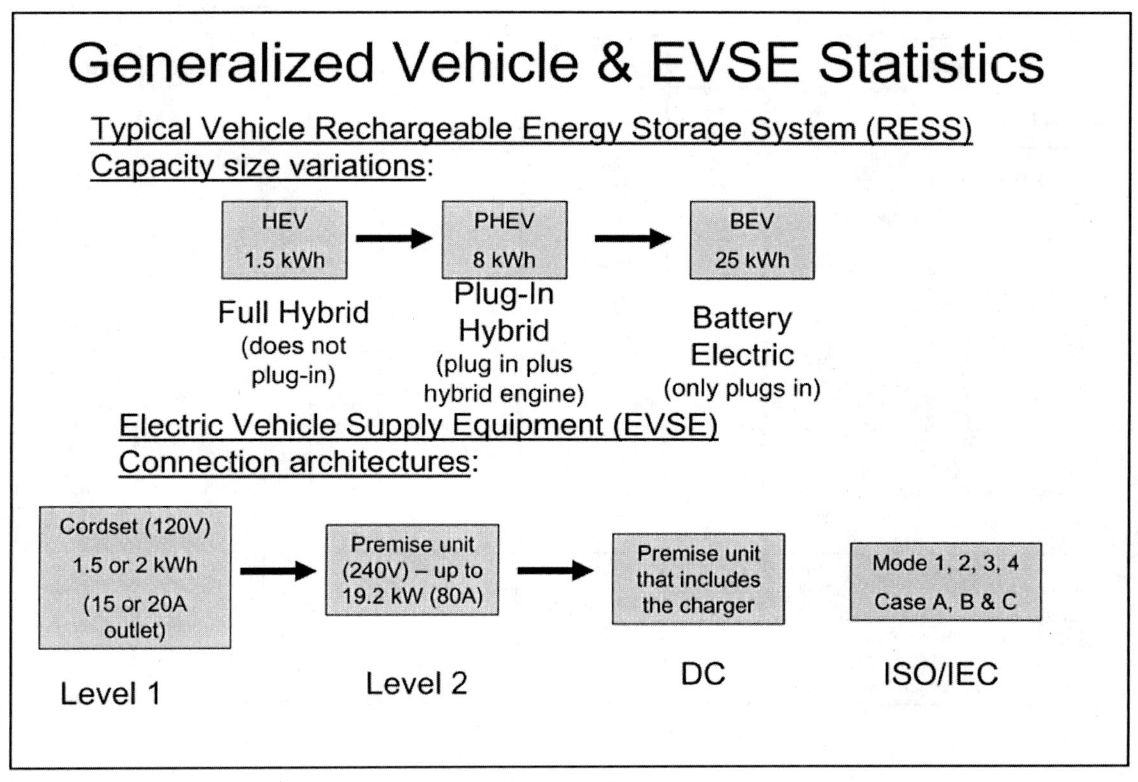

Figure 2. PEV and EVSE Variations

	Notes		Power Delivered	Charger		Typical Charge Times			
				On board	Off board	PHEV	BEV	High end BEV	
	Mobile EVSE	Level 1 (120V)	1.4 kW (16 AWG)	X		7 hours	15 - 17 hours	37 hours (1.5 days)	Same vehicle connector (Level 1 & 2 is the same, DC is being proposed as the same for these power levels)
Standard	Stationary EVSE but simple	Level 2 (240V)	3.3 Kw (16 AWG)	X		3 hours	6 - 7 hours		
	Onboard charger matches EVSE output (7.68 kW) but not being considered for initial vehicle design (large and heavy addition)	Level 2 (240V)	7 kW (8 AWG)	X		1 hour 20 minutes	3 hours	6.5 hours	
	Onboard charger matches BEV	Level 2 (240V)	19.2 kW (8 AWG)	X				2.5 hours	
Standard upgrade (same connector)	Level 2 EVSE but includes a 20 kW off-board charger (may double the cost of the EVSE)	DC (Std)	19.2 kW (8 AWG)		X	1/2 hour	50 - 60 minutes	2.5 hours	
Next Step (updated or added 2nd connector)	More expensive and complex system. Added (2nd connector) to vehicle. Option considered for BEV, not PHEV, therefore lower vehicle volumes).	DC (Fast)	45 - 60 kW (4 AWG - 150A EVSE to PEV)		X	7 - 10 minutes	30 minutes	50 minutes	2nd vehicle connector required
	Even more expensive and complex system. Added a different (2nd connector) to vehicle. Option is still considered for BEV, not PHEV, therefore lower vehicle volumes).	DC (Rapid)	120 - 140 kW (4/0 - 360A EVSE to PEV)		X	Not allowed	7 minutes	15 minutes	Different 2nd vehicle connector required

PHEV assumes 0% (initial) SOC to 100%
BEV assumes 20%(initial) SOC to 80%

Figure 3. Typical Charge Times

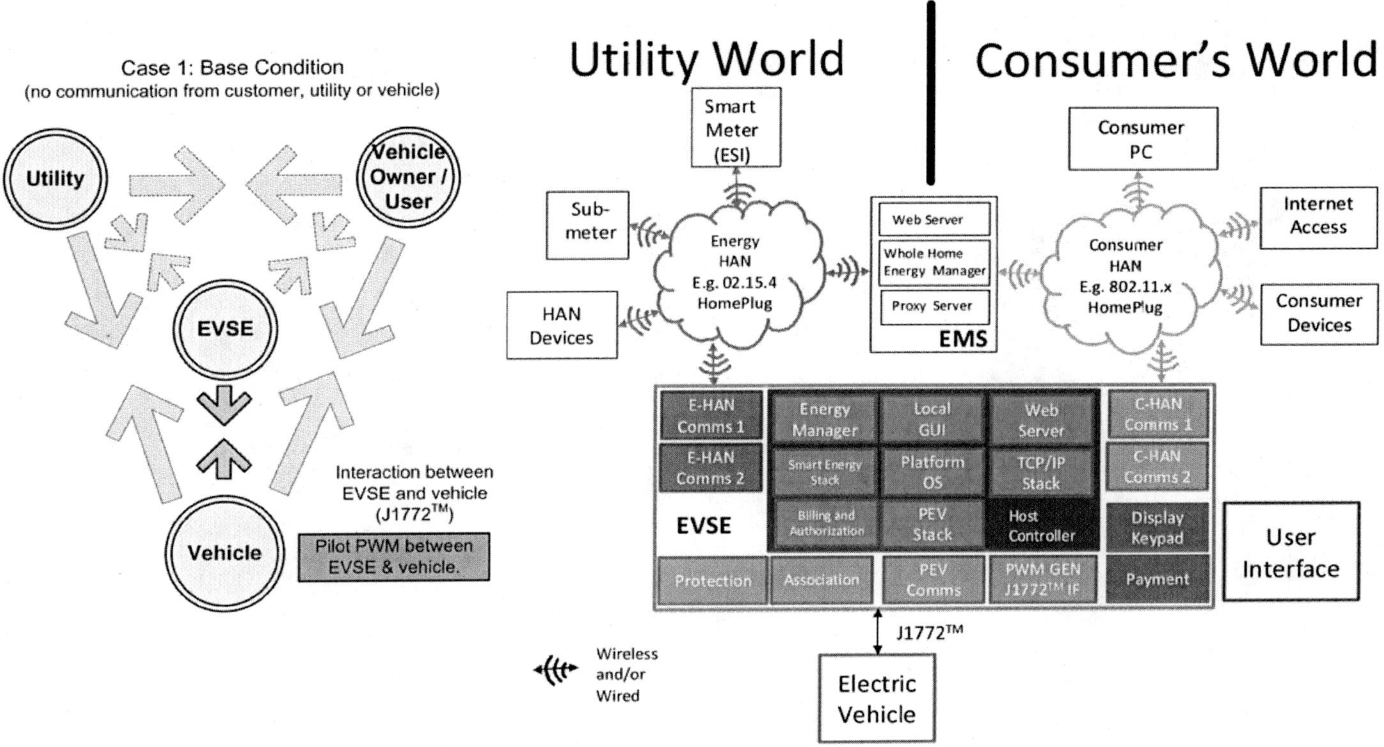

Figure 4. Communication Paths and Modularization of the EVSE

Technologies

2009-01-0168

Intelligent Vehicle Technologies that Improve Safety, Congestion, and Efficiency: Overview and Public Policy Role

Eric C. Sauck
University of Michigan

ABSTRACT

At the forefront of intelligent vehicle technologies are vehicle-to-vehicle communication (V2V) and vehicle-infrastructure integration (VII). Their capabilities can be added to currently-available systems, such as adaptive cruise control (ACC), to drastically decrease the number and severity of collisions, to ease traffic flow, and to consequently improve fuel efficiency and environmental friendliness. There has been extensive government, industry, and academic involvement in developing these technologies. This paper explores the capabilities and challenges of vehicle-based technology and examines ways that policymakers can foster implementation at the federal, state, and local levels.

INTRODUCTION

MOTIVATION - The American driving public has rapidly increased road use over the past decade. According to the Federal Highway Administration, between 1996 and 2006, the number of vehicle miles traveled increased by 21%, from 2,497,901 million to 3,033,753 million. Several conditions can be partly attributed to this growth: the number of traffic fatalities has remained constant, despite better safety technologies, congestion has increased, and the environmental impact of road transportation has increased. These issues translate into a significant economic loss.[1]

Safety - Despite the 17% *decrease* in traffic fatality rate—measured in deaths per 100 million vehicle miles traveled (VMT)—between 1996 and 2006, the total number of traffic fatalities has remained relatively constant at approximately 42,000 deaths per year (see Figure 1). To put this number in perspective, it is equal to one fully-loaded 747 jetliner crashing every four days. In 2004, motor vehicle collision was the number-one cause of death for Americans between the ages of 2 and 34.[2,3]

The Department of Transportation's National Highway Transportation Safety Administration (NHTSA) estimates the economic cost of motor vehicle crashes in 2000 to be $230.6 billion, including property, medical, productivity, and other losses.[3]

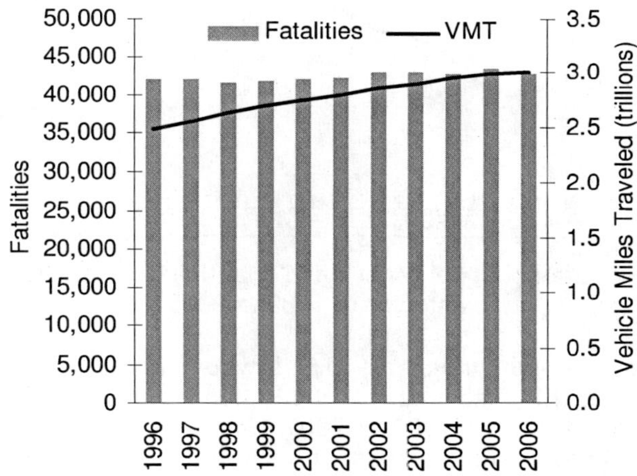

Fig. 1: U.S. motor vehicle related fatalities and vehicle miles traveled from 1996 to 2006[1]

Congestion - While the total number of vehicle miles traveled increased 21% between 1996 and 2006, the number of roadway lane-miles barely grew by only 3%. Furthermore, the U.S. Department of Transportation (DOT) projects that the use of combined road and rail will increase by 250% by 2050, while roadway lane-miles will increase by only 10%. Americans are experiencing the effects of this every day in traffic congestion and delay, which has risen since 1982, according to studies by the Texas Transportation Institute (TTI) (see Figure 2). The difference will have to be resolved by advanced technology, transit, and operations management.[1,4]

The TTI 2007 Urban Mobility Report estimates that, in 2005, congestion in U.S. cities has caused people to lose 4.2 billion hours of their time and to waste 2.9 billion gallons of fuel, equating to an economic loss of $78 billion.[4]

Fuel efficiency and environmental friendliness - When traffic is congested, fuel is wasted. This unnecessary fuel burn has a direct correlation to unnecessary greenhouse gas (GHG) emissions in congestion, which have been rising from 1982 to 2005 (See Figure 2).

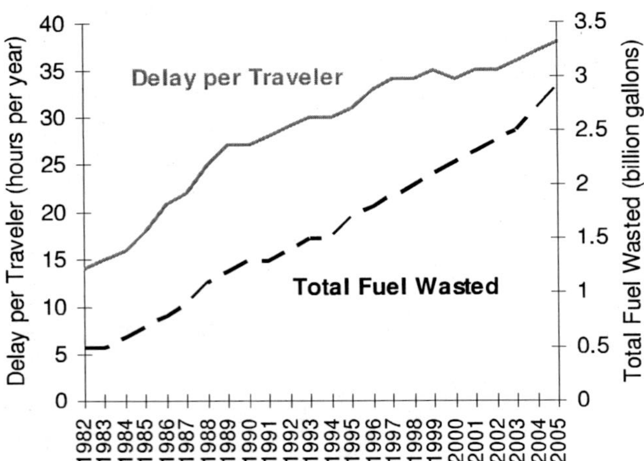

Fig. 2: Annual delay per traveler and annual total fuel wasted from 1982 to 2005[4]

In all light-duty travel, the U.S. Environmental Protection Agency (EPA) estimates that GHG emissions have gone up by 19% from 1990 to 2003. The EPA attributes this growth to both an increase in vehicle miles traveled and a limited improvement in fuel economy associated with an increase in the proportion of light-duty trucks to passenger cars. During the same period, GHG emissions from heavy-duty vehicles, including freight-carrying trucks, increased by 57%.[5]

The economic cost of GHG emissions to society is difficult to quantify. The consumption of foreign oil has national security and market collusion implications. Increased greenhouse gases have become accepted as a man-made cause of global warming with longer term impacts on society, including increased health problems.

SOCIETAL DEMANDS - In recent years, there has been a popular demand in the U.S. to decrease GHG emissions from automobiles and to decrease transportation costs due to high fuel prices. Consequently, government-mandated Corporate Average Fuel Economy (CAFE) standards were increased through the Energy Independence and Security Act of 2007. The new rules require each manufacturer's fleet of light-duty automobiles to attain a combined average of 35 miles per gallon by 2020. These new CAFE standards will require new ways to approach fuel economy; traditional methods like improving drivetrain efficiency will likely not be enough to feasibly satisfy CAFE requirements.[6]

Improvements in drivetrain efficiency have traditionally been counteracted by, among other things, additional safety features, which usually add to vehicle mass. Moreover, increases in EPA-rated fuel economy can be nullified by congestion, since vehicles should ideally travel between 40 and 60 mph (and at constant speed) to achieve optimal fuel economy. More massive vehicles, along with more frequent stop-and-go traffic cycles, are especially detrimental to fuel economy, and therefore increase GHG emissions.[7]

Because of these trade-offs, it is especially desirable and timely to implement technology that addresses all three, sometimes opposing, issues: safety, congestion, and efficiency.

SCOPE - Although there have been many developments in transportation infrastructure technology—such as signal coordination, roadway surfacing, and intersection design—this paper will focus on vehicle-based technology and its interface with infrastructure.

VISION ZERO - The goal of vehicle safety is to adopt systems that prevent collisions from occurring altogether. This so-called "Vision Zero" will require advanced *active* safety devices that entail significant technical, political, organizational, and societal challenges.

The achievement of Vision Zero will have many long-term implications for our concept of a vehicle. When vehicles no longer crash, and when anti-crash systems are proven to be fail-proof, automakers will be able to remove passive safety devices—devices such as air bags, crumple zones, and eventually, seat belts—thereby saving mass, fuel economy, affordability, and complexity.

ROADMAP: CURRENT, SHORT-, AND LONG- TERM INTELLIGENT VEHICLE TECHNOLOGY

The automotive industry has been developing safety technology at a rapid pace. For active safety systems that also mitigate congestion, the potential capabilities, benefits, and associated challenges will be briefly discussed.

CURRENT TECHNOLOGY: DRIVER WARNING AND ASSISTANCE - The market currently offers an array of driver warning and driver assistance aids to consumers.

Driver warning aids are meant to alert the driver in unsafe situations. Driver assistance aids help the driver to perform driving tasks more safely.

Current driver warning aids include the following:

- *Forward collision warning (FCW)* uses forward-facing laser or radio waves to identify an imminent crash; depending on the system, it can warn the driver, pre-load the brakes, close the windows, and tighten the seat-belts. In some cases, it can apply the brakes, as well.
- *Lane departure warning (LDW)* uses a forward-facing camera to identify lane markings and warn the driver using visual, audio, and haptic (touch) feedback.
- *Blind-spot warning (BSW)* uses cameras and radar to recognize vehicles in the driver's blind spot. If the driver begins to merge when a vehicle is in the way, the system will warn the driver via a warning light or chime.

Current driver assistance aids include the following:

- *Adaptive cruise control (ACC)* uses laser or radio waves to determine the distance, speed, and acceleration difference between the subject vehicle and a vehicle preceding it to keep a safe following distance while maintaining a preset speed whenever possible. It automatically applies throttle and braking as necessary. Some systems are capable of functioning at all speeds, while others work only above a minimum speed.
- *Lane-keep assist (LKA)* usually employs a forward-facing camera to identify lane markings, in conjunction with active steering or brake assist to maintain the vehicle in its lane.
- *Self-park system* uses cameras and short-range ultrasonic sensors to identify a parking space and automatically guide the vehicle into it by controlling the steering, acceleration, and braking.

The intended benefits of these systems are to increase driver comfort, convenience, and safety. However, ACC also acts to diminish congestion.

Safety - It is clear that an ACC-equipped vehicle can react many times quicker than a human driver, and that the control system will not overreact like humans in changing traffic conditions. These advantages may prove to have a multiplier safety effect; that is, smoother driving of the ACC-equipped vehicle will make driving in surrounding vehicles safer.

Safety benefits of driver aids in the context of NHTSA-reported related factors in fatal accidents are highlighted in Table 1.

Congestion - According to simulations done by California Partners for Advanced Transit and Highways (PATH), ACC can moderately increase single-lane roadway capacity from the current 2050 vehicles per hour with manual control to 2200 vehicles per hour with ACC. It is important to consider that roadway capacity is affected by many variables and that these numbers are intended only for comparison purposes.[8]

Challenges - With the introduction of these driving aids, there is a danger that the driver will lose attentiveness or gain a false sense of trust in the vehicle in the event of an accident. Also of concern is whether the driver will react predictably to the warning systems. Human factors concerns will be discussed further in following pages.

Another challenge is that these features are currently available mainly on premium vehicles; historically, safety features have "trickled down" to more mainstream vehicles when the cost of the technology has decreased. However, even on premium vehicles, safety features are often bundled with luxury and convenience items as option packages for marketing and cost reasons.

SHORT-TERM TECHNOLOGY: ADDING VEHICLE-TO-VEHICLE COMMUNICATION FUNCTIONALITY - Vehicles will soon be able to communicate with each other using a system called Dedicated Short-Range Communications (DSRC). The Institute of Electrical and Electronics Engineers (IEEE) is finalizing a standard, 802.11p, for wireless inter-vehicle communication. It is similar to wireless computer networking, and it uses a 5.9 GHz frequency band allocated by the Federal Communications Commission.[9]

A vehicle equipped with DSRC is capable of sharing information—position, velocity, acceleration, and other data, like braking capability—with other nearby vehicles over the secure, "ad-hoc" network. When integrated with ACC, the system becomes *cooperative* adaptive cruise control (CACC). Some of the potential improvements in safety, congestion, and fuel efficiency and environmental friendliness are described below.

Safety - The introduction of vehicle-to-vehicle communication (V2V) using DSRC could potentially eliminate certain types of crashes when combined with driver aids. It can mitigate eight of the top ten related factors in traffic fatalities identified by NHTSA (see Table 1).

Congestion - Full market penetration of CACC can further improve single-lane capacity over ACC alone to 4550 vehicles per hour, according to PATH simulations. This capacity is highly dependent on the preset following distance between vehicles, which can be increased or decreased based on industry consensus and/or on driver comfort. Figure 3 shows the simulation results for various market mixes of manual control, ACC, and CACC. Note that high market penetration of CACC is necessary to achieve significant increases in roadway capacity.[8]

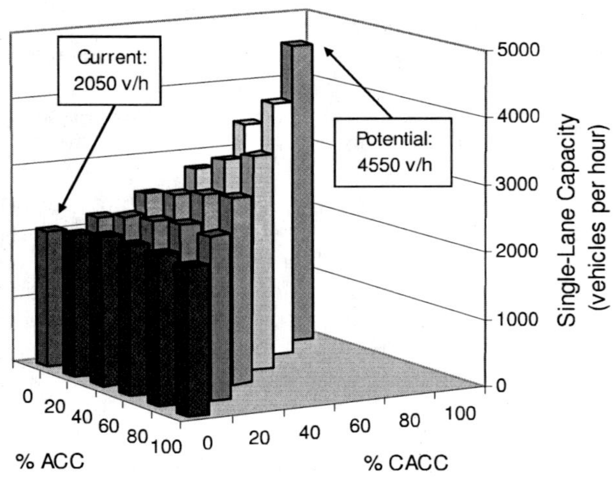

Fig. 3: Effect of ACC and CACC mix on lane capacity (reproduced with permission from author)[13]

In another model, similar benefits were found: a traffic simulation by the Netherlands-based TNO research institute found a more than 15% decrease in delays when one-half of the vehicles were equipped with CACC.[11]

Fuel efficiency and environment - While CACC can "smooth out" velocity changes in traffic and therefore improve fuel economy by maintaining more consistent speeds, the fuel efficiency benefits of V2V extend to steady-speed cruise. In a "platoon," a closely-spaced string of vehicles, aerodynamic drag force is reduced. Using CACC, the vehicle spacing can be safely shortened enough to improve fuel efficiency at highway speeds.

The effect is especially significant in freight trucking. According to the Department of Energy's 21st Century Truck Partnership, aerodynamic drag consists of 53% of the non-engine energy losses of a heavy truck at 65 mph.[12]

Challenges - Although V2V affords clear benefits to congestion, safety, fuel efficiency, and the environment, the added benefits would be limited at low market penetration of the technology, as shown in Figure 3. One possibility to more quickly attain the advantages of V2V is to retrofit all existing vehicles with DSRC transceivers, which would allow them to send their position, velocity, and acceleration status to vehicles with driver aids. When retrofit vehicles broadcast their actions, other vehicles with driver aids can help their drivers to avoid colliding with the retrofitted vehicle.

Technical challenges remain in creating efficient algorithms and processing power to handle the vast amounts of transmitted and received data. Field trials are also required to more accurately predict the effect of CACC systems in mixed (DSRC- and non-DSRC-equipped) traffic.

LONG-TERM TECHNOLOGY: ADDING VEHICLE-INFRASTRUCTURE INTEGRATION FUNCTIONALITY - Beyond V2V with CACC, the next stage in intelligent vehicle technology is vehicle-infrastructure integration (VII), where strategically-placed roadside equipment with DSRC sends data to and receives data from vehicles. It can also send it to third parties, like a traffic management office for usage statistics or a communications provider for car-based Internet access.

Safety - There are many potential safety benefits to equipping roadways, intersections, and signals with DSRC capability. In addition to extending the functionality of CACC-type systems, VII can allow drivers to receive in-car (or out-of-car) warnings of impending traffic signal violations, curve speed warnings, notices of upcoming traffic congestion and re-routing guidance, and weather alerts.

Congestion - Sophisticated road management functions can be automatically controlled with VII. Controllers can optimize highway on-ramp metering and signal priority for maximum traffic throughput based on real-time conditions.

Furthermore, VII can allow road managers to collect traffic flow information. If traffic flow indicates a problem, managers can pinpoint the problem area and quickly deploy emergency response, road maintenance, or snow removal crews.

Another aspect of VII that can potentially improve congestion is open-road tolling. Without having to stop at a toll both (a system still in use in many states), drivers will pay precisely for the road they use.

Fuel efficiency and environment - One way to improve fuel efficiency with added VII capability is to integrate three-dimensional, or topological, road maps, precise GPS location, and ACC. In a 2006 study by researchers at Linköping University in Sweden, heavy trucks were equipped with a sort of road-predictive cruise control. Based on road slope, an on-board computer selected speeds and transmission gears for optimal fuel efficiency. In their simulation along a 127-km stretch of Swedish highway, trucks reduced fuel consumption by 2.5% without adding to travel time.[13]

Topographic maps already exist for the U.S., through Geographical Information System (GIS). Road maps can be overlaid onto the topological maps. If corrections need to be made to the overlay, then vehicles equipped with GPS and DSRC can potentially upload the new road geometry values to a database. This database can be distributed to vehicles via VII and used by their ACC systems to further improve fuel efficiency.[14]

Challenges - Funding will be a major hurdle in implementing VII. Federal Highway Administration researchers estimate the cost to equip all intersections around the country with DSRC to be on the order of billions of dollars, not including expansion to other parts of roadways, like dangerous curves or construction zones. The VII Coalition, comprised of the U.S. DOT, state DOTs, and automobile manufacturers, estimates the cost to be $5-8 billion to create and $100 million per year to maintain a national VII infrastructure.[15,16]

Moreover, roadway infrastructure is the domain of states and municipalities, which receive funding from the federal government only for select highways. The decentralized nature of highway management can diminish the influence of the federal government in establishing a uniform VII system across the United States.

Technical issues must also be resolved. Particularly, engineers must find a way to implement V2V in vehicles in the near term while ensuring compatibility with VII in the long term. This requires allocating extra computer power for future enhancements of V2V and VII.

VII has increased potential, compared with V2V alone, to address factors in fatal accidents, as shown in Table 1.

Related factor and percent in fatal accidents[10]	Driver aid	+ V2V	+ VII
Failure to keep in proper lane or running off road: 28.5%	LDW, LKA	•	•
Driving too fast for conditions or in excess of posted speed limit or racing: 21.3%		•	•
Under the influence of alcohol, drugs, or medication: 12.7%		•	•
Inattentive (talking, eating, etc.): 7.9%	LDW, FCW		
Failure to yield right of way: 7.3%	BSW	•	•
Overcorrecting/oversteering: 4.6%			
Failure to obey traffic signs, signals, or officer: 4.2%		•	•
Swerving or avoiding due to wind, slippery surface, vehicle/object/person in road, etc.: 3.7%		•	•
Operating vehicle in erratic, reckless, careless, or negligent manner: 3.6%		•	•
Vision obscured (rain, snow, glare, lights, building, trees, etc.): 2.7%	ACC	•	•

Table 1: Safety potential of driver aids, V2V, and VII to affect the top 10 fatal accident factors

HUMAN FACTORS CHALLENGES

The automation of vehicles will likely be an evolutionary process. The market is already seeing automated collision warning and lane departure warning systems leading to automated control-assist devices like active cruise control and lane keep assist. In the distant future, this process will lead to full automation.

DRIVER ROLE CONCERN - Between now (warning and control-assist) and the distant future (full automation), difficulties will arise in clearly defining the driver's role and in assuring the driver understands a vehicle's capabilities. Both of these aspects can adversely impact safety.

In fact, automakers have already begun to diverge in their vehicles' capabilities. There is concern about how a driver who is familiar with one type of system can adapt to driving a vehicle with another, similarly-named system.

Furthermore, in light of increasing automation, experts are considering the likelihood that the driver could lose focus on the driving task.

Stakeholders will need to address these issues as a system. As a 2001 California PATH report states, "the roles of the driver and the automation system will need to be defined so that, when combined, all of the essential safety-critical functions are performed at least as well as they are today."[8]

REMOVING THE DRIVER - By definition, human factors issues can be eliminated by completely removing the driver from the system. Although this would be a worthy goal to pursue, the Department of Defense is taking the lead in seeking autonomous (driverless) vehicles; its aim is to decrease battlefield fatalities. The resulting technology should benefit civilians as well as military personnel.

Specifically, the Defense Advanced Research Projects Agency (DARPA) held racing competitions in 2004, 2005, and 2007 to encourage industry and academic groups to develop autonomous (driverless) vehicles for the military to use in place of conventional vehicles in high-danger situations. In 2007, the race was held for the first time in an urban environment. In the DARPA Urban Challenge, contestant vehicles "simulated military supply missions while merging into moving traffic, navigating traffic circles, negotiating busy intersections, and avoiding obstacles."[17]

The number of finishers for each year's competition showcases the rapid development of fully automated vehicles:

- The 2004 Grand Challenge, held on a 142-mile desert course, finished 0 of 15 finalists.
- The 2005 Grand Challenge, held on a 132-mile desert course, finished 4 of 23 finalists.
- The 2007 Urban Challenge, held on a 60-mile mock-urban course, finished 6 of 11 finalists.[17]

Many experts consider full automation the only way to remove human error from the driving experience. Although the progress made by science and engineering in this area is impressive, most academic, industry, and government experts in attendance for the 2008 IEEE Intelligent Vehicles Symposium predicted that fully-automated, mass-market vehicles would only be available only after 2030, 2040, or later.[18]

MARKET ACCEPTANCE CONCERN - There exists a valid concern over how consumers will accept driver aids, V2V, and VII as on-board helpers to the task of driving. California and Michigan are two states where real-world intelligent vehicle tests involving human participants have been conducted. The participants' feedback about the experience is valuable in determining how the public will react to intelligent vehicles.

California PATH platoon demonstration - An eight-vehicle platoon was demonstrated in 1997 on freeways in San Diego, California. In this demonstration, riders were driven by fully-automated vehicles, and the vehicles safely maintained short separation. Despite the "tailgating" effect of the 21-foot separation that made riders uneasy at first, "most of them quickly adapt and develop a sense of comfort and security because of the constantly maintained separation." [19]

UMTRI pilot test - In a 2008 University of Michigan Transportation Research Institute (UMTRI) report for the DOT, 18 subjects drove vehicles equipped with several driver aids like FCW, LDW, BSW in a pilot test. Following the test, the subjects were asked to subjectively rate the systems on a +2 to -2 scale. The report states, "The mean usefulness score is 1.33 and the mean satisfaction score is 0.75, both of which indicate positive feelings towards [driver aids]."[20]

As part of the same study, UMTRI performed a pilot test for heavy trucks equipped with similar driver aids. Following the test, three of the five truck drivers generally liked the systems, while two of the five disliked the systems due to false alarms. The rate of false alarms will steadily decrease with further development, so this issue should not be a problem with future driver aids.[20]

V2V and VII systems will have much more data to process than current driver aids, so avoiding false alarms will remain an important issue throughout development. Fortunately, many of the driver alerts generated by V2V and VII can be routed through the same user interface as driver aids, which the subjects of the UMTRI study liked.

PUBLIC POLICY ROLE

The U.S. government has the potential to positively impact the development and implementation of intelligent vehicle technologies. Through mandates, rules and regulations, tax incentives or penalties, and subsidies, government has a wide array of options to affect intelligent vehicle technology.

CURRENT INVOLVEMENT - Federal, state, and local government has played a role in several stages of development, from research to field trials, of intelligent vehicle technologies.

Research - The federal government has assisted industry in high-risk research. In a 2008 interview of the administrator for the DOT Research and Innovative Technology Administration, Paul Brubaker, by ITS International, he said, "Government should support basic and applied research, then get...out of the way and let the private sector and localities get on and do things." [14]

Prize competitions - Programs such as the DARPA Urban Challenge for initiating development in automation technology are mutually beneficial to government, industry, and academia. Besides the obvious recognition and awards that winning teams get, government gains an advantage in defense technology, and industry gains know-how and a skilled pool of potential new-hires from academia. Industry also gains solid reassurance that the government is committed to purchasing products and services stemming from this technology in the future.

Technology transfer - In cases where government groups and private industry seek some of the same capabilities, but where each has previously conducted independent development, it can be prudent to exchange knowledge between interested organizations.

Technology transfer has already occurred in the realm of intelligent vehicles. The U.S. Department of Defense, the Department of Transportation, and the Department of Commerce have held a Joint Military/Civilian Seminar On Intelligent Vehicle Technology Transfer. The event has been unclassified and open to all interested parties. At the third, 2008 seminar, leaders of various industry and government projects gave 20 presentations over a two-day period.

Field trials - The federal, state, and local governments have a track record of sponsoring field operation test of near-term advanced technology.

From 2006 to 2008, the VII Coalition, including the U.S. DOT, Michigan DOT, and Oakland County Road Commission, has created the Developmental Test Environment (DTE) in Detroit, Michigan. The DTE demonstrates the proof-of-concept of VII, and 57 sites in Oakland County have been equipped with roadside equipment to communicate with vehicles over DSRC. The DTE will prove the technical viability of the VII system architecture; the DTE also will prove the applications viability of VII to support safety, mobility, and private/commercial services. If the results of DTE are satisfying to industry, they may begin to incorporate VII into their future product plans. The Michigan DOT is also using the DTE to prove operations aspects of VII, such as snow removal and road maintenance.[16]

At the state level, the Michigan Economic Development Corporation is sponsoring the Connected Vehicle Proving Center along with several industry partners. The center will allow developers to share costs and coordinate testing in expensive facilities and in public roadways. Similar joint efforts between federal, state, and local agencies are taking place in 13 other states across the U.S.[16]

STANDARDS DEVELOPMENT - A major challenge in developing a vast new intelligent transportation system is getting agreement from disparate parties in the automobile industry, academic institutions, and from state and federal government research agencies and regulatory bodies.

The government, by executive decree, is obligated to use voluntary consensus standards developed by the private sector whenever practicable. According to the Executive Office of Management and Budget in its Circular No. A-119, this applies to all agencies of the federal government. Although voluntary consensus means that all parties will be in agreement, this approach can take much longer than mandates used in other countries.[21]

REWARDING IMPLEMENTATION - Implementing new technologies on a wide scale can entail high initial costs that can make them unattractive to consumers. Government can stimulate sales by providing incentives in the interim deployment stage, until the technology becomes established.

Performance-based incentives - Government could carefully create performance criteria for awarding tax credits or subsidies to the customer.

For driver aids alone, the federal government could subsidize the cost of safety features like CACC, LDW, and FCW. The incentive would be vehicle-specific and be based on the capability of its driver aids. In cases where vehicle manufacturers offer safety features bundled with luxury amenities, government could impose a rule to separate the safety options from amenities like leather seats and entertainment systems.

For V2V, the government could financially assist those who seek to retrofit their vehicle with DSRC or those who purchase a new vehicle equipped with DSRC. As a comparable precedent, the government-mandated 2009 switch from analog to digital television comes with a subsidy. Called the TV Converter Box Coupon Program, it allows all U.S. households to receive two $40 coupons toward a digital-to-analog converter for old, analog television sets (these converters cost approximately $60). Similarly, government could offer coupons for retrofitting vehicles with DSRC transceivers while mandating that new vehicles come equipped with them.[22]

Another example is the Federal government's New Energy Tax Credits for Hybrids, which varies the tax credit according to the mileage performance of the hybrid vehicle and phases out the incentive after sales of a model reach 60,000 units. Similarly, the incentive could be greater for more capable V2V systems than for less capable systems, and the incentive could be gradually phased out once a "critical mass" of V2V-equipped vehicles are sold. After a certain portion of the market possesses the technology, economies of scale and desire to offer competitive features could drive down cost and thereby increase market penetration.

For VII, the customer is not the car-buying public, but rather the state or local government that is considering an infrastructure upgrade to adopt the VII standard. Performance-based incentives could depend on the safety, congestion, and environmental effects of the planned infrastructure over the current condition. This would allow states and local governments the freedom to decide which roads to equip with VII first.

Marketing strategies - For all driver aids, V2V, and VII, the federal government can provide public awareness of what the safety technologies mean for them.

Crash *avoidance* ratings - Through NHTSA's New Car Assessment Program (NCAP). NCAP has provided star ratings (based on a five-star scale) for front and side impacts. These ratings, along with the Insurance Institute for Highway Safety (IIHS) crash test ratings, have influenced consumer purchasing habits.

More recently, NHTSA has begun evaluating and assigning stars for roll-over safety. In addition to awards for a vehicle's crashworthiness, NHTSA and IIHS could evaluate the vehicle's performance in avoiding crashes altogether. Since driver aids are available now, NHTSA and IIHS could begin awarding stars and ratings for today's technology, and then increase requirements for high ratings as new technologies emerge.

Addition to fuel economy ratings - Since V2V and VII also affect fuel economy of the equipped car as well as having a multiplier effect on other equipped- and non-equipped vehicles, the EPA could add a numerical value of the fuel savings next to the standard miles-per-gallon rating. For example, if a 40-mpg vehicle lowers consumption by 10% when using CACC, its new EPA fuel economy could read "40 mpg + 4 Intelligent Vehicle mpg)."

ACCOUNTING FOR UNINTENDED CONSEQUENCES - Given the potential benefits of V2V and VII systems, it may be easy to forget to consider the side effects caused by their implementation. Careful government action can help to mitigate the negative effects on the state and local levels.

Increased vehicle use and tolling - Advancements in fuel economy standards and reduction of congestion might influence vehicle use in relation to other modes of transportation. However, past CAFE increases have not significantly decreased overall fuel consumption in the U.S., because it lowered transportation fuel costs and subsequently increased VMT. Taking this history into account, 35 mpg by 2020 may bring with it the unintended consequence of increasing travel and *not* achieving its intended goal of decreasing overall consumption.

VII holds one possible answer in open road tolling. State and local governments can enact usage fees for driving on the most heavily congested roads, using DSRC for collection at cruising speed. The driver would pay for exactly the amount of road driven. The proceeds from these tolls would be ideally suited for implementing more VII capability around the state or municipality, in effect creating a self-perpetuating system after an initial investment in heavily congested areas.

The federal government grants authorization for states to enact tolling on Interstate highways. However, much of the public often opposes such measures, because tolls become revenue-generating sources for other state spending purposes.

Therefore, careful stewardship of these tolls would be required to ensure that they reflect the cost of the infrastructure, and not other government programs. This can be ensured by enacting policies that limit the use of toll income to further investments in road infrastructure.

Safety problems and investigation - Many lessons have been learned following the ill-fated introductions of automatic seat belts and high-powered first-generation airbags, and engineers are now thoroughly testing every piece of technology that goes onto a vehicle, especially safety equipment. However, there is a chance that a critical algorithm, component, or safeguard will be overlooked. For this reason, NHTSA has a complaints database that can be accessed by concerned members of the public.

If an accident of national importance does occur, the National Transportation Safety Board (NTSB) has the mandate to perform a thorough and impartial investigation and recommend actions to the appropriate organs of government and industry. However, it can take many years for safety problems to manifest themselves, and to do so generally requires extensive market penetration of the problematic technology. Needless to say, this means of addressing safety issues reactively is the least desirable option. Also, it can take days to years for NTSB recommendations to be implemented by the responsible party.

CONCLUSION AND RECOMMENDATIONS

Among new transportation technologies, intelligent vehicles provide an attractive mix of benefits to safety, congestion, fuel efficiency, and environmental friendliness. The improvements afforded by current, near-term, and long-term intelligent vehicle technologies are real, and they are being proven by current pilot programs. Challenges introduced by increased automation and complex systems integration are being resolved by engineers around the world.

However, the success of intelligent vehicle technologies ultimately depends on the actions of a few—and often non-unified—key players in the public policy arena. Fortunately, it is the public who choose (indirectly, in some cases) the policymakers. It is our duty as citizens to make sure our voices are loud enough, and it is the duty of the policymakers to listen.

For a timely and efficient transition to intelligent vehicle technologies, government should do the following:

- Continue to sponsor intelligent vehicle competitions, technology transfer, and field trials
- Add to NHTSA star ratings to reflect the active safety benefits of V2V—the deployment of which should coincide with the first launch of the technology
- Add to EPA fuel economy ratings to reflect the energy savings of V2V—the deployment of which should coincide with the first launch of the technology
- Offer performance-based incentives to car buyers during the introduction of V2V by offering subsidies for retrofitting DSRC to existing vehicles or for buying V2V-equipped new vehicles.
- Offer performance-based federal incentives and disincentives to state and local governments during the implementation of a standardized VII system
- Approve state and local governments to use VII-based toll collection on Interstate highways, with funding restricted to roadway projects and further VII implementation.

ACKNOWLEDGMENTS

The author wishes to thank the following people for their wise advice and invaluable assistance: Mr. Timothy Mellon, SAE International Government Affairs Director; Dr. Jeffrey King, 2008 WISE Faculty Member in Residence and Assistant Professor at the Missouri University of Science and Technology; Mr. James Bryce, 2008 WISE Intern, University of Missouri; and Ms. Brooke Buikema, 2008 WISE Intern, Calvin College.

REFERENCES

1. "Highway Statistics," HM-60 and VM-2, Federal Highway Administration, Washington, DC, 1997-2007.
2. Boeing Commercial Airplanes - Boeing 747-8, http://www.boeing.com/commercial/747family/747-8_facts.html, accessed July 25, 2008.
3. "Traffic Safety Facts, 2006 Data: Overview," pp. 1-3, DOT HS 810 809, National Highway Transportation Safety Administration, Washington, DC, 2008.
4. Schrank, David, Tim Lomax, "The 2007 Urban Mobility Report," Texas Transportation Institute, http://mobility.tamu.edu, 2007.
5. "Greenhouse Gas Emissions from the U.S. Transportation Sector, 1990–2003," pp. 8-11, EPA 420 R 06 003, United States Environmental Protection Agency, Washington, DC, 2006.
6. "Energy Independence and Security Act of 2007," Title I, Public Law 110-140, United States of America, 2007.

7. Davis, Stacey, "Transportation Energy Data Book," p. "7-22," Oak Ridge National Laboratory, National Technical Information Service, Springfield, VA, 2001.
8. Shladover, Steven, et al, "Development and Performance Evaluation of AVCSS Deployment Sequences to Advance from Today's Driving Environment to Full Automation," pp. 72-77, California Partners for Advanced Transit and Highways, Berkeley, CA, 2001.
9. Jiang, Daniel and Luca Delgrossi, "IEEE 802.11p: Towards an International Standard for Wireless Access in Vehicular Environments," pp. 2036-2040, Vehicular Technology Conference, 2008, IEEE, Singapore, 2008.
10. "FARS/GES 2006 Data Summary," p. 19, DOT HS 810 819, National Highway Transportation Safety Administration, Washington, DC, 2008.
11. van Arem, Bart, et al, "Design and evaluation of an Integrated Full-Range Speed Assistant," p. 44, TNO Traffic and Transport, The Netherlands, 2007.
12. "21st Century Truck Partnership: Roadmap and Technical White Papers," p. 2, 21CTP-0003, Department of Energy, Washington, DC, 2006.
13. Hellström, Erik, Anders Fröberg, Lars Nielsen, "A Real-Time, Fuel-Optimal Cruise Controller for Heavy Trucks Using Road Topography Information," 04-03-2006, SAE International, Warrendale, Pennsylvania, 2006.
14. "Driven Man," pp. 16-17, ITS International magazine, January/February 2008.
15. Meeting with Raj Ghaman, Travel Management Team Leader, Federal Highway Administration Office of Operations R&D, July 10, 2008.
16. VII Coalition, http://www.vehicle-infrastructure.org.
17. DARPA Urban Challenge, http://www.darpa.mil/grandchallenge/, Accessed July 20, 2008.
18. E-mail communication with Steven Shladover, California PATH and Keynote Speaker in IEEE Intelligent Vehicles 2008 Symposium, June 30, 2008.
19. "Vehicle Platooning and Automated Highways," California Partners for Advanced Transit and Highways, http://www.path.berkeley.edu/PATH/Publications/Media/FactSheet/VPlatooning.pdf.
20. Green, P. J.Sullivan, O. Tsimhoni, J. Oberholtzer, M.L. Buonarosa, J. Devonshire, J. Schweitzer, E. Baragar, J. Sayer, "Integrated Vehicle-Based Safety Systems (IVBSS): Human Factors And Driver-Vehicle Interface (DVI)Summary Report," pp. 86-93, DOT HS 810 905, U.S. Department of Transportation, Washington, DC, 2008.
21. Circular No. A-119, Executive Office of Management and Budget, http://www.whitehouse.gov/omb/circulars/a119/a119.html.
22. TV Converter Box Coupon Program, https://www.dtv2009.gov.

CONTACT

Eric Sauck is pursuing a Bachelor of Science in Mechanical Engineering and a Master of Automotive Engineering, both at the University of Michigan. He has worked for several automotive suppliers and manufacturers. In the summer of 2008, he was sponsored by SAE International to participate in Washington Internships for Students of Engineering (www.wise-intern.org) to gain exposure to the public policy process.

Questions, comments, and discussion should be directed to esauck@umich.edu.

DEFINITIONS, ACRONYMS, ABBREVIATIONS

Light-duty vehicle: Light-duty vehicles are defined as vehicles with a gross vehicle weight rating (GVWR) of less than 8,500 lbs. They include passenger cars, sport-utility vehicles, minivans, pickup trucks, and motorcycles.

ACC: Active cruise control

BSW: Blind Spot Warning

CACC: Cooperative active cruise control

DARPA: Defense Advanced Research Projects Agency

DOT: Department of Transportation

DSRC: Dedicated short range communications

EPA: U.S. Environmental Protection Agency

FCW: Forward collision warning

GHG: Greenhouse gas

GPS: Global Positioning System

IEEE: Institute of Electrical and Electronics Engineers

IIHS: Insurance Institute for Highway Safety

LKA: Lane keep assist

LDW: Lane departure warning or lateral drift warning

mpg: Miles per gallon

NCAP: National Highway Transportation Safety Administration New Car Assessment Program

NHTSA: National Highway Transportation Safety Administration

PATH: California Partners for Advanced Transit and Highways

TTI: Texas Transportation Institute

V2V: Vehicle-to-vehicle communications

SAE TECHNICAL PAPER SERIES

2010-36-0280
E

Eco Navigation with Vehicle interaction

RICARDO YOSHIKAZU TAKAHIRA

FILIADA À
SAE International

XIX Congresso e Exposição Internacionais de Tecnologia da Mobilidade
São Paulo, Brasil
05 a 07 de outubro de 2010

Este trabalho técnico/científico recebeu apoio financeiro do **Conselho Nacional de Desenvolvimento Científico e Tecnológico - CNPq – Brasil**

AV. PAULISTA, 2073 - HORSA II - CJ. 1003 - CEP 01311-940 - SÃO PAULO – SP

2010-36-0280

Eco Navigation with Vehicle interaction

Ricardo Takahira
SAE Associate

New Business / Program Manager @

Magneti Marelli Sistemas Automotivos Ind. E Com. Ltda,

Electronics Division

ABSTRACT

On board Electronic Solutions – In Vehicle Navigation with interactive interface within powertrain ECU's. Eco route, eco driving concepts. Navigation device integration with infra structure for real time traffic information. Driving education feedback focusing on better fuel consumption, so consequently lower emissions. On Going Technologies and Road Map CO_2 percentage reductions got by components or functions added to the vehicle architecture. Trend of use in EV (Electrical Vehicles) or HEV (Hybrid Electrical Vehicles), new ECU's targeting this type of vehicles, architecture looking better autonomy or driving interaction with infra structure for refueling/recharging. Smart Grid and navigation oriented to electrical Vehicles.

INTRODUCTION

We just started our dependency as drivers on the GPS technology, nowadays is common to see these navigation devices in our vehicles. For the old ones that not came with this from the factory a lot of Aftermarket PNDs-Portable Navigation Devices are available.

AVAILABLE ECO TECHNOLOGIES

The existent automotive solutions may have already reached the maximum of CO_2 emission reduction. The devices and ECUs on board can reduce fuel consumption and emission but the gain got actually by dollar / Euro spent may not be interesting to get the percentage of CO_2 emission reduction nowadays.

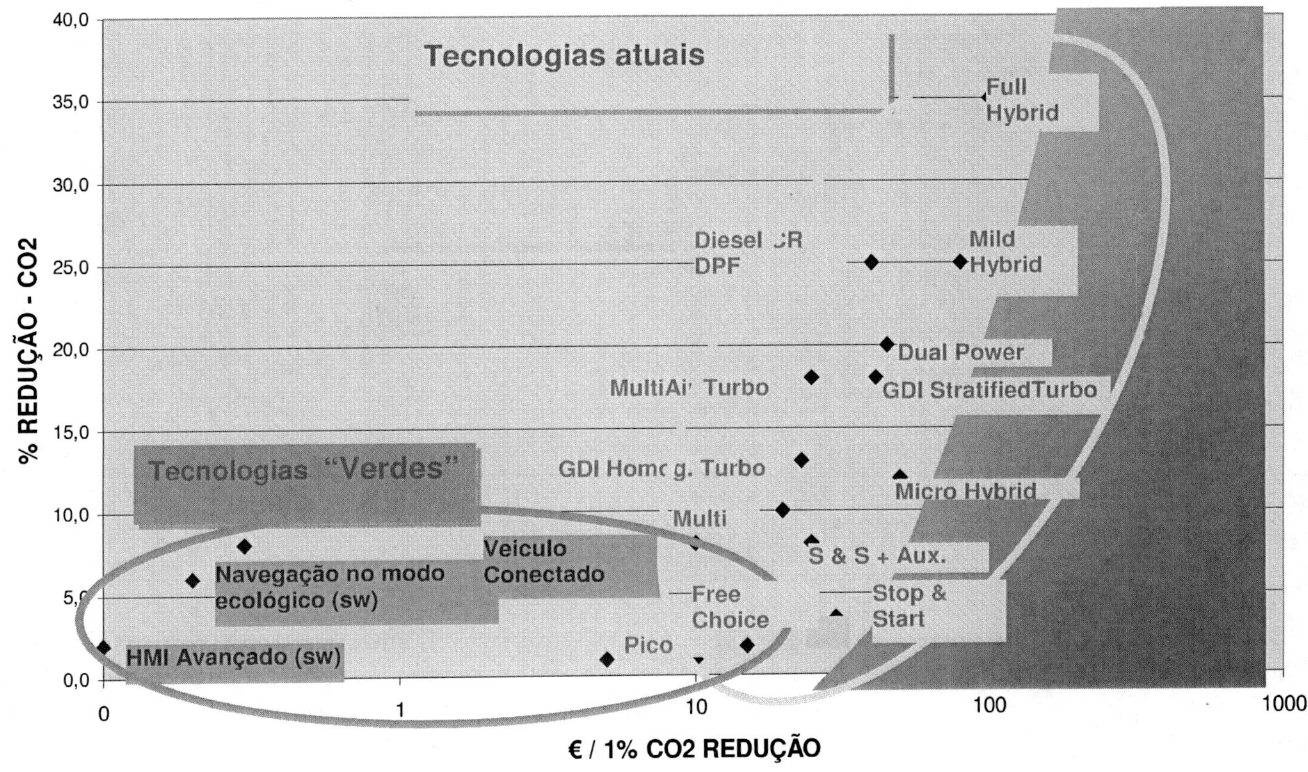

NORTH AMERICA AND EUROPEAN EMISSON REGULATON

Each year the motor industry is being pushed to make cars that pollute less. Ethanol is not an economical solution for most of the countries which are forced , without the possibility to adopt a green fuel to go hybrid or electric.

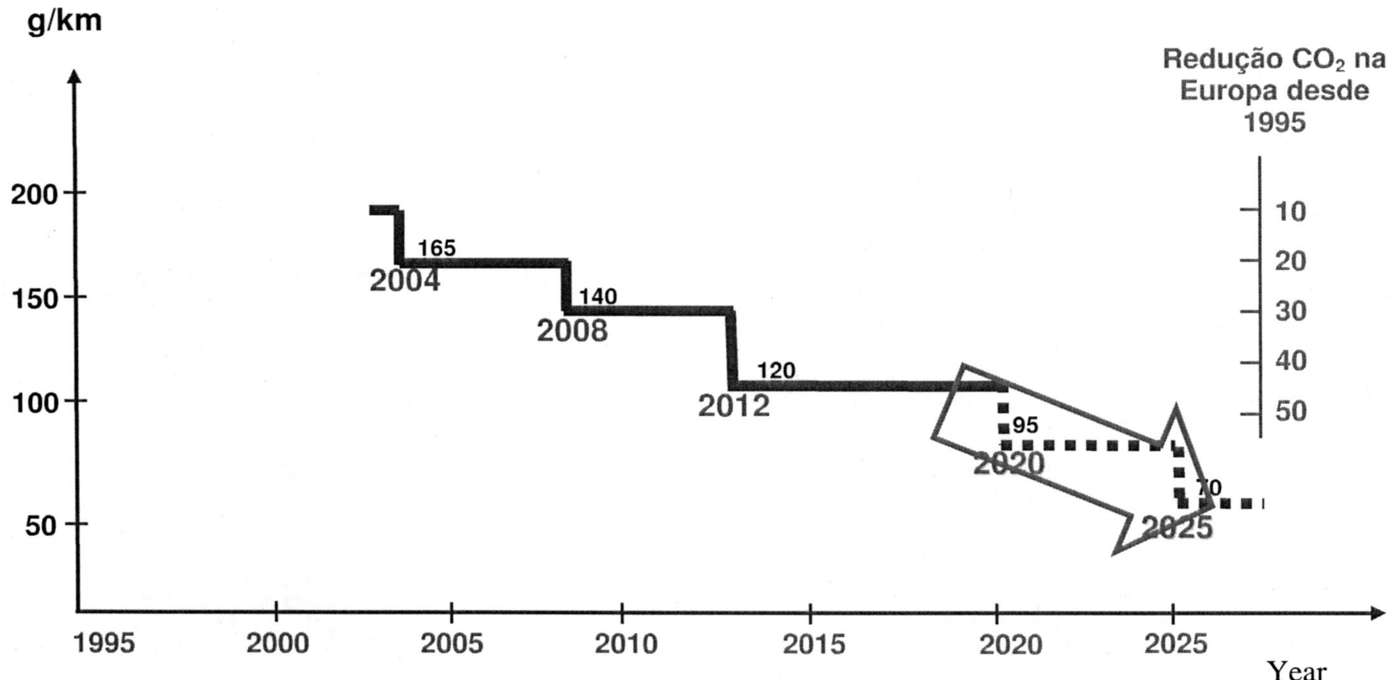

To reduce CO2 emission on the current car and engine may the only way is interfere on drivers' habits even with advices or direct interaction. Other possibility is to use infrastructure support and thinking on the navigation technologies The next step of this evolution is the interaction of maps more exactly horizon prediction with car data

ECO DRIVING CONCEPT

The next step of this evolution is the interaction of maps more exactly horizon prediction with car data information, like current gear used, velocity and engine rotation. Targeting fuel consumption reduction is possible using a dedicated software to give suggestion for the driver with is the best ecological way to conduct your vehicle reducing fuel consumption so consequently emission. This concept is defined as ECO DRIVING, which consist using a HMI or voice giving feedback to the driver to change your way of driving, shifting down or up the manual gear, speeding up or reducing velocity or engine rotation acting on gas pedal, optimizing according the road slope or geographical shaping the engine response versus fuel consumption.

Is up to the drive to follow up the recommendations but if he does the ECO MERITY will be increasing meaning that the driver is doing what is recommended focusing a better way to drive.

ECO ROUTE

Is already normal the use of RDS TMC, over FM modulation, regular FM Broadcasting, free of charge or billed as one shot payment, lifetime use, monthly payment renewing a license fee to receive and decode traffic information in real time - RTTI. And interacting with software navigation indicating which roads are congested, has constructions, flood and incidents or may blocked suggesting a deviation or alternate route manually or automatically. The LBS - Location Based Systems, the ones using GPS + GSM Modem can supply dynamic data when installed into cars with a good quality because we know which type of vehicle, passenger car, truck, bus or motorbike is being monitored. Of course the driver identity is filtered granting the privacy required. The same channel that send individual information can be used to receive a compiled data worked over a representative dynamic traffic data, and must be filtered when sent based on vehicle location to reduce amount of data with is relevant for the vehicle position. FM Broadcast automatically does it due typical range of 100Km.

The next wave is to use Internet Browsers connected trough GSM 3G modems built-in on In Dash Computers - PC Car. Accessing web pages or remote routers that has such kind of traffic flow information for interactions suggesting or changing dynamically the route based on RTTI.

Starting from point A to point B, there are others ECO decisions to make based on MAP Parameters: Slope, climbing or descending may is worst than get around a mountain instead when fuel consumption or travel time is considered.

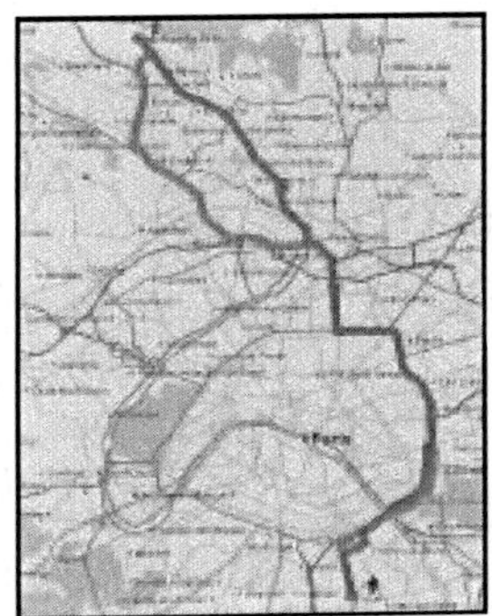

	Normal Route	Green Route	Fast / Green Ratio
Time	76 min	76 min	0%
Fuel	0.65 Gal	0.62 Gal	-5%
CO2	5.7 Kg	5.4 Kg	-5%

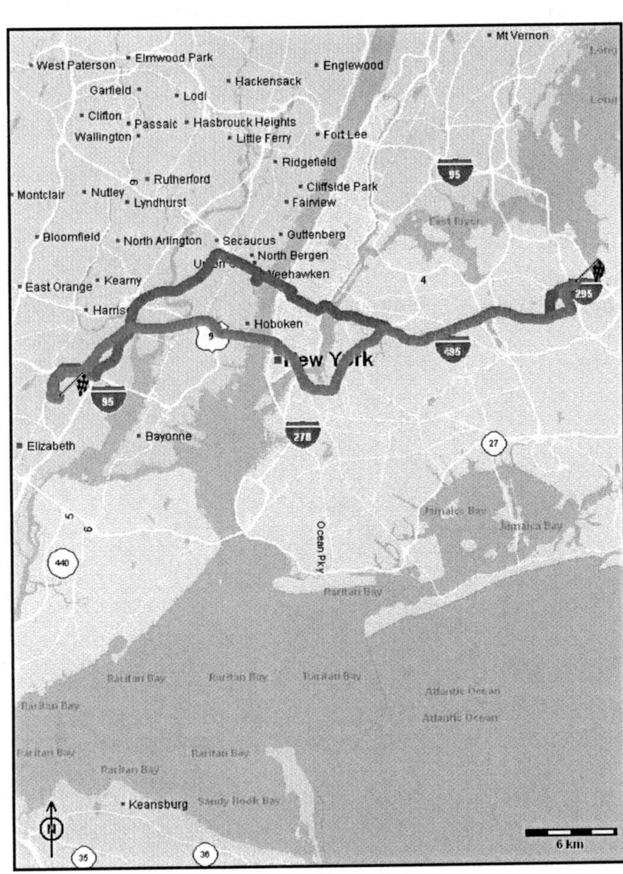

ECO NAVIGATION

When ECO DRIVE plus ECO ROUTE are put together into the system we can call the solution ECO NAVIGATION, when the driver is oriented and the environment information achieves the vehicle and the driver

	Normal Route	Green Route	Fast / Green Ratio
Time	43 min	38 min	-11%
Fuel	0.78 Gal	0.67 Gal	-14.1%
CO2	6.8 Kg	5.8 Kg	-14.7%

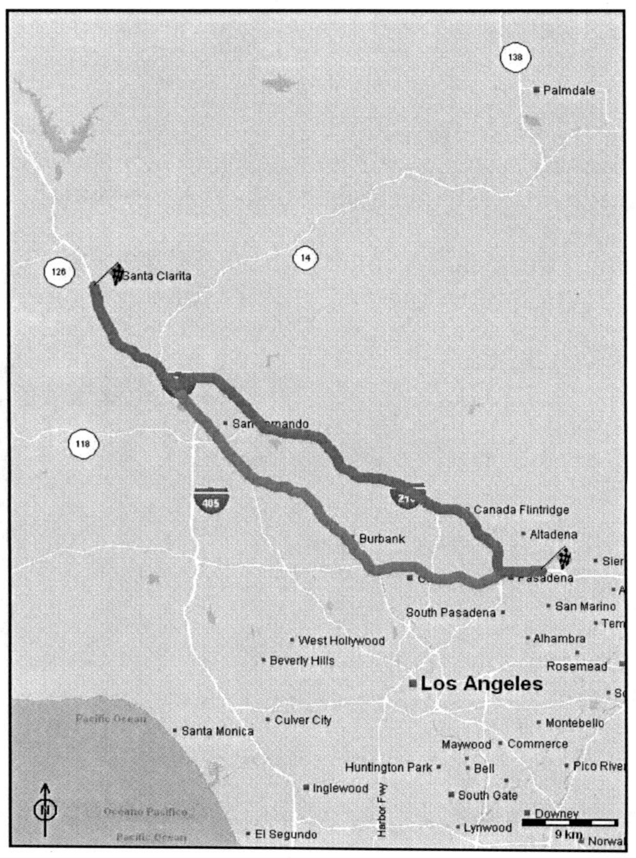

In the case the car has some automation content is possible to interfere setting or changing speed when adaptative cruise control or changing gear using automatic shift gear or AMT - Automatic manual transmission such as Free Choice ™ Magneti Marelli

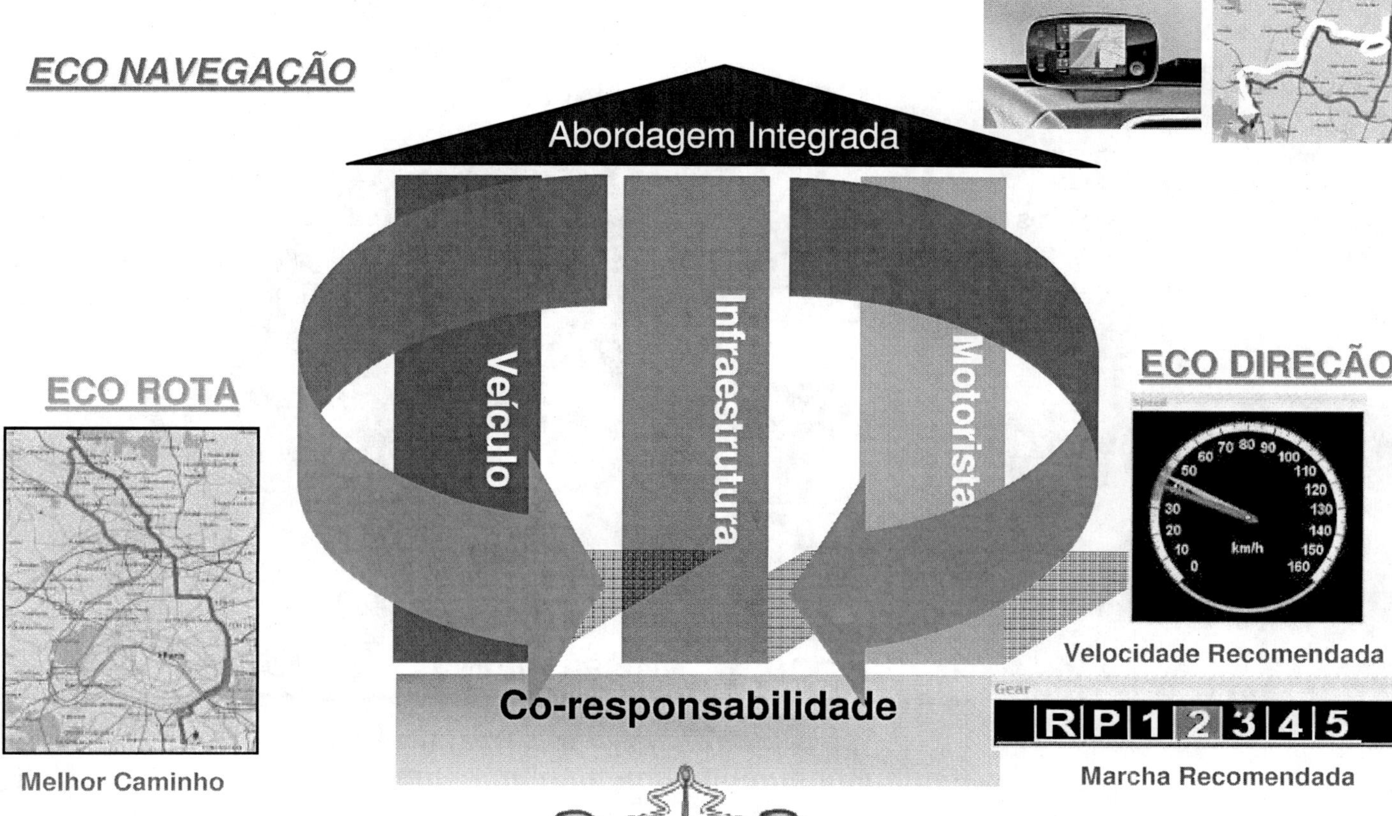
ECO NAVEGAÇÃO
Abordagem Integrada
Veículo
Infraestrutura
Motorista
ECO ROTA
ECO DIREÇÃO
Co-responsabilidade
Melhor Caminho
Velocidade Recomendada
R P 1 2 3 4 5
Marcha Recomendada
Carro -> Infrestrutura -> Carro

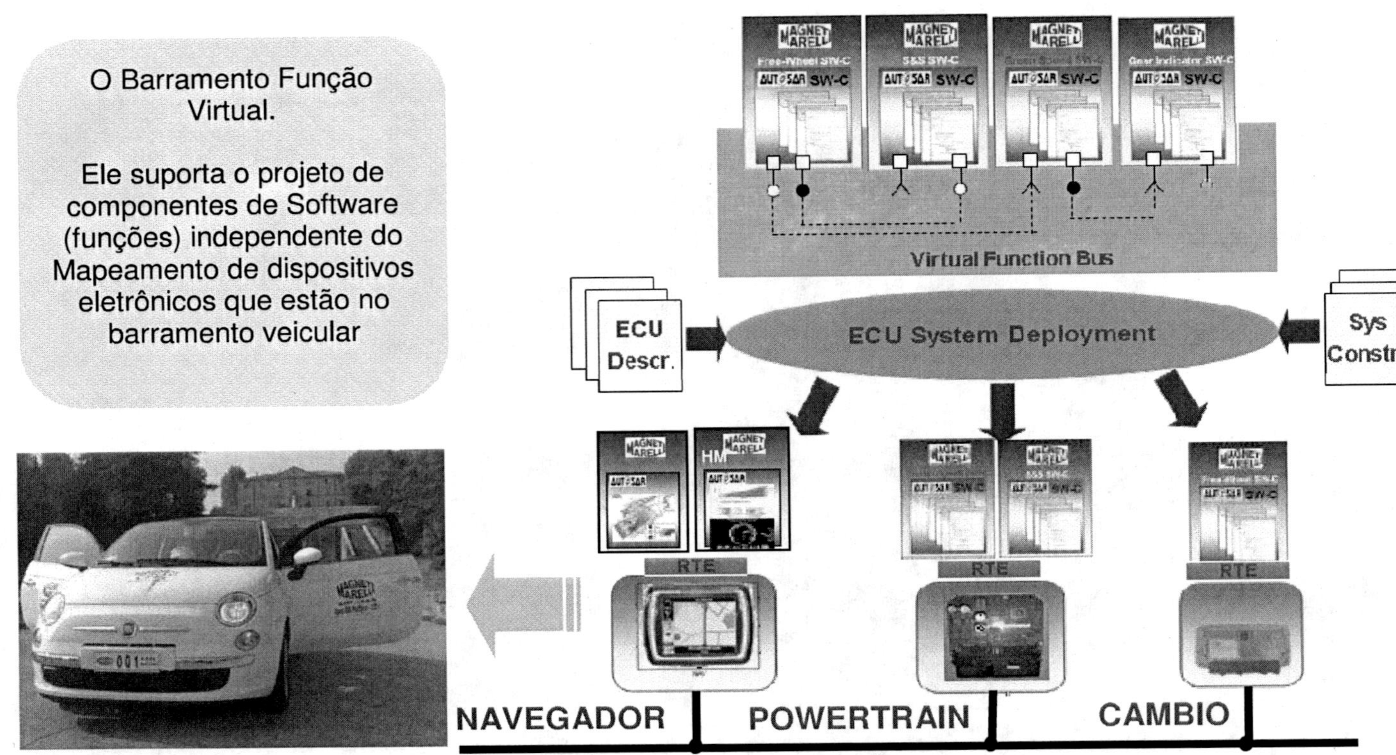

ACTIVE GREEN DRIVING

When the vehicle has minimum automation and the ECO Driving Software is capable to send commands, better say recommendation thought CAN system to engine and electronic gear shift ECUs. The ECO system advice can be performed despite of driver will. In this case we say we have an AGD - Active Green Driving solution where the software is able to monitor external information, use map information, monitor vehicle parameters and indicate the best way to drive interfering directly on the powertrain system looking for the best approach for performance x ecology.

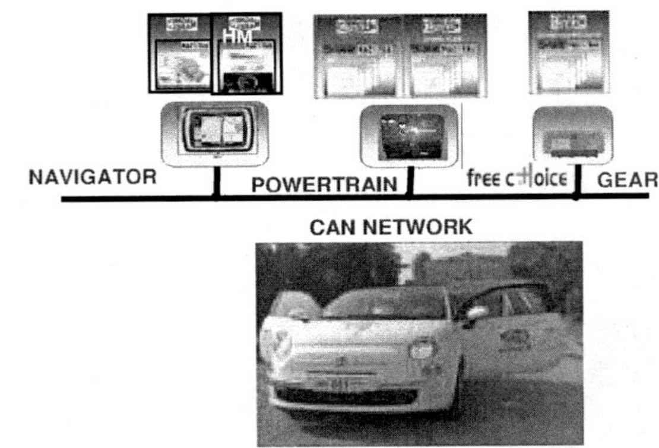

	Fast Route	Green Route	Fast / Green Ratio
Time	25 min	29 min	+16%
Fuel	1.42 L	1.31 L	-7%
CO2	3.35 Kg	3.1 Kg	-7%

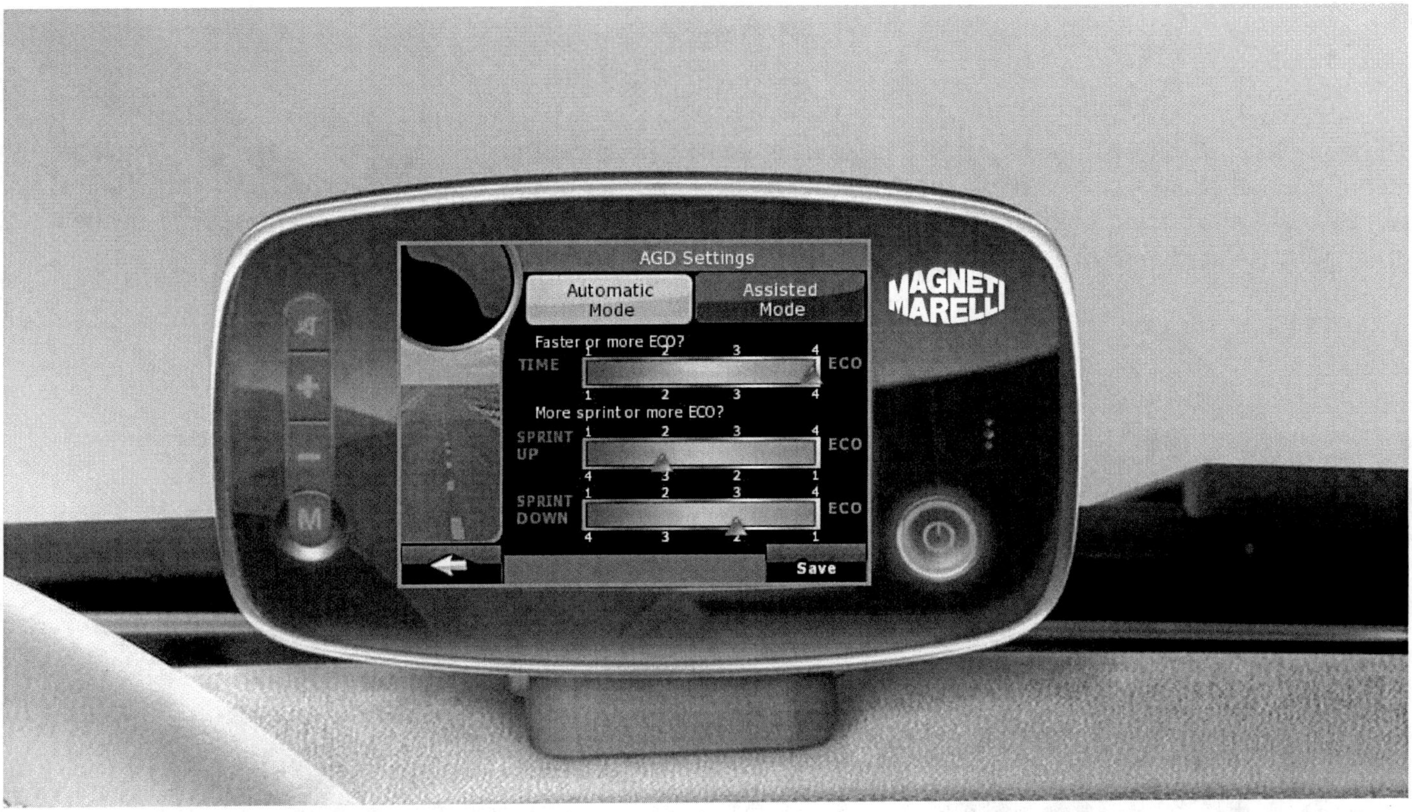
AGD Settings
Automatic Mode
Assisted Mode
Faster or more ECO?
TIME
ECO
More sprint or more ECO?
SPRINT UP
ECO
SPRINT DOWN
ECO
Save
MAGNETI MARELLI

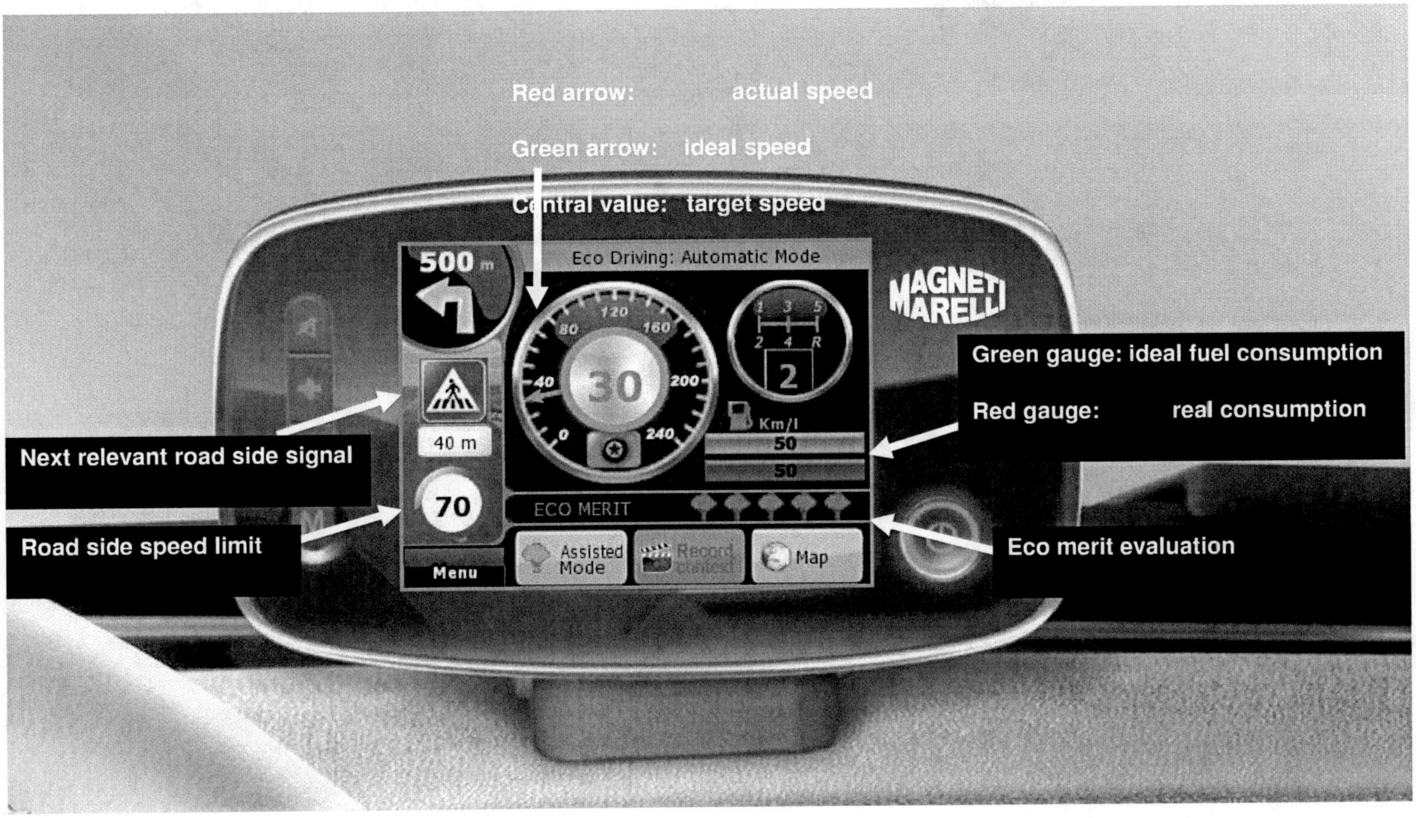
Red arrow: actual speed
Green arrow: ideal speed
Central value: target speed
Eco Driving: Automatic Mode
500 m
40 m
70
Menu
30
Km/l
50
50
ECO MERIT
Assisted Mode
Map
MAGNETI MARELLI
Green gauge: ideal fuel consumption
Red gauge: real consumption
Next relevant road side signal
Road side speed limit
Eco merit evaluation

The complete solution is not restricted to devices and ECUs mentioned, with adoption of new contents like reconfigurable clusters, message center displays, PC cars or other devices which can support a very good HMI and display information combined with more and more car to infrastructure services and maybe in the near future Car to Car solution.

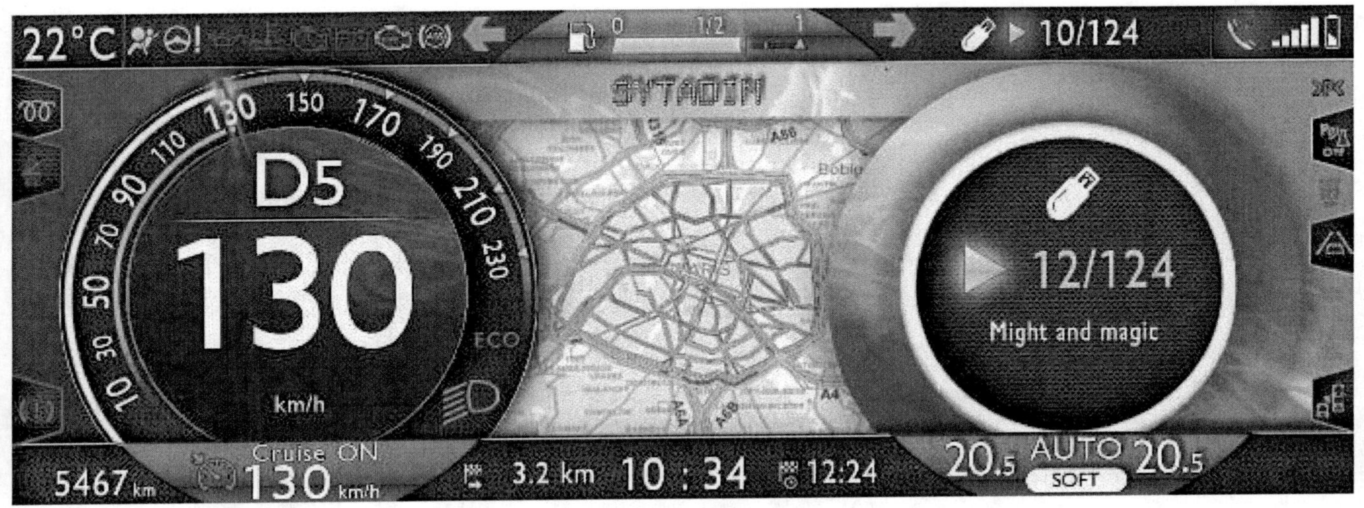

VEHICLE TO INFRESTRUCTURE SOLUTION

So LBS and Tracking - Telematics devices called T-Box can supply communication, positioning, flow data for traffic information

Sistemas de Navegação

- Interface Homem-Máquina HMI
- Atributos de mapas

Parametros do Motor

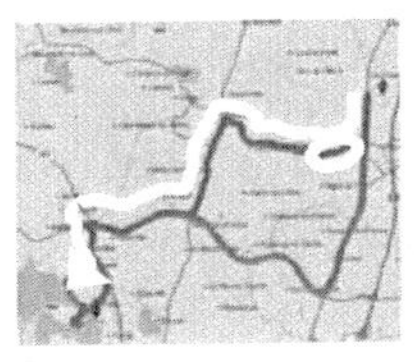

- TMC
- Navegação t-dependente
- V2X

- Altitude/ aclives declives
- Curvatura da Via
- Sinais de Transito

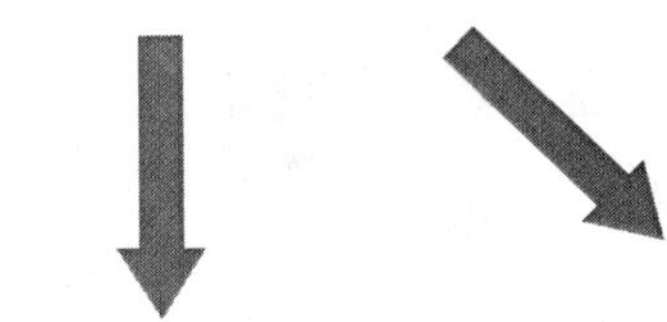

Marcha Recomendada

Comportamento do Motorista

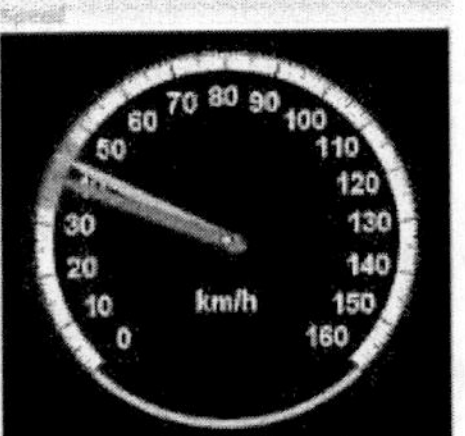

Velocidade Recomendada

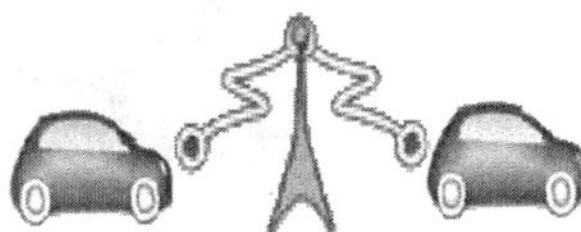

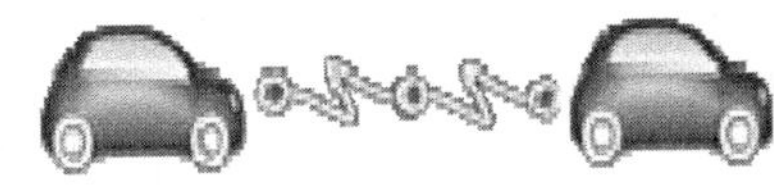

Portable or embedded solution holds the map information on their memories being able if a interface communication is available to receive, treat and use traffic info, map update, accident warning and with image technology add drive safely technologies like night vision, rear parking aid, 360 degrees car monitoring as Bird view.

PC Cars or In Dash Computers, allow the web and internet access, so dynamically information cam be used into the cars equipped with communication modules and Rich HMI. Infotainment contents are strongly necessary to support the new wave of hybrid and Electrical Cars because a simple trip using these cars must be monitored to assure the customer driver will be able to arrive, recharge, replace batteries and then return back to your origin or destination.

AMT - AUTOMATIC TRANSMITION

A big step into Eco navigation solution oriented to AGD it's AMT, not a fully automated gear shift solution but an automatzed gear shift low cost solution, giving to the system the possibility of autonomous gear change without the actuation of the driver if he allows it activating the ECO Driving system. Once the AGD is not desired just turn it off and the vehicle will not assume the way for fuel saving consumption. A small hydraulic circuit controlled by a ECU is basically the Free Choice $_{TM}$ System. It cam be applied better for small and medium vehicles but the light and heavy truck may get more advantage of it due used hour, kilometers drove in these kind of vehicles

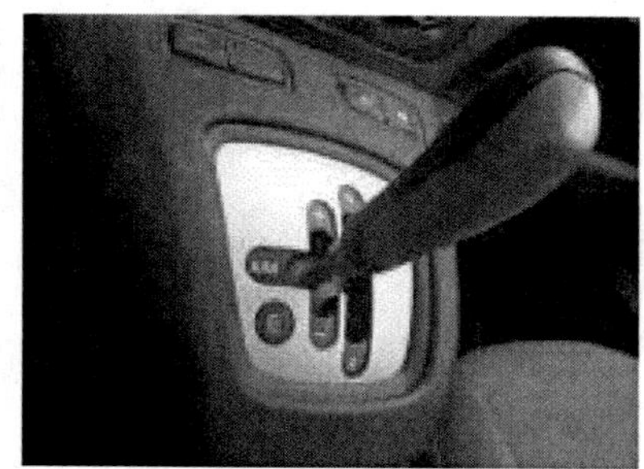

free choice

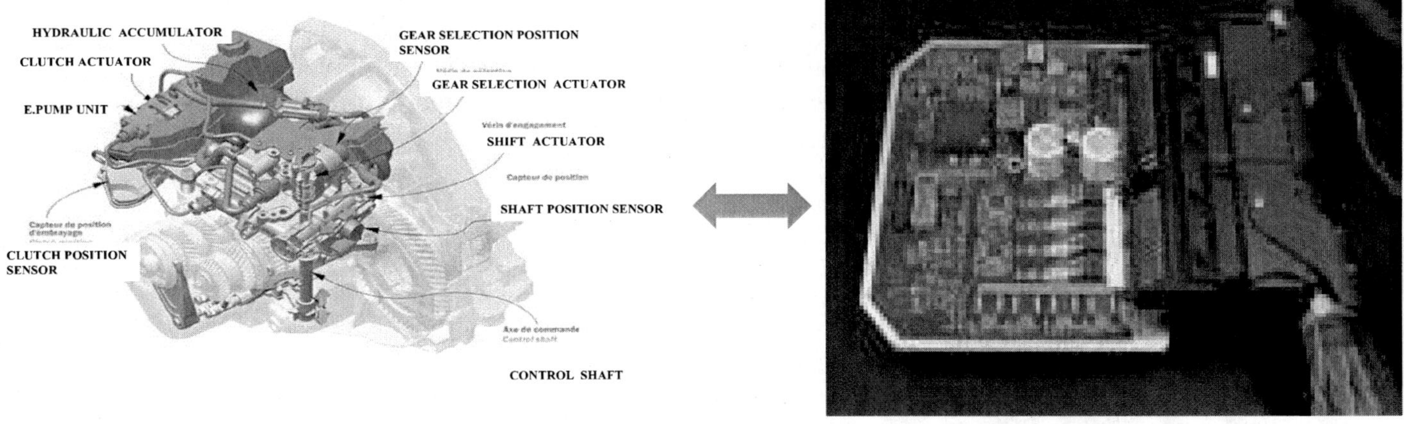

HYDRAULIC ACCUMULATOR
CLUTCH ACTUATOR
E.PUMP UNIT
GEAR SELECTION POSITION SENSOR
GEAR SELECTION ACTUATOR
Vérin d'engagement
SHIFT ACTUATOR
Capteur de position
SHAFT POSITION SENSOR
CLUTCH POSITION SENSOR
Axe de commande
CONTROL SHAFT

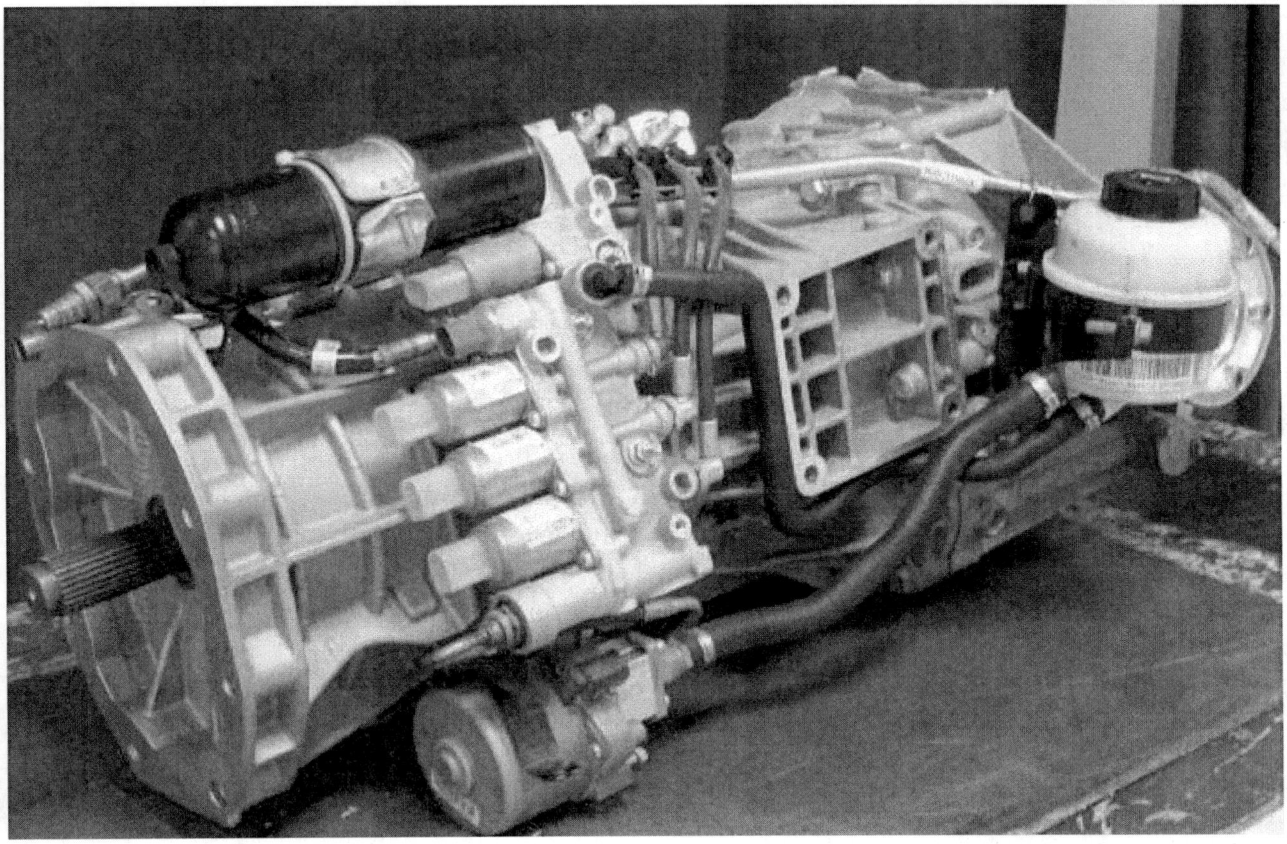

ELECTRICAL & HYBRID VEHICLES

Of course from the point of view of environment the Hybrid and pure electrical vehicles are more efficient and has less emission, but they still cost more than the conventional Diesel or Otto, gasoline, ethanol engine powered. And specially in Brazil and other countries were is possible to use a renewable fuel power and new discovery of oil reserves, like pré-sal with a large amount of investment available from the governmental sources is difficult to convince just based on ecological reasons to change immediately to EV & HEV.

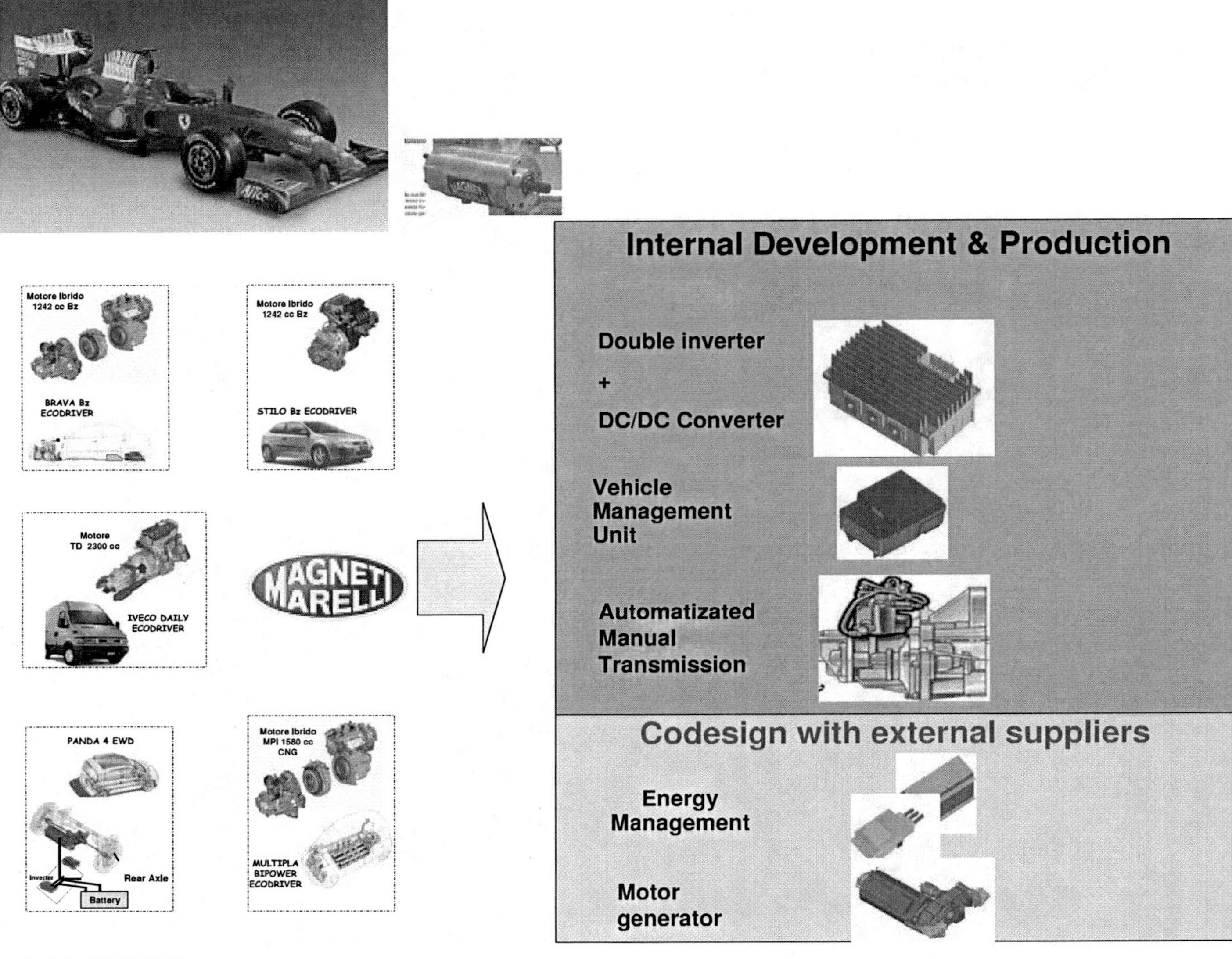

SMART GRID

Talking about V2I or V2V is better are ecological correct to speak V2X, because with EV, HEV that are ecologically correct will connect on all system mentioned before and more, is mandatory to have the vehicle connected to electrical distribution infrastructures, because in the public charging spots the identification for billing is mandatory and the planning for battery change on switching stations to reserve and schedule stop replacement based on your travel plans.

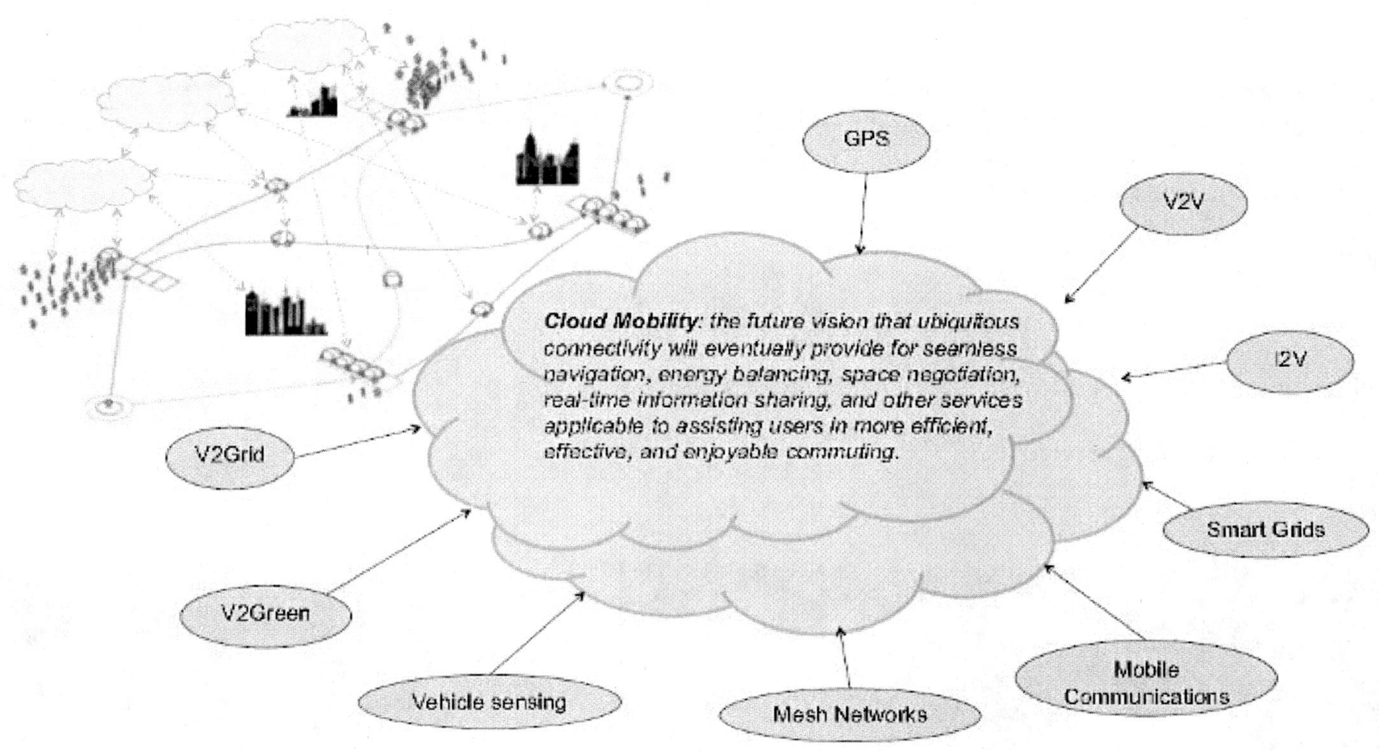

A lot of regulations and standardizations will be required to really turn the electrical vehicles part o of SMART GRID

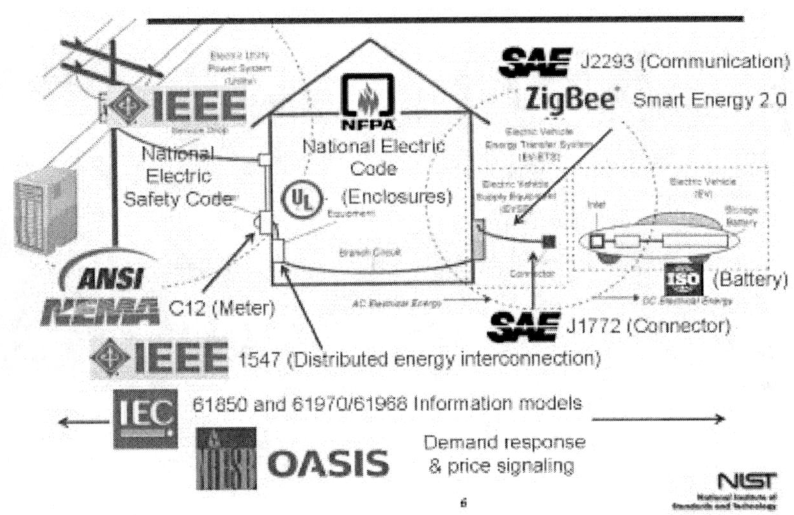

The Electrical exists, the cost is a constrain but the real constrain is the Infrastructure: Several issues must be solved and the solution must be according to the each local player and regulations:

- On Board Management system;

- Billing;

- Cables for slow and fast charging;

- Navigation systems for travel planning and logistics refuel or charging;

- Car to Infrastructure service ;

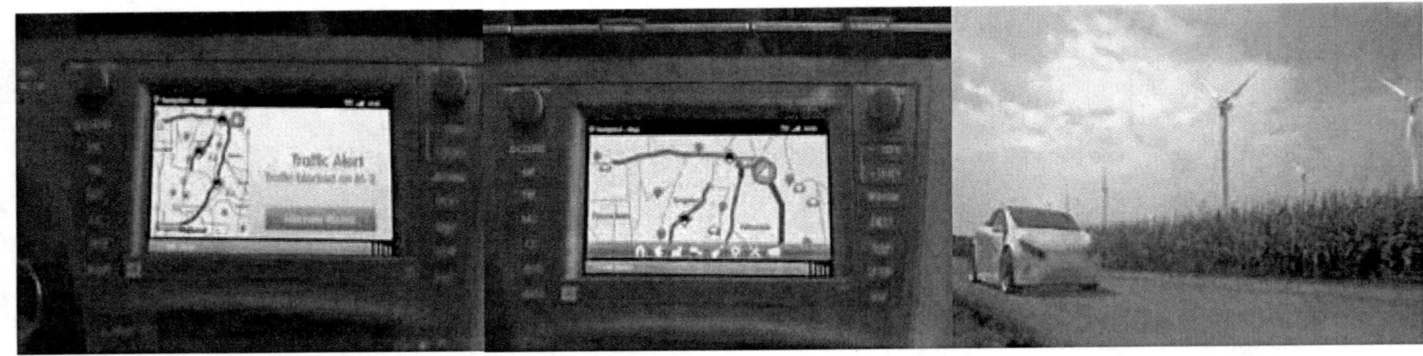

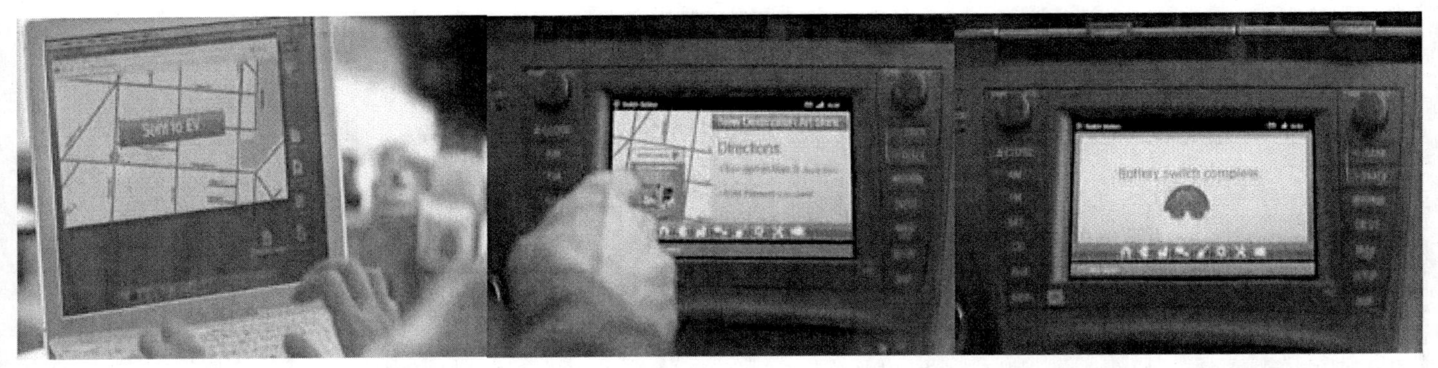

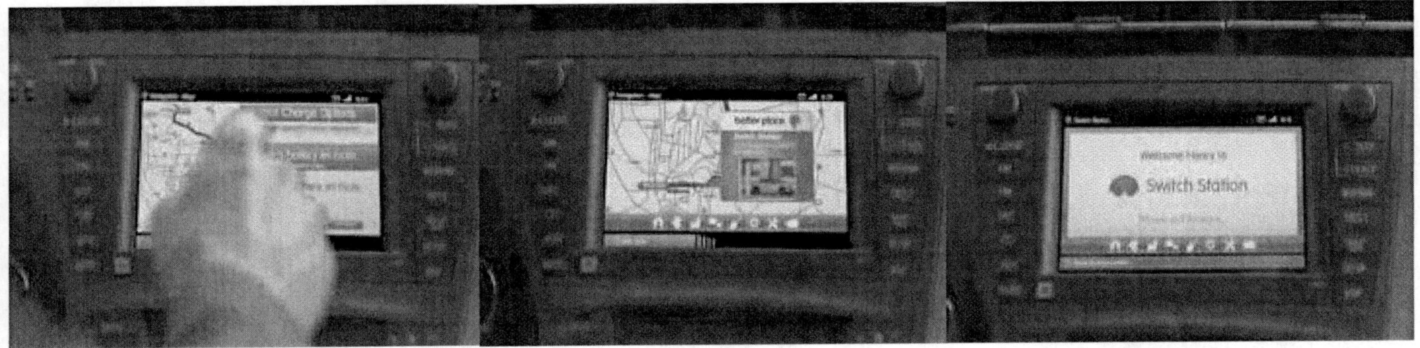

VEHICLE TELEMATICS CONTENTS ORIENTED TO HEV & EV.

The future is now and is impossible to imagine the next generation of cars without the Telematics solution into our vehicles; this is already them for several engineering meetings and events. The EV PROJECT, Better Place and a lot of car sharing programs make it evident on their solution. Logistics of the vehicle when travelling, billing for the electrical power with identification, battery replacement on switching stations. Must have a heavy support of interactive embedded systems and Infrastructure.

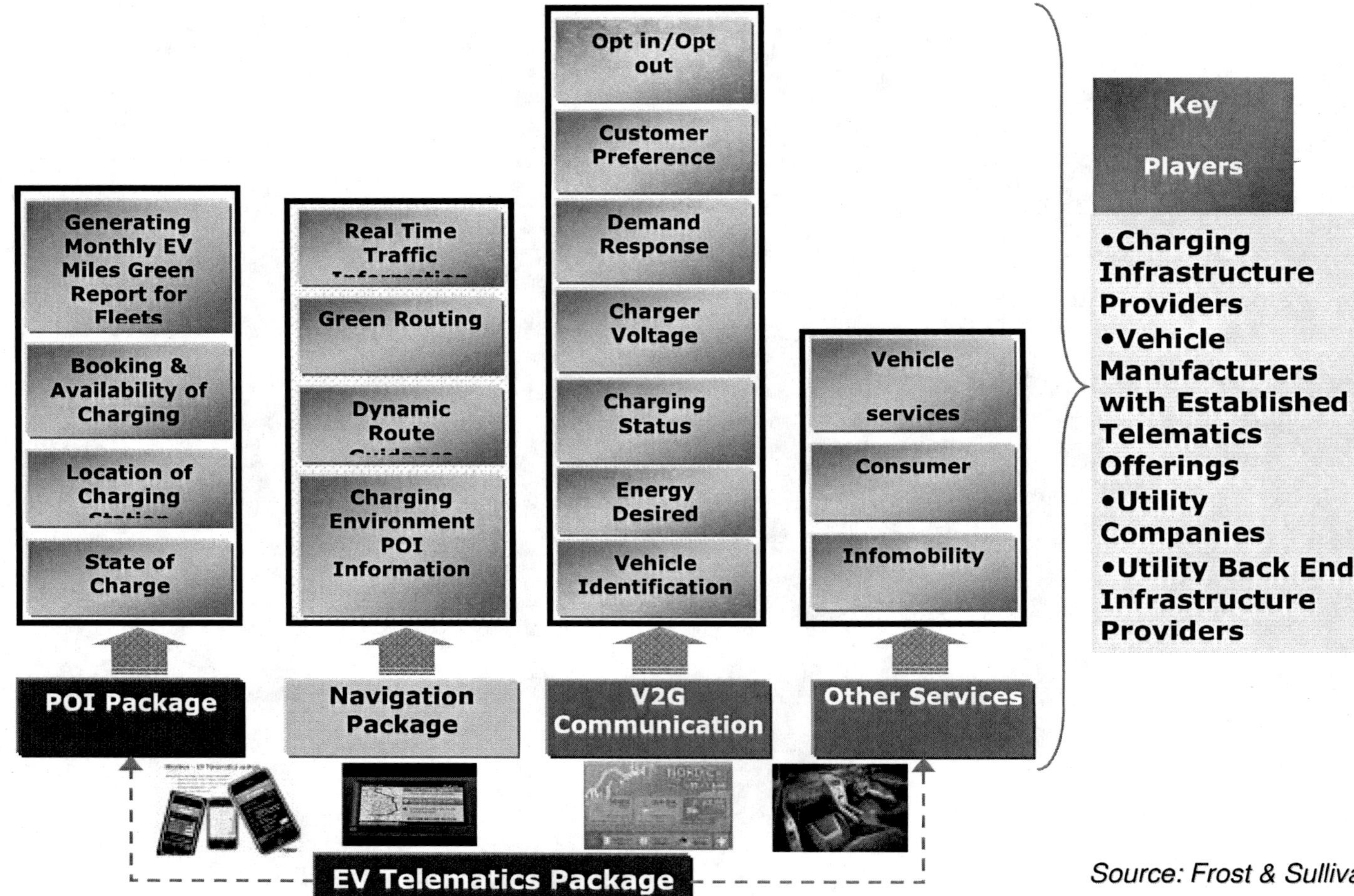
Generating Monthly EV Miles Green Report for Fleets
Booking & Availability of Charging
Location of Charging
State of Charge
Real Time Traffic
Green Routing
Dynamic Route
Charging Environment POI Information
Opt in/Opt out
Customer Preference
Demand Response
Charger Voltage
Charging Status
Energy Desired
Vehicle Identification
Vehicle services
Consumer
Infomobility
Key Players
•Charging Infrastructure Providers
•Vehicle Manufacturers with Established Telematics Offerings
•Utility Companies
•Utility Back End Infrastructure Providers
POI Package
Navigation Package
V2G Communication
Other Services
EV Telematics Package

Source: Frost & Sullivan

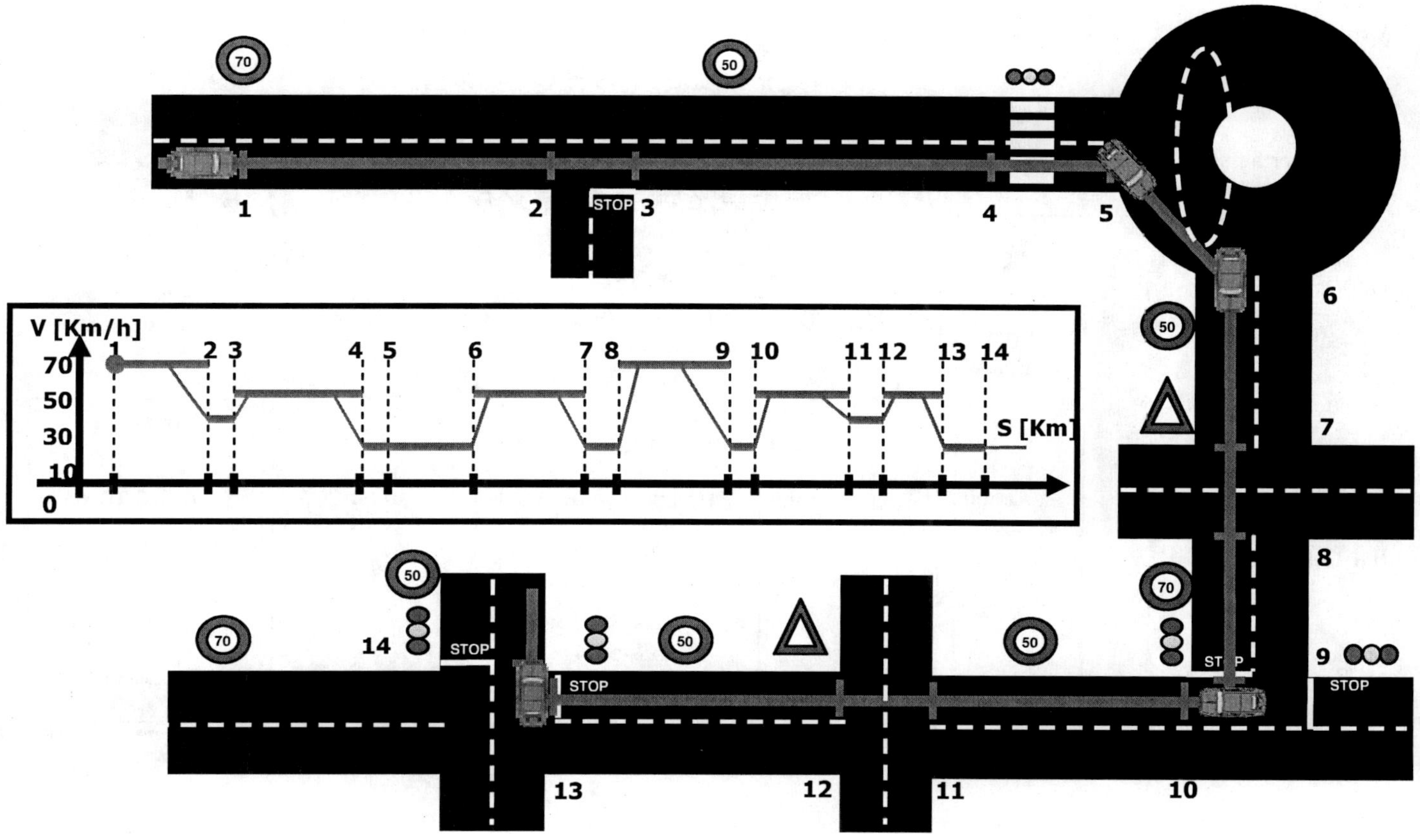

ITS AND ON BOARD ELECTRONICS.

After all technology were being used into the cars is time to cities and road infraestrute receive investment turning the city infrastructure smart enough to interact with cars, like opening traffic lights for crossing incoming vehicles, synchronizing traffic light signals, change speed, monitoring traffic using cameras, floor loop sensors for velocity speed ticketing etc.

TECHNOLOGICAL VALUE

Despite of several non connected solution are feasible the real world expect for really integrated and connected solutions, if not take in consideration the engine size, optimun use of torque curve and get information on Maps parameters there is real and effective gain on CO_2 emission reduction.

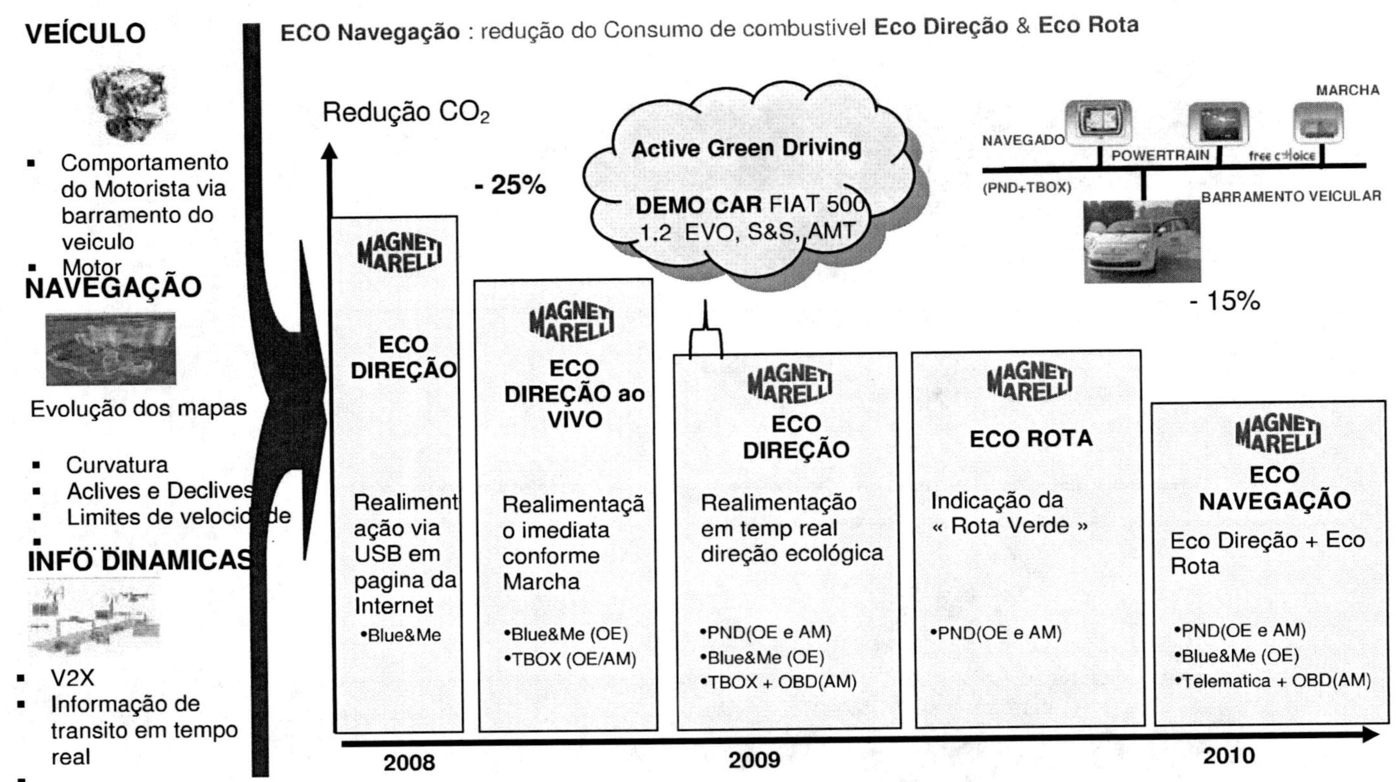

SUMMARY/CONCLUSION

No matter what good condition we have, we must look at future pursuing green technologies that will preserve the enviroment, even helping to clean, not degradete more than we have done right now.

REFERENCES

1. Internal Material Magneti Marelli - Powertrain & Electronics Division
2. Smart Grid Forum - Sao Paulo Aug 2009
3. Telecom Smart Grid - Sao Paulo June 2010
4. Electrical Car Seminar - INEE CPFL - Campinas Nov 2009
5. Frost & Sullivan - Telematics for Electrical Cars - Telematics Detroit - Michigan - June 2010

CONTACT INFORMATION

Ricardo Takahira
New Business Program Manager
Magneti Marelli Sistemas Automotivos Industria e Comercio Ltda.
Electronics Division
Phone: +55 19 2118 6509
Mobile: +55 19 8111 1570
Fax: +55 19 2118 6500
ricardo.takahira@magnetimarelli.com

ACKNOWLEDGMENTS

To my coleghes from Powertrain and Electronics division giving me support and information to really understand about Telematics, ITS, Smart Grid and future on EV & HEV.

DEFINITIONS/ABBREVIATIONS

ECU - Electronic Control Unit

CAN - Control Area Network

ITS - Intelligent Transport Systems

V2X:

V2I - Vehicle to infrastructure

V2V - Vehicle to Vehicle,

V2G - Vehicle to Grid

EV - Electrical Vehicles

HEV - Hybrid Electrical Vehicles

POI - Point Of Interest

APPENDIX

Powerpoint presentation

SAE TECHNICAL PAPER SERIES

2010-36-0310
E

Comparative Analysis of Automatic Steering Technologie and Intelligent Transportation System Applied to BRT

LEOPOLDO RIDEKI YOSHIOKA
MAURICIO MICOSKI
RENATO DUARTE COSTA
EDSON RODRIGUES
JOSÉ ROBERTO CARDOSO

FILIADA À
SAE International

XIX Congresso e Exposição Internacionais de Tecnologia da Mobilidade
São Paulo, Brasil
05 a 07 de outubro de 2010

Este trabalho técnico/científico recebeu apoio financeiro do **Conselho Nacional de Desenvolvimento Científico e Tecnológico - CNPq – Brasil**

AV. PAULISTA, 2073 - HORSA II - CJ. 1003 - CEP 01311-940 - SÃO PAULO – SP

2010-36-0310

Comparative Analysis of Automatic Steering Technologies and Intelligent Transportation System Applied to BRT

Leopoldo Rideki Yoshioka
Jose Roberto Cardoso
USP - University of Sao Paulo

Mauricio Micoski
Renato Duarte Costa
Edson Rodrigues
COMPSIS Computadores e Sistemas Ind. Com. Ltda

ABSTRACT

In the recent years the urban transport system known as BRT (Bus Rapid Transit System) is gaining importance due to the growing demands from alternatives to rail systems. However, unlike rail systems, the performance of BRT depends on the driver's ability to perform accurate docking maneuvers on the bus stop platform and to travel in narrow bus lane quickly and safely. In this scenario, the automation of the bus through the technology of automated steering shows up as a viable alternative, with excellent prospect of operational performance improvement. This article shows how sensing technologies (including magnetic and optical), computational intelligence and electromechanical actuator can transform standard bus in automatically guided vehicle. In addition, it discusses the importance of integrating automated vehicle guidance system (AVGS) with Intelligent Transport System (ITS) to increase operational performance and safety of the BRT.

RESUMO

Nos últimos anos o sistema de transporte urbano conhecido como BRT (Bus Rapid Transit –Corredor Expresso de Ônibus) vem ganhando importância em função da crescente demanda por alternativas aos sistemas sobre trilhos. Entretanto, diferentemente dos sistemas sobre trilhos, o desempenho do BRT depende da habilidade do motorista em realizar manobras de acostamento nas paradas com precisão e trafegar em vias estreitas com rapidez e segurança. Diante deste cenário, a automatização do ônibus por meio da tecnologia de guiagem automática mostra-se como alternativa viável, com excelente perspectiva de ganho de eficiência operacional. O presente artigo mostra como as tecnologias de sensoriamento (magnético e ótico), inteligência computacional e atuador eletromecânico permitem transformar um ônibus comum num veículo guiado automaticamente. Além disso, discute a importância de se integrar o Sistema de Guiagem Automática (SGA) com o Sistema Inteligente de Transporte (ITS) para aumentar o desempenho operacional e a segurança do BRT.

INTRODUCTION

This article presents a systematic analysis of the implementation of the Automated Vehicle Guidance System (AVGS) and Intelligent Transportation System (ITS) in the Bus Rapid Transit Systems (BRT). It is based on the experiences obtained through the development of advanced transportation system for Sao Paulo city's BRT - Expresso Tiradentes - and research project conducted with support of FINEP - Brazilian Studies and Projects Support Agency.

Currently, there exists uncovered demand for medium capacity passenger transportation system (15,000 to 30,000 passengers / hour / direction) [1-3]. Confirming this fact, data from the IBGE (Brazilian Institute of Geography and Statistics), shows that there are 81 cities in the Brazil with over 300,000 inhabitants and 14 of them are over one million inhabitants [4]. Thus, there is an urgent need for efficient urban passenger transportation system solution with respect to the technical and financial standpoint.

There is no doubt that one of best solution for population mobility in large urban centers is the Metro (subway) that has exceptional attributes of efficiency and quality. However, due to the high cost of both deployment and operation, only six cities in Brazil have this system, in other words only a small percentage of the population are benefited. There are other transport modals such as Light Rail Vehicle (LRV) and Monorail, whose implantation costs are lower than Metro, nevertheless the required investments are considered very high for the reality of the country. Of course, the bus modal is the most widespread, being the most accessible, but due to limitations in transport capacity and poor service quality has been unable to meet the demand adequately. Given this scenario it is important to develop transportation solution that can be applied broadly in the Brazilian cities (low cost requirement) that meet the demands (high capacity requirement) with quality (speed, regularity and comfort requirements) and that are attractive (modernity, technology and visual identity requirements) [5].

This article shows how sensing (magnetic and optical), computational intelligence and electromechanical actuator technologies can transform a conventional bus in automatically guided vehicle. In addition, discusses about importance of integrating the AVGS with ITS to increase the performance and safety of the BRT.

AUTOMATIC STEERING TECHNOLOGIES

OPERATIONAL CONCEPT

The AVGS was developed to replace driver's vehicle steering control action. It consists of sensors, signal processors, onboard computer and eletromechanical actuator. It can be installed in any vehicle. It is able to perform vehicle positioning and alignment in the roadway, automatically, with accuracy and repeatability. It allows the vehicle to perform bus stop approach and docking maneuver accurately (with one centimeter margin) and quickly. The gap between the platform side and the vehicle can be set by software configuration for five to ten centimeters, allowing the passengers embarkation/disembarkation to be done in less time, with comfort. The vehicle can also travel in narrow busway quickly and safely [6,7].

As shown in the Fig. 1 pictures sequence, the vehicle can operate in manual or automatic mode and the driver is still present, being also responsible for control of speed, stops and starts. The set of pictures of Fig. 2 shows the details of the maneuver of precision docking at bus stop platform.

FIGURE 1 - Pictures sequence showing the driver passing from manual to automatic steering mode (Sao Paulo City's BRT - Expresso Tiradentes).

FIGURE 2 - Pictures sequence showing details of precision docking at bus stop platform. (Sao Paulo City's BRT, Mercado Station - Expresso Tiradentes).

SYTEM DESCRIPTION

The AVGS consists of four main segments: (1) Position Sensing, (2) Signal Processing, (3) Steering Control, (4) Steering Actuator. It is presented in the following a brief description of each of these segments.

1. Position Sensing

This is a fundamental part of the AVGS. From the position sensing information, the exact lateral position of the vehicle on the roadway is determined. There are basically five types of positioning references applicable to the AVGS [6]:

- **Discrete Magnetic Marker**: by measurement of intensity of the magnetic field generated by discrete magnetic markers;
- **Magnetic Adhesive Tape**: by measurement of intensity of the magnetic field generated by magnetic adhesive tape;
- **Optical**: by video camera capturing the lane marking painted on the busway;
- **DGPS**: by capturing differential GPS signals; and
- **Wire Current Loop**: by measurement of electromagnetic field of wire current loop.

Evaluating the applicability for the AVGS, with respect to each type of positioning reference – based on the criteria of safety, robustness, flexibility, durability, deployment and maintenance costs – come to the conclusion that both optical and discrete magnetic marker are the most appropriate [7].

Examples using discrete magnetic references can be seen in MGS (PATH, California, USA [8]), APTS (Phileas, Eindhoven, Netherlands [9]), IMTS (Toyota, Nagoya, Japan [10]) and SGM (SPTrans, Sao Paulo, Brazil [11]). The OPTIGUIDE (Siemens, Rouen, France [12]) and SGO (Compsis, Sao Jose dos Campos, Brazil [13]) are using optical sensing. The LMAG (Laboratory of Applied Electromagnetic of Sao Paulo University [14]) is developing magnetic guidance techniques since 2009.

The Figure 3 shows the picture of ferrite material magnetic marker, while Fig. 4 shows a lane marking painted on the busway (Espresso Tiradentes, Sao Paulo).

FIGURE 3 - Magnetic Markers (ferrite magnets with dimension: 25x100 mm).

FIGURE 4 - Lane Marking painted on the busway (lane width: 100 mm) - Sao Paulo City's BRT - Expresso Tiradentes.

2. Signal Processing

This segment is responsible for extracting vehicle lateral deviation from the signals captured by the sensing system. Let us describe in the following the signal processing for Magnetic Marker and Optical Sensing.

Magnetic Marker Sensing: the signal processing technique is applied in order to determine the peak position of the magnetic field profile generated by the discrete magnet. From this information the lateral deviation of the vehicle is estimated [15-18]. The Fig. 5 illustrates a three dimensional view of the magnetic field profile generated by a sequence of magnetic marker. One can observe that the peaks follow the polarity (north or south) of the magnets, which can be used to encode information like track geometry.

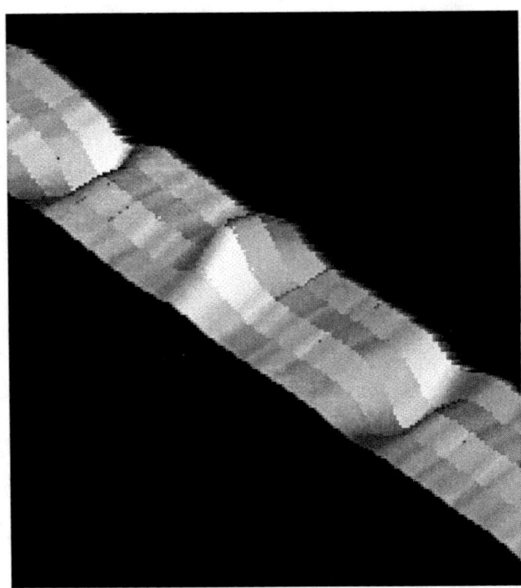

FIGURE 5 - Three dimensional view of the magnetic field profile generated by sequence of magnets.

Optical sensing: the image processing technique is applied to determine the lateral deviation of the vehicle. The reference lane marking painted on the road is captured by a video camera and each image frame is analyzed by the image processor. The block diagram of Fig. 6 shows the software architecture of image processor.

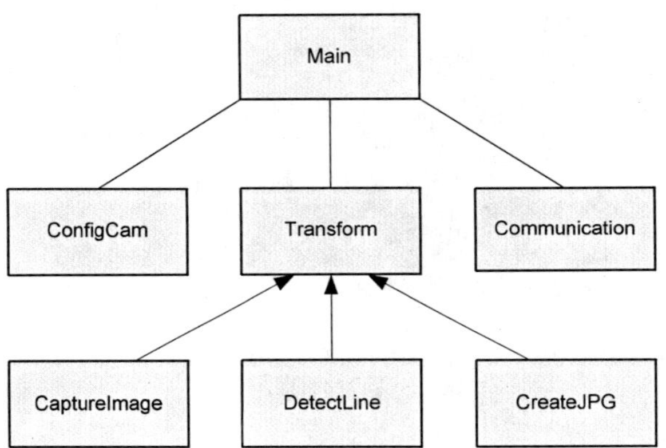

FIGURE 6 - Software Architecture of the image processor.

Description of image processor modules:

- **Main:** it contains the main( ) function that is called after a processor reset. This module defines the sequencing of calls from other modules;
- **ConfigCam**: it is specialized module to configure camera parameters;
- **Transform**: it sets the default interface for all modules that use the captured image. The image is stored in a memory region with the format BUFFIMG and it is accessed via parameter passed in the call of each module. They are defined three types of transformations: CaptureImage, DetectLane and CreateJPEG.
- **CaptureImage**: it fills the memory with image data of the new frame acquired from the camera.
- **DetectLine:** it locates the edges of the image objects and defines the parameters of the corresponding lines.
- **CreateJPEG:** it converts the captured image to JPEG format data file and sends to the serial port.

- **Communication**: sends the parameters of the lane marking edges to the serial port.

3. Steering Control

The control segment is responsible for maintaining the correct lateral position of the vehicle on the road. From the information of the vehicle lateral deviation determined by signals processing segment, the control generates the actuator command to apply the correct steering angle to correct the lateral deviation of the vehicle [11,19,20]. They are described in the following, some of the components used in the control system.

Kalman Filter [21]: The vehicle's states are defined from the following variables:

- x: lateral distance measurement from de magnetic track to the vehicle, in the perpendicular direction to the vehicle body;
- y: longitudinal position of the vehicle measured along the road;
- Ψ: angle between vehicle body and the road;
- α: front tire steering angle, which is the angle between the vehicle body direction and the front tire pointing direction;
- ρ: curvature of the positioning reference to be followed. The curvature is defined as angle variation of the tangent line of the positioning reference curve in relation to the traveled distance.

When the magnetic sensor is used, the read variables are x, y, Δα and ρ (where the value of ρ is obtained from the design of magnetic track and the measure of the longitudinal position of the vehicle). The Kalman filtering is used to estimate the values of x, α and Ψ.

When the optical sensor is used, the read variables are x, Ψ, y and Δα (determination of ρ depends on the implementation of an efficient method for marking the optical path). The Kalman filtering is used to estimate the values of x, ρ and α.

Vehicle State Estimator: The state estimator reads the information from sensors. Applies them to the Kalman filter and decides how to use the results. An important decision of this module is to define if the current state is reliable or not.

- In the case of magnetic sensing, the approximation between the forecast and the state actually measured is used in this decision.
- In the case of optical sensing, an additional element to be considered is the quality factor of the captured image.

When the quality factor falls below the minimum, the position of the optical range is used only if there is close approximation with the prediction. In the case of high quality factor, the restriction can be lower for the data to be accepted. In any case, if the agreement between prediction and measurement is greater than the limit by a predetermined distance, the system will consider that optical reference is lost and forcing the return to the reference search state.

Control Algorithm: The function of the control algorithm is to calculate the steering angle to be applied to compensate the lateral deviation at one point ahead. When the reference comes from the optical sensor, it is necessary to consider two differences in the use of magnetic sensor signal: the separation between samples and the use of track curvature.

- In the case of magnetic sensor, the separation between the samples is given by the distance between the magnets, usually of two meters. As the vehicle speed varies, the time between samples also vary, which brings stability problems at high speeds.
- In the case of optical sensor, the separation between samples is given by the captured image frame rate, usually 15 or 30 frames per second. Thus, the time between samples does not change with vehicle speed, which minimizes the problem of stability.

The use of the curvature of the reference (magnetic or optical path) depends if the system is able to determine precisely the longitudinal position of the vehicle.

- In the case of magnetic sensors, this is done by creating binary codes based on the polarity of magnets installed. Observe that the simple count of the magnets provides longitudinal location information. With the position information, the control law can take into account the current curvature and the curvature ahead, anticipating the steering at the entrance of curves.
- In the case of optical sensor, it is necessary to establish identifiable markers for video and whose detection is reliable and suitable for available processing capacity. Due to these constrains, the initial version of control system was based on optical sensor used only local curvature, calculated from the Kalman filtering and not considering the curvature ahead.

Fig. 7 shows the block diagram of steering control system.

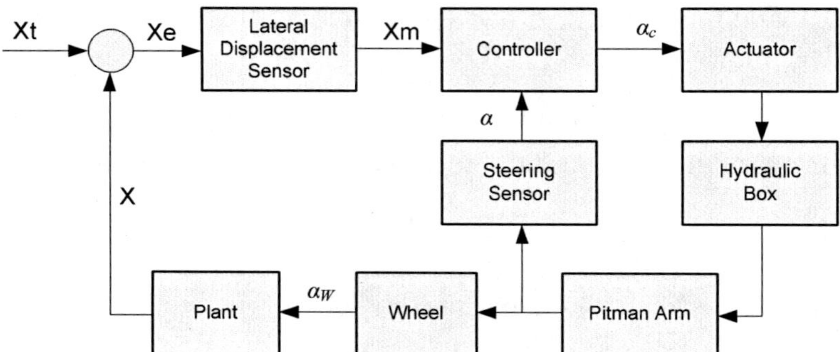

FIGURE 7 - Functional block diagram of the steering control system.

4. Steering Actuator

The actuator is an electro-mechanical component. Transform the output of the steering control system in mechanical drive of the steering system in order to provide the appropriate steering for directional wheel of the vehicle. Produce the correction of lateral deviation of the vehicle to keep in the correct road trajectory. It consists of controller, servo-motor and mechanical coupling with the steering system of the vehicle. The controller is responsible for communicating with the guidance computer that processes the steering control algorithm. The servo motor in conjunction with the mechanical coupling produces the mechanical motion required to drive the steering system of the vehicle. The schematic diagram of Fig. 8 illustrates the components of the actuator, while the picture of Fig. 9 shows its assembly configuration used for laboratory tests.

Every fifty milliseconds the guidance computer calculates a new angle position to be taken by the steering bar. This information is forwarded via CAN network to the servo motor controller. The controller drives the motor in closed loop. This causes the angular position of the axis follow very precisely the values determined by the Guidance Computer. The motor shaft is mechanically coupled to a reduction box, which amplifies the torque capacity from 2Nm to 20Nm over the steering bar, enough to control the steering in all situations. The output shaft of the reduction box is connected to the clutch, which is the control element coupling the servo motor to the steering bar. The clutch is controlled electrically by the button Auto/Manual located on the driver's dashboard. When the system is in Auto mode, the clutch transmits the movement of the output shaft of the reduction box to the steering bar. When the system is in Manual mode, the clutch disengages the two axles, and steering bar rotates freely relative to the reduction box output. As illustrated in Fig. 8, the clutch is connected to the bus driving through a belt attached to a pulley installed on the output shaft of the clutch and another one installed in the steering bar.

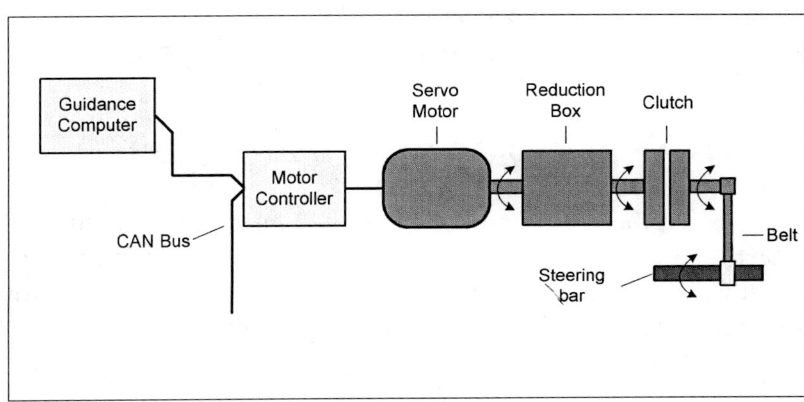

FIGURE 8 - Schematic diagram of the steering actuator.

FIGURE 9 - Actuator assembly configuration for laboratory tests (IP&D Laboratory at UNIVAP).

AVGS INFLUENCE ON BRT PERFOMANCE

As seen in the previous section AVGS increases the accuracy of vehicle guidance (with about 1 cm margin in bus stops and 5 cm margin along the track), providing significant improvements in operational performance of BRT [6,7], as shown below.

- Reducing the passenger embarkation and disembarkation time. It is estimated that time reduction is 1-3 seconds per passenger.
- Increase accessibility for users with visual or motor disabilities, children and elderly.
- Possible elimination of ramps for wheelchair access;
- Possibility of operation in narrow busways, enabling the deployment of exclusive lanes in urban centers.
- Reducing the construction cost of the bus corridor. It is estimated that the width of the roadway may be reduced from 3.50 m to 2.90 m.
- Reducing the time lost due to the approaching and the departure in the bus stops.
- Increase the comfort of passengers due to the standardization of the bus trajectory (smooth movements).
- Reducing the driver workload, who with the automated guidance can concentrate on controlling the acceleration and braking.

INTEGRATION OF AVGS AND ITS FOR BRT OPERATION

In the previous section has presented the technology of automated guidance as an important alternative to increasing the operational performance of bus corridors. In this section, a technology component known as ITS or Intelligent Transport System will be present. Furthermore, we will show that integration of these both technologies allows BRT to reach higher service level, providing greater mobility to the population.

The ITS is a combination of information technology and communication technology destined to management and control of transportation systems. It provides instant access to traffic information, integration with the vehicles and a wide range of services for authorities, operators and passengers. Also, provides increasing the decision power in order to plan actions more wisely and efficiently.

There are several ITS technologies that can help the improvement of the BRT services. Here, they are analyzed five of them, enough to show the effects of application of ITS in the operation of the BRT. They are the followings:

- Telematic;
- Intelligent traffic light control;
- Intelligent display;
- Video monitoring;
- Control Center.

Telematic: consists of on-board computer installed in the vehicle with ubiquitous connection at the control center and driver's terminal with display and keyboard that allows the exchange of information between the driver and operator of the control center. Sensors collect information on speed, engine rotation speed and other vehicle conditions allowing remote monitoring the vehicle operating conditions. Also, vehicle location data are processed at the control center in real-time, generating information on waiting times for users.

Intelligent traffic light control: communication devices and vehicle queue sensors on the road allow centralized monitoring and controlling of the traffic lights at intersections. From the traffic information and preferred vehicle approach communication, the central computer calculates, every second, the traffic signal timing needed to optimize the flow of vehicles.

Intelligent display: from the electronic message panels, the user can check the waiting time for the arrival of the next vehicle. The panels communicate with the control center computer through data networks, receiving in real time the messages to be displayed. Also, electronic message panels can be placed in the corridor side to warn drivers about the traffic situation ahead.

Video monitoring: the image resources have become increasingly important for monitoring the operation. Risk situations of users and vehicles can be detected remotely and generate alerts and allowing preventive actions quickly. Currently, digital video cameras with IP interface and high compression capabilities allow them to be monitored extensively the critical points of the terminals, bus stops and corridor. Advanced tools like video based incident detection software may detect dangerous situation instantly, and it can work 24x7 uninterruptible.

Control Center: it is where all information about the operation of the system converges: traffic conditions, location of bus, delays, images of critical points, alerts, alarms and more. Has a wide data network that is interconnected with all subsystems. It requires high computing power to process all collected data and present information in real time to operators and users. The collected data feed a database and the specialist application software, automatically, generates statistical analysis reports and performance index of the operation.

The Fig. 10 and Fig. 11 show the example of ITS application in Sao Paulo City's BRT.

FIGURE 10 - ITS Technology: Control Center - Sao Paulo City's BRT - Expresso Tiradentes.

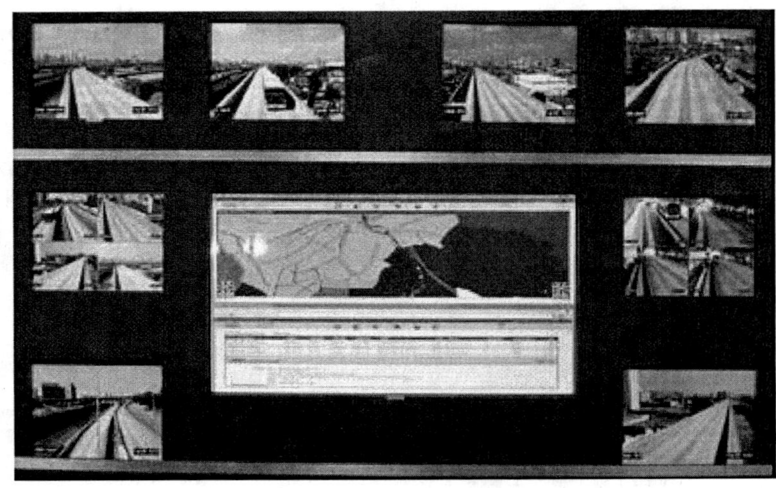

FIGURE 11 - ITS Technology: Video Wall - Sao Paulo City's BRT Control Center - Expresso Tiradentes.

CONCLUSIONS

The automated guiding technology for passenger transport is already successfully applied today in various places in the world. Brazil, through the pioneering work of the Sao Paulo City's BRT - Expresso Tiradentes - and the Compsis's development of optical guiding technology supported by FINEP, has this technology.

ITS technology is widely applied in Brazil in urban control systems and in concession highways control systems.

Brazil is today a world reference in the development of BRT or express bus corridors. Has know-how in vast field of technology of various technologies in public passenger transport, as shown by several studies and real applications in this area. However, it suffers from the problem of universalization of the experience, due to the shortage of financial resources to invest in rail transportation network, particularly outside major urban areas, cannot meet the demand for mass transportation with traditional bus corridors.

It was shown that AVGS and ITS applied to the BRT transportation systems could allow many Brazilian cities to increase their transport capacities and qualities of the existing corridors to meet their growing demands.

As can be seen in this work, the two technologies, AVGS and ITS, are complementary and when applied together, enable the transport solutions, where alternatives on rail like the Metro and LRV (light rail) are expensive against the necessity.

REFERENCES

1. Ferraz, A. C. P., Urban Public Transport. Rima Editors, Sao Carlos, SP, 2004.
2. Writes, S. , Urban Transportation Systems (1st ed.). McGraw-Hill Professional Publishing, New York, USA, 2002.
3. Vuchic, V. R., Urban Transit - Systems and Technology, John Wiley & Sons, 2007.
4. IBGE - Statistical data from Brazil, April 2010.
5. NTU - National Association of Urban Transport. Issue No. 143, July 2009.
6. Shiladover, S. E. et al, Lane Assist Systems for Bus Rapid Transit, Volume I: Technology Assessment, California PATH Rearch Report, University of California at Berkeley, November 2007.
7. Zhang, W. et al. Lane Assist Systems for Bus Rapid Transit, Volume II: Needs and Requirements, California PATH Rearch Report, University of California at Berkeley, November 2007.
8. MGS / PATH-Berkeley: http://www.path.berkeley.edu/PATH/research/magnets/
9. APTS-Phileas: http://www.apts-phileas.com/
10. IMTS-Toyota: http://www.gizmohighway.com/transport/expo_bus.htm
11. Micoski, M., Yoshioka, L.R., Costa, R.D., Teixeira, A.S., Jr. et al., "Automatic Pilot for Buses: A Brazilian Reality," SAE Technical Paper 2008-36-0088, doi:10.4271/2008-36-0088.
12. OPTIGUIDE-Siemens: https://www.swe.siemens.com/france/web/en/sts/newspress/releases/Pages/OptiguideCastellon.aspx
13. SGO-Compsis: http://www.compsis.com.br/uni_onibus.php?id=311&idpai=149
14. LMAG - Applied Electromagnetism Laboratory - Polytechnic School of USP: http://www.lmag.pea.usp.br/
15. Chan, C.Y. A System Review of Magnetic Sensing System for Ground Vehicle Control and Guidance. California PATH Program, Berkeley, 2002.
16. Chan, C.Y.; Tan, H.S. Evaluation of Magnetic Markers as a Position Reference System for Ground Vehicle Guidance and Control. California PATH Program, Berkeley, 2003.
17. Tan, H.S.; Bougler, B. Experimental Studies on High Speed Vehicle Steering Control with Magnetic Marker Referencing System. California PATH Program, Berkeley, 2000.
18. Micoski, M., "Lateral Displacement Calculation Algorithm for a Magnetic Guidance System," SAE Technical Paper 2007-01-2833, 2007, doi:10.4271/2007-01-2833.
19. Guldner, J., Tan, H.S.; Patwardhan, S. On Fundamental Issues of Vehicle Steering Control for Highway Automation. California PATH Program, Berkeley, 1997.
20. Ogata, K., Modern Control Engineering, Prentice Hall, New Jersey, ISBN-13: 978-0130609076, 2001.
21. Welch, G., Bishop, G., An Introduction to the Kalman Filter. Department of Computer Science - University of Noth Carolina at Chapel Hill, 2006.

INFORMATION ABOUT AUTHORS

Leopoldo Rideki Yoshioka: Graduated in Electronic Engineering from Aeronautical Institute of Technology, Brazil, Master and Doctor from Tokyo Institute of Technology, Japan. lryoshioka@lps.usp.br.

José Roberto Cardoso: Graduated in Engineering and Electricity from Sao Paulo University, USP. Master and Doctor from USP. jose.cardoso@poli.usp.br

Mauricio Micoski: Graduated in Electronic Engineering from Aeronautical Institute of Technology, ITA, Brazil, Expert in Software Engineering from Campinas University, UNICAMP and Master in Electronic and Computer Engineering from ITA. mauricio.micoski@compsisnet.com.br .

Renato Duarte Costa: Graduated in Electronic Engineering from Aeronautical Institute of Technology, ITA, Brazil, and Master in Electronic and Computer Engineering from ITA. renato.costa@compsisnet.com.br .

Edson Rodrigues: Graduated in Communications from ESPM, Sao Paulo, Brasil. Master in Planning and Regional Development Management from Taubate University, UNITAU, Sao Paulo, Brazil. edson.rodrigues@compsisnet.com.br

ACKNOWLODGMENTS

To FINEP (www.finep.gov.br) from financial support of research project "Embedded System and Software for Automatic Steering of Buses in Express Corridors", contract No. 01.07.06.57.00.

To Professor Doctor Antonio Teixeira Junior of Vale do Paraiba University, UNIVAP, Sao Jose dos Campos, Brazil, from support to Magnetic Guidance Project.

DEFINITIONS/ABBREVIATIONS

AVGS	Automatic Vehicle Guidance System
BRT	Bus Rapit Transit
DGPS	Differential GPS
FINEP	Study and Project Support Agency, Rio de Janeiro, Brazil
GPS	Global Positioning System
IP&D	Institute of Research and Development at UNIVAP, Brazil
ITS	Intelligent Transportation System
LRV	Light Rail Vehicle
UNIVAP	University of Vale do Paraiba, Sao Jose dos Campos, SP, Brazil
USP	University of Sao Paulo, Brazil
SGO	Optical Guidance System

Development of HMI and Telematics Systems for a Reliable and Attractive Electric Vehicle

2011-01-0554
Published
04/12/2011

Shoichi Yoshizawa, Yoichiro Tanaka, Masahiro Ohyamaguchi, Satoshi Kitazaki, Kouichi Kuroda, Shinpei Sato, Tetsu Obata, Yuumi Hirokawa, Masayasu Iwasaki and Kenji Maruyama
Nissan Motor Co., Ltd.

doi:10.4271/2011-01-0554

ABSTRACT

This paper describes the HMI, navigation and telematics systems developed specifically for the Nissan LEAF electric vehicle to dispel drivers' anxieties about operating an EV. Drivers of EVs will need to understand various new kinds of information about the vehicle's operational status that differ from conventional gasoline-engine vehicles. Additionally, owing to the current driving range of EVs and limited availability of charging stations, drivers will want to know acccurate the remaining driving range, amount of power and the latest information about charging station locations. It will also be important to ensure that people unfamiliar with EVs will be able to operate them easily as rental cars or in car-sharing systems without experiencing any inconvenience. These needs have been met in the Nissan LEAF mainly by prioritizing displayed information, adopting a combination main meter-navigation system display and providing a two-way communication capability along with real-time information.

INTRODUCTION

In December 2006, Nissan announced a medium-term environmental action plan called the Nissan Green Program 2010. In addition to continued efforts for improving the efficiency and fuel economy of gasoline-fueled engines, this program also calls for the development of hybrid and plug-in hybrid cars, fuel-cell vehicles and electric vehicles (EVs) as part of the company's medium to long-term vision. Full-scale production of the Nissan LEAF EV was launched in September 2010 followed by the start of sales activities.

Electric vehicles (EVs) represent a new type of automobile in which drivers will need to understand various new kinds of information presented about the vehicle's operational status that differ from what is ordinarily shown in conventional gasoline-engine vehicles. Such information includes, for example, the battery state of charge (SOC), the output state of the drive motor, and sometimes the state of the charging system. It will probably take drivers a certain amount of time before they become accustomed to and readily understand these new types of information displayed about the operational status of an EV. However, because the displayed information may also include warnings, drivers will have to be able to understand the presented information instantly. It will be necessary to ensure that even first-time drivers of EVs can understand the displayed information immediately and know what to do if a problem occurs, so that they are not inconvenienced in any way.

Another factor to be considered is that EVs have a shorter driving range than conventional vehicles owing to the present level of battery capacity. Consequently, when traveling longer distances or when the battery SOC starts to become not enough, drivers may constantly worry about how they should drive or about the remaining battery charge so as not to run out of electricity. The availability of charging facilities is also an issue. The number of charging stations is expected to increase rapidly under the green policies being promoted by national governments, local municipalities and businesses, and EV charging facilities are continually being installed in new locations. Two types of battery chargers are now generally available -a quick-charge type and an ordinary type. Drivers will have to know the locations of charging stations before they depart, and if they should need to charge the battery en route, it will be helpful to provide them with the latest information on nearby charging stations.

Typical examples include use as rental cars and in car-sharing systems, even people who are not normally used to operating an EV will want to start driving immediately toward their destination in ordinary city traffic without being concerned about travel time as they look for their route. Naturally, people must be able to operate EVs smoothly and easily right away without experiencing any unnatural feeling or inconvenience.

With that aim in mind, driving tests using EV development mules and simulations using digital mockups were conducted to identify thoroughly any elements that might cause anxiety in drivers in various driving situations. The results were carefully analyzed to identify major anxiety-producing factors, and effective measures were determined for dealing with them. Three approaches were taken in this regard. First, display positions were determined in terms of the priority and grouping of information so that drivers can easily see and comprehend the presented information. Second, the displays of the main meter and navigation system were linked in a combination system, and pull-type and push-type information displays were adopted so that drivers can easily notice the displayed information that is intended to prompt suitable action by them. Third, thoroughgoing driver support is provided through the telematics system, including a two-way communication capability via a mobile phone or the Internet and the provision of real-time information to facilitate remote operation.

These functions have been implemented on the Nissan LEAF electric vehicle by adopting digital color LCD twin meters and dedicated EV navigation and telematics systems. The result is a reliable and attractive HMI display system that is designed to dispel drivers' anxieties about operating an EV.

DETERMINATION OF DISPAYED INFORMATION AND DISPLAY POSITIONS

This work began from the advanced development stage of the HMI display system. The discussions extended through the launch of the Nissan LEAF development project and continued until conclusions were reached in the digital planning phase when decisions were made about the information items to be displayed for expressing the vehicle's operational status. Subsequently, in the physical engineering phase that included driving tests, extensive public road tests of the Nissan LEAF were conducted in Japan, the U.S. and Europe. The purpose was to identify once again any factors that might cause driver anxiety and to confirm the effectiveness of the measures designed to overcome them.

IDENTIFYING POTENTIAL WORRIES AND THE INFORMATION DRIVERS DESIRE ALONG WITH PURSUING EASY-TO-UNDERSTAND DISPLAYS AND EASY TRANSMISSION OF INFORMATION

It was found that the remaining driving range is the biggest worry. It was also learned that drivers want to know the locations of charging stations and desire clear information on what has happened when some type of system failure occurs and what they should do about it. It was concluded that easy-to-understand displays could be achieved by presenting information in simple terms without any unnatural feeling.

Easy transmission of information was treated in the context of the methods used to present displayed information. The reason for taking these approaches is that drivers' attention can vary greatly because they are listening to music, talking with passengers or thinking about something else while driving.

CONCEPT OF INFORMATION ZONES FOR DETERMINING DISPLAY POSITIONS

Three areas in the cockpit, designated as Zone A, Zone B and Zone C in Fig. 1, were considered as locations for displaying information.

Zone A: HUD and Upper Meter Location

The information displayed in this zone has a high degree of priority and should be presented in the driver's effective field of vision. Displaying information here in the driver's forward view enables drivers to obtain essential information effortlessly at all times. However, dashboard layout limitations make it difficult to secure sufficient display space.

Zone B: Main Meter Location

Information on the vehicle's operational conditions is displayed here so that it can be viewed instantly at a glance. Drivers are accustomed to this traditional location for meters and gauges. However, information is displayed at a lower viewing angle than the driver's effective field of vision for looking ahead of the vehicle. Consequently, in terms of visibility, drivers unconsciously experience a certain workload in viewing information in this zone compared with Zone A.

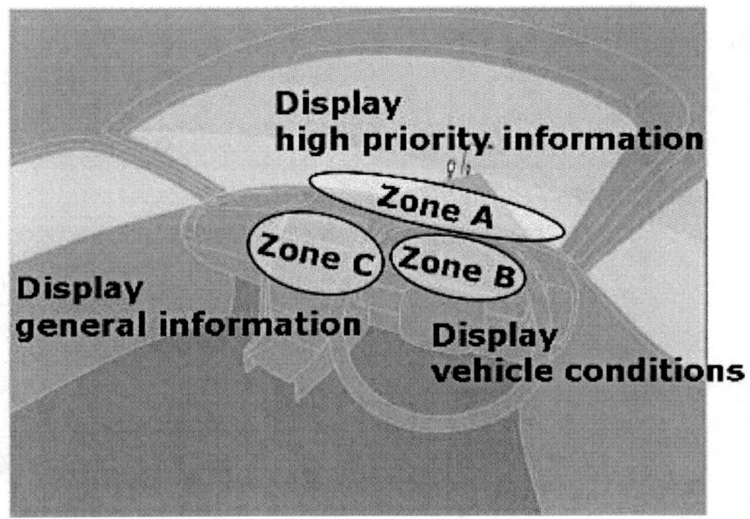

Figure 1. Cockpit zoning concept for displaying information

Zone C: Navigation System Display Location

General information of low priority is presented in this zone where the navigation system display is usually located. This information can also be seen by the front-seat passenger and even sometimes by rear-seat occupants. Ordinarily, the information displayed here includes navigation maps and the controls of the audio system and air-conditioning system. The navigation display screen is markedly larger in size than the displays in the other two zones, so it has the potential for expanding the amount and kinds of information displayed. However, it is not normally used to present equipment failure messages because that might cause passengers unnecessary anxiety. One exception here is a message to the driver to stop the vehicle immediately.

ATTRACTIVE INFORMATION DISPLAYS DEVELOPED FOR THE NISSAN LEAF

The information to be displayed in the Nissan LEAF and the display positions were determined on the basis of the foregoing zone concept.

Upper Meter Location

It was decided to display the driving speed, outside temperature and eco-driving advice here. Drivers often check this information because it is of high interest to them.

Main Meter Location

It was decided to position the power meter, SOC gauge, driving range, the battery temperature gauge and warnings here. This information was carefully selected as being items that drivers need to know regarding the operational status of the EV.

Navigation Display Location

Information items unique to an EV were added to the conventional navigation system functionality. These include a display of a map display of the driving range, a map display of charging station locations and a screen for setting various EV functions.

The digital color LCD twin meters adopted for the Nissan LEAF are shown in Fig. 2. Under ordinary circumstances, the driver only needs to look at the upper meter. Eco-driving advice is also shown visually by the eco-indicator. If the vehicle develops a problem, the master warning lamp provided in the upper meter illuminates to tell the driver that a warning message is displayed on the main meter.

UNIQUE EV WARNING LAMPS AND DISPLAYED INDICATIONS

There are ten warning indicatons lamps altogether that are unique to this EV (Figure 3). Two warning lamps were newly added to the Nissan LEAF as a result of re-examining all the warning indications developed previously for Nissan EVs. One is a "Head lamp warning light" indicating owing to the adoption of LED head lamps. The other is a "Ready" indicating that the vehicle is ready to drive. In our previous EVs, a "Ready" indicating was displayed. Since it is planned to market the Nissan LEAF globally, it was decided to use a symbol mark indicator in place of the word indicator.

Upper Meter

Lower Meter

Figure 2. Digital color twin meters

	Color	Symbol	Meaning
12-volt battery charge warning light	Red		12-volt battery needs to be charged
Plug in indicator	Green		Charge connecter is connected
Ready operation indicator light	Green		Vehicle is ready to drive.
Power limitation indicator light	Yellow		Motor power is limited.
EV system warning light	Yellow		Failure of EV system.
Electric shift warning light	Red		Failure of electric shift
Head light warning light	Yellow		Failure of LED head lights
Regeneratve brake warning light	Yellow		Failure of cooperative regenerative brake
Electric parking brake warning light	Red	PARK (USA) (P) (Eu Japan)	Failure of electric parking brake
Low battery charge warning light	Yellow		Electric energy is getting low

Figure 3. 10 dedicated EV warning indications

UNIQUE EV DISPLAY CONTENTS

The SOC meter shows the remaining battery energy as a ratio of the total battery capacity (Figure 4). The principle adopted here is the same as that of mobile phones and notebook PCs. The narrow indicator along the right side is a distinctive point in that it shows the battery's total capacity at any given moment. This information enables the driver to know the state of battery degradation.

The power meter indicates the output level of the drive motor and replaces the tachometer in a conventional gasoline-engine vehicle (Figure 5). It also indicates the amount of power produced by cooperative regeneration during vehicle deceleration and braking. The driver can check the amount of power being consumed or regenerated in real time while driving.

The eco-indicator is provided to encourage drivers to drive in aneco-friendly manner (Figure 6). The shape of the outside ring indicates the present state of eco-driving in real time, thereby providing guidance to the driver about proper acceleration and braking, air-conditioner settings and other aspects. In addition, the eco-trees show the cumulative level of eco-driving per trip, which is intended to motivate drivers to improve their eco-driving style.

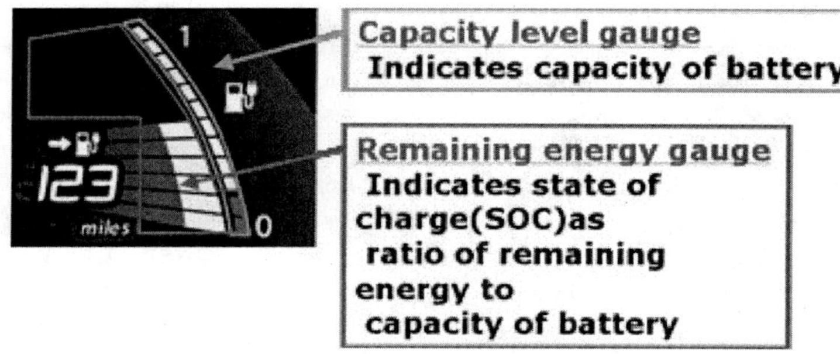

Figure 4. SOC Meter

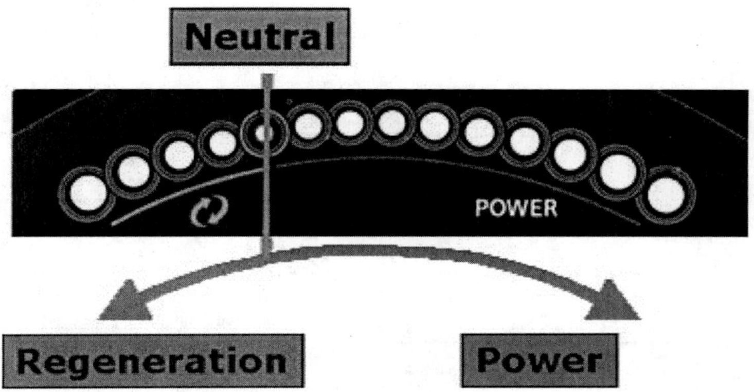

Figure 5. Power meter

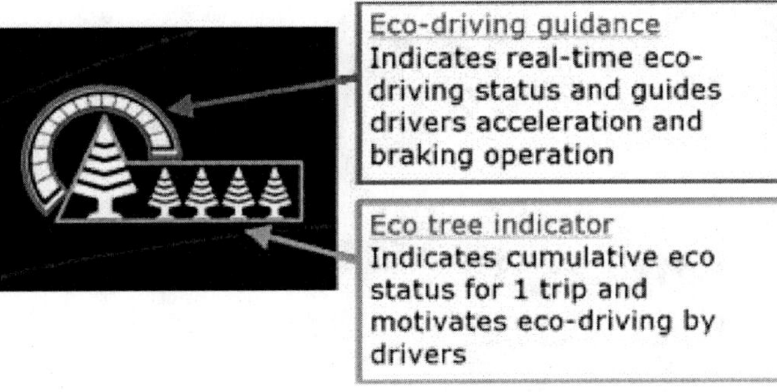

Figure 6. Eco-driving gui dance and eco tree indicator

The outside ring of the eco-indicator shows the real-time eco-driving status (Figure 7). The 15 levels of the indicator change when the vehicle stops, accelerates or decelerates. It functions to tell the driver intuitively the vehicle's power consumption in every operational state.

☆: Guide
0: Not guide

	Stopped	Accelerating/ Cruising	Braking
Climate control	☆	☆	☆
Idling time length	☆	0	0
Accelerating	0	☆	0
Braking	0	0	☆

Figure 7. Guidance details in different modes

COMBINATION MAIN METER-NAVIGATION SYSTEM DISPLAY

The foregoing discussion has described the displayed information and its grouping in three different zones. This approach to presenting information enables the driver to readily obtain easy-to-understand information about the vehicle's operational status. The Nissan LEAF adopts a combination main meter-navigation system display that ensures critical information is conveyed to the driver without fail. This linked presentation of information has two attractive aspects in the form of emotional value and functional benefits.

The emotional value of this combination display technology was evaluated in a driver survey using the three parameters of advanced technical impression, knowledge system impression and impression of connectivity with the vehicle. Interesting results were obtained in that all three parameters received a positive evaluation in the range of 60-70%. Evaluation results for the functional benefits revealed that in the case of a navigation system display without an audible alarm, drivers were slow to respond to a message displayed on the screen. In contrast, drivers responded more quickly when an indication was also simultaneously shown in the main meter. This result confirmed the effectiveness of the combination display. Moreover, drivers' response time was markedly improved by the addition of an audible alarm or a voice announcement (Figure 8). [1] [2] [3]

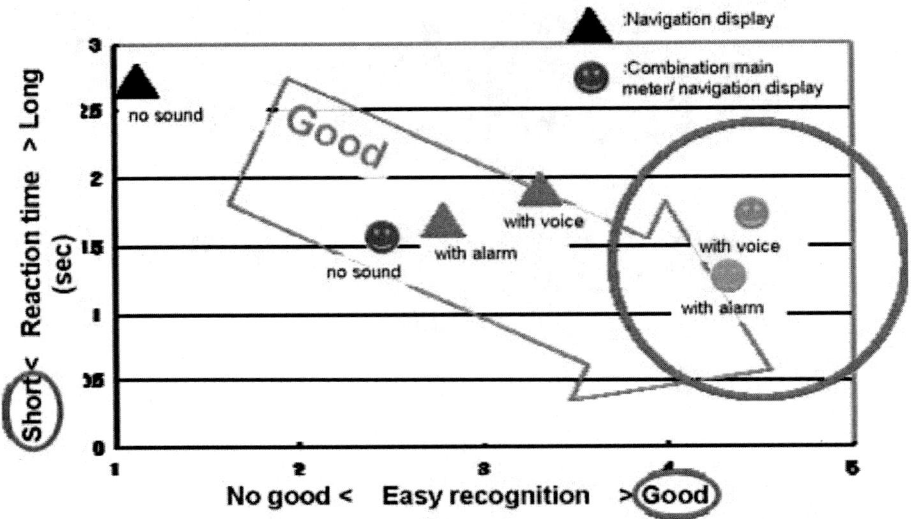

Figure 8. Survey results for effectiveness of combination display

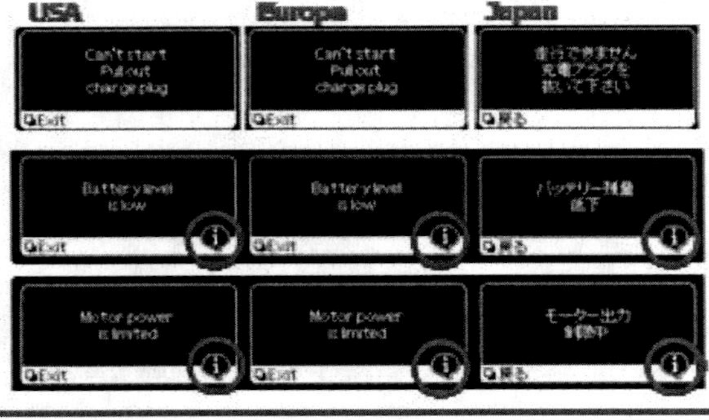

Figure 9. Multi-function display Combination displays for warnings

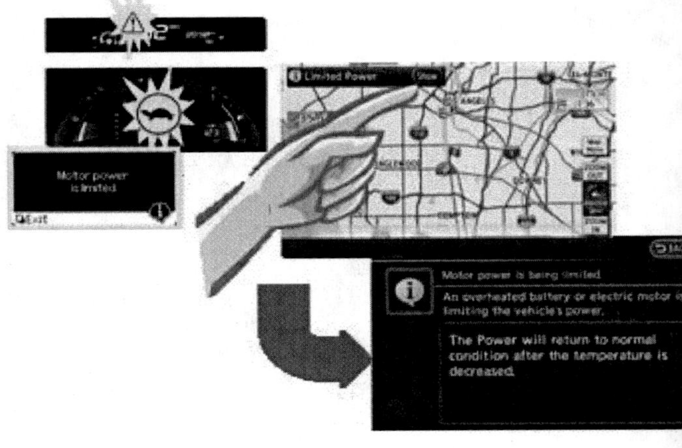

Figure 10. Limited power warning

Alarm indications that should be presented to the driver considerately and without fail were narrowed down on the basis of three perspectives: degree of necessity, frequency and availability of some alternative means. Two indications were selected-low battery level and power output limitation due to four reasons (Figure 10, 11, 12, 13). In each case, the corresponding warning indicator is illuminated in the main meter as a push-type display (Figure 9) and simultaneously an icon is illuminated to tell the driver that supplementary information is shown on the navigation system screen. If necessary, the driver can obtain more detailed information by touching the screen (pull-type display). The displayed information not only provides a supplementary explanation of the nature of the warning, it also tells the driver what to do next. Navigation guidance is also presented, and the system will connect the driver to a service operator in the event that the driver is overwhelmed by the situation.

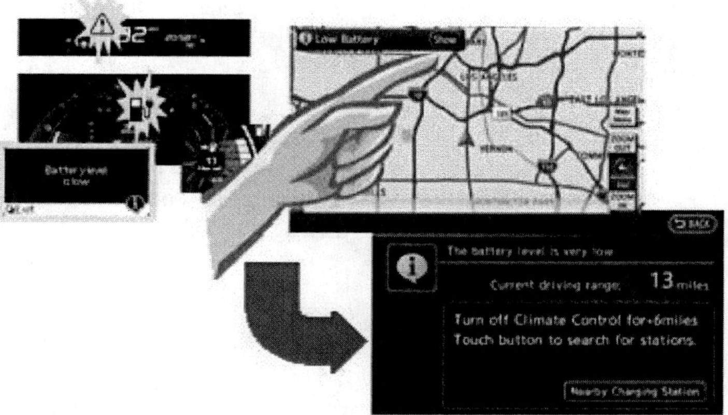

Figure 11. Charging station map and advice when SOC is low

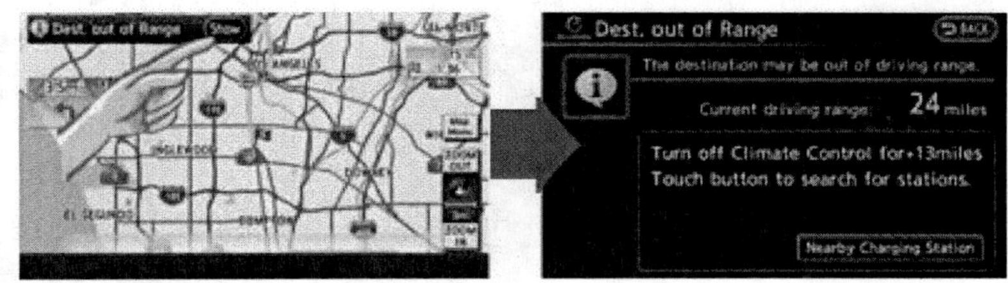

Figure 12. Charging station map and advice when vehicle can't reach destination with

Figure 13. Charging station map and advice when driving range meter shows "--

UNIQUE EV NAVIGATION AND TELEMATICS FUNCTIONS

The foregoing discussion has explained the display technologies embodied in the Nissan LEAF to dispel drivers' anxieties about EVs and to enable them to concentrate on driving with full trust in the vehicle. This section explains the EV navigation and telematics functionality that is also provided to support the operation of this EV. This has been achieved by adding three new dedicated functions that are both essential and attractive for driving an EV.

(1) Map display of estimated driving range: The estimated driving range is shown conceptually on a map as concentric circles with the vehicle's present position as the origin. If the destination has been entered in the navigation system, the driving range from that location can also be displayed in the same way. Useful information is displayed visually and intuitively. Since drivers will probably use the driving range display frequently, this information can also be displayed immediately by pushing a switch on the steering wheel pad (Figure 14).

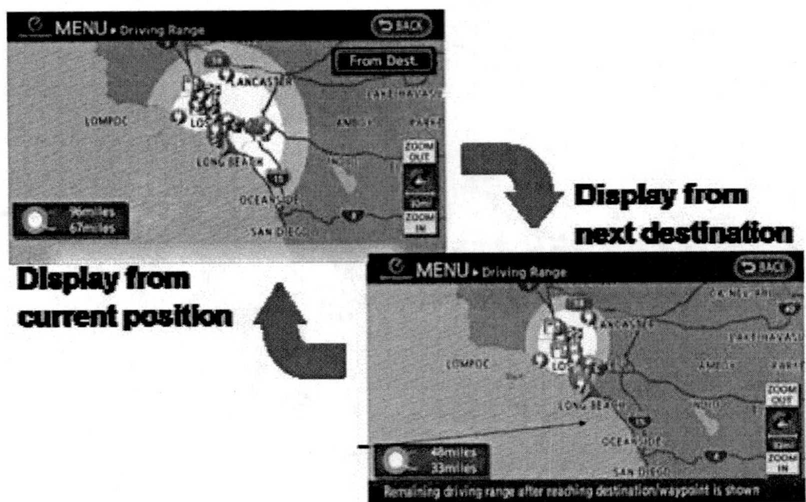

Fgiure 14. Driving range display

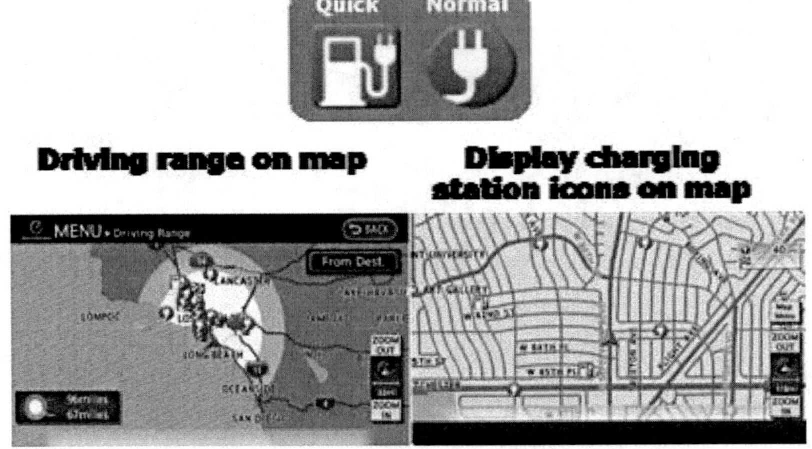

Figure 15. Map showing charging station icons

(2) Support for judging the need for battery charging and for using charging stations: Besides helping the driver judge whether the battery needs charging, information is also presented regarding the latest charging station installations. The locations of ordinary-type and quick-type chargers are shown on a navigation screen map. Detailed information about the charging stations is also displayed, including the connector type, number of connectors and charging fees. The latest information on charging stations can be updated either automatically or manually (Figure 15).

(3) Additional remote functionality: Drivers also have 24-hour connectivity to a telematics center via their mobile phone, enabling them to access various handy functions (Figure 16). Specifically, they can charge the battery or activate the air-conditioning system in advance through remote control, obtain an email message when charging is completed or also confirm the battery SOC. This two-way communication capability with the vehicle via a mobile phone enables drivers to check the vehicle's condition anywhere, anytime along with charging the battery or operating the air-conditioner remotely (Figure 17).

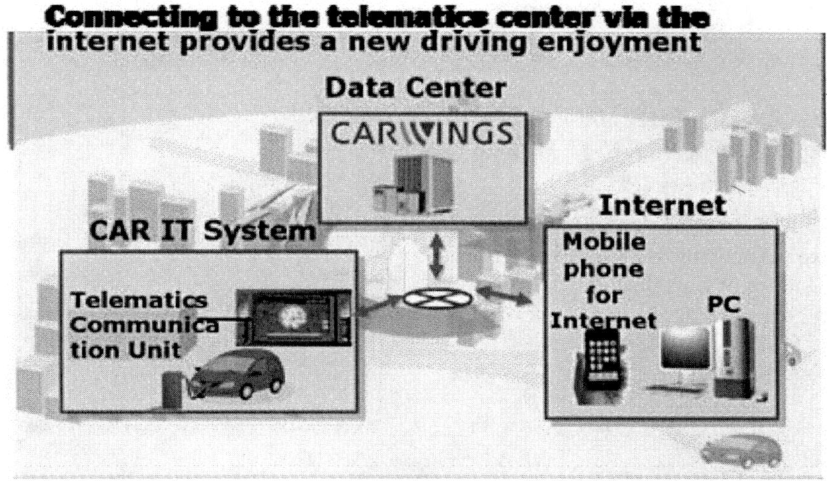

Figure 16. Benefits of telematics

Figure 17. Benefits of telematics

SUMMARY/CONCLUSIONS

This paper has described some of the convenient functions that have been achieved in the Nissan LEAF by adopting digital color LCD twin meters and dedicated EV navigation and telematics systems. The adoption of the many new functions explained here will enable drivers to begin driving the Nissan LEAF without any prior special preparations, trusting the vehicle just as they would conventional automobiles and charging the battery themselves in the course of using the vehicle. The HMI display system is designed to support comfortable, enjoyable driving with smooth, powerful performance provided by the electric drive motor. There is the additional attraction of zero-emission mobility made possible by charging the battery at home or elsewhere without being dependent on gasoline stations.

REFERENCES

1. Yasunori, Maruyama et al. (2008), A study of useful information for driver assistance, JSAE, No. 146-08 Page 5-10(2008.10.22)

2. Hiroshi, Takahashi et al. (2000), A study on intervention timing of the automated driver assistant system to driver manual operations. From the point of trust between human and machine, IEIC Technical Report Vol.100 No. 214(SSS2000 7-13) Page21-26(2000.07.25)

3. Deguchi, Yoshitaka et al. (2003), HEV Charge/Discharge control system based on car navigation information, JSAE, No.29-93 Page1-4(2003.05.21)

CONTACT INFORMATION

Address: 560-2, Okatsukoku, Atsugi-shi, Kanagawa 243-0192, Japan

Nissan PV Product Development Division No. 1

shoichi-y@mail.nissan.co.jp

Name: Shoichi Yoshizawa

DEFINITIONS/ABBREVIATIONS

EV
Electric vehicle

SOC
State of charge

Applications

Commercial Business Viability of IntelliDrive[SM] Safety Applications

2010-01-2313
Published
10/19/2010

Robert White, Tao Zhang, Paul Tukey and Kevin Lu
Telcordia Technologies

David McNamara
MTS LLC

ABSTRACT

This paper presents modeling, analysis, and results of the business viability of a set of IntelliDrive [1] safety applications in a free market setting. The primary value drivers for motorists to adopt the IntelliDrive system are based on a set of safety applications developed and analyzed by the US DOT. The modeling approach simulates IntelliDrive on-board equipment adoption by motorists based on the value of the safety applications. The simulation model uses parameters that are based on adoption rates in a similar dynamical system from recent history and incorporates feedback loops such as the positive reinforcement of vehicle-to-vehicle applications value due to increased adoption. This approach allows the analysis of alternative IntelliDrive business approaches, deployment scenarios, and policies. The net present value of the IntelliDrive system to the nation is computed under alternative scenarios.

INTRODUCTION

The Promise of IntelliDrive is Improved Safety and Mobility

IntelliDrive is a U.S. Department of Transportation (US DOT) initiative to develop and demonstrate technologies for using wireless communications technologies to improve transportation safety, mobility, and sustainability. The nation-wide deployment represents a significant investment by a diverse set of stakeholders, both public and private. As described by the US DOT outreach website, "IntelliDrive aims to enable safe, interoperable networked wireless communications among vehicles, the infrastructure, and passengers' personal communications devices." IntelliDrive will ultimately enhance the safety, mobility, and quality of life of all Americans, while helping to reduce the environmental impact of surface transportation.

IntelliDrive as the Vehicle Infrastructure Integration (VII) was formally announced at the ITS America's 2003 Annual Meeting. The US DOT gathered stakeholders from state DOTs, automobile manufacturers, and others to create a working group representing the public and private interests and to create a consensus over deployment. Many at that time expected that VII would be a federally funded project of the magnitude of the US Interstate Highway System; it would be a "wireless network" that covered our nation's highways and intersections. Since that time, several successful test beds and demonstrations have been funded, but no consensus or plan for a nation-wide deployment has emerged. Today, IntelliDrive remains largely a federally funded set of research projects.

As IntelliDrive matures beyond the research phase, a crucial issue becomes a viable business case for the nation-wide deployment of IntelliDrive systems and applications. Studies have shown that IntelliDrive applications, when widely deployed, can provide significant economic benefits [1]. However, it is also widely recognized that some important IntelliDrive applications, such as vehicle safety applications based on vehicle-to-vehicle communications, will provide benefits to drivers only after a high percentage of all the vehicles are equipped with the same applications, which can take many years and require heavy investments. Similarly,

[1]The IntelliDrive[SM] logo is a service mark of the U.S. Department of Transportation (US DOT).

traffic signal phase and timing (SPAT) applications based on Dedicated Short-Range Communications (DSRC) will generate significant benefit only when a large number of the dangerous intersections are equipped with DSRC communications capabilities.

Collaboration and Investment possible with a Viable Business Model

Further complicating matters is the fact that supporting IntelliDrive applications requires collaboration among many industry sectors, private and public. For example, automotive manufacturers need to install communications and applications capabilities on vehicles. Network operators and transportation agencies need to collaborate to deploy roadside network infrastructures such as DSRC equipment at intersections. Network providers, software providers, and automotive manufacturers need to collaborate to establish the enabling infrastructure required for vehicle communications, such as the public key infrastructure for supporting security for vehicle communications. Device makers will deploy IntelliDrive applications on after-market devices. As illustrated by these examples, supporting any set of IntelliDrive applications will require multiple parities to deploy different pieces of an integral system.

Given that many IntelliDrive applications will provide significant value only when either a large number of vehicles or a large infrastructure network is deployed, stakeholders have been reluctant to jump in with their respective investments. Automotive manufacturers want to see roadside infrastructures be deployed before deploying onboard equipment. Parties involved in deploying infrastructure networks don't want to invest in the deployment and wait through a long uncertain period of time before their deployment can generate economic benefits.

The experience of the authors is that before an endeavor of this magnitude is undertaken -- installation and on-going maintenance of a vast transportation system -- key business questions must be addressed:

• Who will benefit?

• Who can invest?

• What and when is the pay back: i.e., what is the business model?

• When and how will the project be launched; i.e., the business plan?

These important questions are typically addressed as part of a business plan which details the cost of deployment and projects the return on investment over-time. We think the current issues with deployment stem from a lack of a viable and credible business plan. The attitude of "build it and they will come" is even more unrealistic in light of the current global financial crisis. An IntelliDrive business plan, including the formal statement of a set of goals, the rationale for why they are attainable, and the financial and operational plans for reaching these goals, is of paramount importance today. Fortunately, many in the community see this need and the dialogue is now underway, with deployment ideas being considered. The authors consider the model detailed by this paper an important element in creating a viable IntelliDrive business plan - assessing and quantifying the benefits in credible financial terms, is the cornerstone.

System Dynamics as a Tool to Model and Quantify Benefits

In this paper, we study the commercial viability of deploying safety applications. We present a business-modeling tool that can be used to answer the fundamental question: driven by commercial markets, what IntelliDrive business models and deployment strategies will be viable and practical?

A major challenge in understanding the business viability of deploying complex systems and applications, such as IntelliDrive systems and applications, is to model the complex interactions of the many factors that impact the business cases. As discussed above, the value to the users depends on the set of applications and how widely they are deployed. Some applications can provide value only after a high percentage of other vehicles are equipped with the same capabilities or when a large roadside network infrastructure is implemented, while other applications may provide value even when a small number of vehicles are equipped with the applications. As the deployment of the IntelliDrive systems and applications grow, the value to the users will grow; recognizing that this user value growth is typically not linear. As the user value grows, more and more users will be motivated to join the system, further increasing the system value. We present in this paper a business case modeling tool that uses system dynamics techniques [2] to model and analyze the interrelations among the many business impacting factors.

The model can support any combination of applications. For the results presented in this paper, we focused on the safety applications in the Volpe study [1] : Signal Violation Warning, Stop Sign Violation Warning, Curve Speed Warning, and Electronic Braking Lights. We further considered the impact of deploying more applications to increase the value provided to the users.

VALUE OF THE INTELLIDRIVE SYSTEM

The John A. Volpe National Transportation Systems Center in the US DOT has published a study that estimated the benefits of a set of VII[2] safety applications based on reduced

Table 1. National benefits and present value per vehicle of VII safety applications from the Volpe study

Safety Application	Type	National Benefit ($B)	Value per Vehicle ($)
Electronic Brake Lights	V2V	14	54
Signal Violation Warning	RSE/Intersection	11	44
Stop Sign Violation Warning	RSE/Intersection	3	11
Curve Speed Warning	RSE or Database	15	58
Total		42	168

crashes and other benefits under the assumption that intersections and vehicles would be equipped over time under an assumed deployment of vehicles [1]. That study calculated a benefit for each of the safety applications as a present value, in which the benefits from reduced future crashes were discounted to the present day as a total for the entire United States. The safety applications and the projected benefits (in billions of 2008 dollars) are shown in Table 1.

In the current analysis, we are interested in studying the effect of the value of the safety applications on the decisions of individual motorists to adopt the system. This requires a value that the motorists would gain by having an On-Board Equipment (OBE) on their vehicles. The average value of the system per vehicle shown in Table 1 is derived by dividing the national benefit (the national present value of the safety applications) by the current number of vehicles in use in the US (250 million).

The "Intersection" type safety applications rely on an IntelliDrive Road-Side Equipment (RSE) being installed in an intersection for an equipped vehicle to receive transmissions that carry warnings and other information. This means that a significant portion of all intersections need to be equipped before equipped vehicles can receive benefits.

"Electronic Brake Lights (EBL) is a Vehicle-to-Vehicle (V2V) application that would provide a warning to the driver in case of the sudden deceleration of a forward vehicle. The OBE of the lead vehicle would send a signal to other vehicles if its longitudinal deceleration exceeds a predetermined threshold, thereby allowing those following drivers to be aware of this deceleration even if their visibility is limited by weather conditions or obstructed by large vehicles."[3] The value of V2V applications depends on the fraction of all vehicles that are equipped to support the application. This is an externality similar to that of a communications network, in which the benefit of joining the network grows as the network size grows.

"The Curve Speed Warning (CSW) application provides an in-vehicle warning to the driver if the vehicle's speed is higher than the recommended speed for the curve. The system can be designed to receive the information from an RSE or to use the OBE and a downloaded navigation map to make an assessment. In the first case, the RSE compares the vehicle speed with the recommended speed and sends a signal to the vehicle if there is a potential danger. In the latter case, the OBE compares the vehicle speed to the recommended speed that is stored with the navigation map data. Road condition data can also be used in this process to fine-tune the speed warning based on weather and other factors."[4]

Other safety and commercial applications envisioned can be added to the system over time. These include electronic payment for tolls as well as goods and services, and private applications that could be installed either by an OEM or in the aftermarket.

The current modeling and analysis envisions the initial OBEs will be available in the aftermarket to provide the safety applications listed in Table 1. Over time, additional applications will become available and increase the value of the system. As the system is adopted by increasing numbers of motorists, OEMs will be incented to offer an IntelliDrive option on new vehicles.

MODEL STRUCTURE

The approach is to model the growth of value to motorists as intersections are deployed with RSE. Additional value will accrue as OBE are deployed on vehicles and the benefits of V2V applications such as EBL increase. Some safety applications such as CSW can be available as soon as the system is initiated. The general model structure is shown in Figure 1.

---

[2] Vehicle Infrastructure Integration (VII) was the prior name given to IntelliDrive by the US DOT.

[3] See reference [1]

[4] See reference [1]

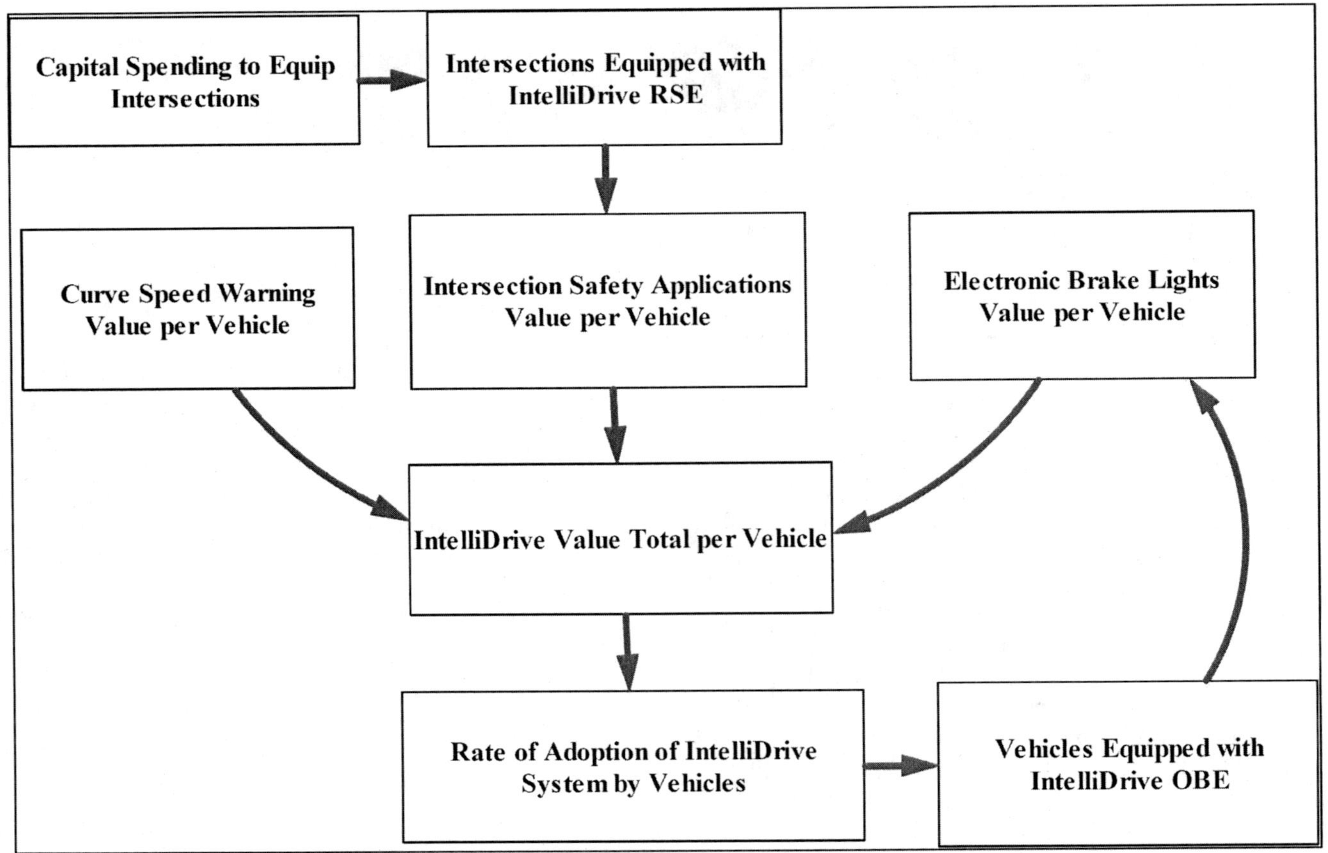

Figure 1. Model influence diagram showing contribution of value flows from safety applications

The model dynamics are based on the underlying assumption that the growth in system adoption is proportional to its average value to the motorist. System value comes from the value of the three groups of safety applications shown in Table 1. The model was implemented in the Vensim system dynamics modeling language [3].

The value of intersection safety applications is driven by the deployment of intersections. Capital spending by the government drives the deployment of RSEs in intersections and other locations. As the number of equipped intersections increases, the value to motorists increases. This value increase, in turn, will motivate more motorists to install the OBE on their vehicles.

The value to motorists of the CSW application can be realized as soon as the system is initialized if this safety application is map driven. The OBE can estimate safe curve speeds from its current position on a map. A warning is issued if the OBE detects that the speed of the vehicle entering the curve exceeds the calculated safe limit. In the current analysis, CSW is an RSE-based application.

The value of V2V safety applications such as EBL is directly tied to the number of other vehicles that are equipped. As the number of equipped vehicles grows, the chances of a vehicle encountering another equipped vehicle increases; this, in turn, increases the value of EBL adding to the value of the entire system.

The question of setting the parameters in the IntelliDrive penetration model is difficult because the system has not yet begun to be deployed and decisions by motorists have not yet been made. Similar systems for which we have a full history of deployment and customer adoption can be used to provide approximate parameters if care is taken in correctly mapping the coefficients. One such system is the E-ZPass electronic tolling system that was originally deployed in New York and New Jersey in the 1990's [4, 5, 6]. We set the functional form in the E-ZPass penetration growth model identically to that in the IntelliDrive penetration growth model. In each case, the penetration growth is proportional to the average value of the system per vehicle. This common structure allows us to use the calibrated E-ZPass coefficients in the IntelliDrive model. By transferring these coefficients we are assuming that motorists' reaction to average value per vehicle is the same

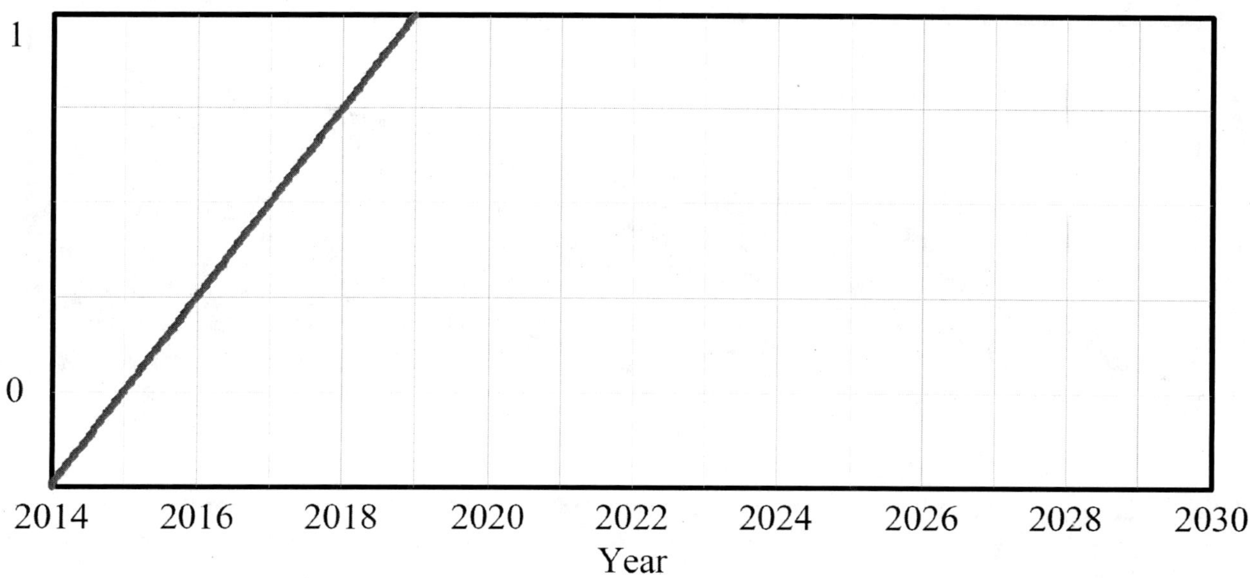

Figure 2. Build-out of RSE locations (the fraction of designated locations that are equipped is on the y-axis)

for both systems. This model calibration is discussed in Appendix 1.

INTELLIDRIVE PENETRATION MODEL BASELINE RESULTS

The baseline scenario for the IntelliDrive Penetration Model (IPM) is based on the Volpe study assumptions in [1]. However, since the business drivers are fundamentally different in the current commercial business model [7] than the model assumed in the Volpe study, some basic assumptions will be different. For example, the Volpe study assumes a deployment schedule of new light vehicles based on a prior US DOT ITS Joint Program Office study [8]. In the current model, the schedule of vehicle OBE adoption is not an input, but is calculated as an output.

Contrasting to the Volpe study that assumed 100% of all vehicles would adopt the system regardless of the value, the ultimate IntelliDrive system penetration in a commercial model cannot be known at this point. Market research can be employed to estimate the market potential as a function of price and features. We have assumed 80% ultimate penetration for the baseline scenario.

The Volpe study assumed a five-year build-out of RSE locations starting in 2014 and continuing through 2018. We interpret this to mean that all 252,000 sites for RSEs identified in the Volpe study are equipped in the five-year rollout. Figure 2 shows the assumed build-out of RSE locations.

The value of the system grows as RSE locations are developed and equipped. The resulting aggregate value from all safety applications is shown in Figure 3. This chart shows value from EBL, which begins to accrue as vehicles are equipped. We included "other" applications in the baseline scenario that add value to the system without identifying these explicitly. Other applications are assumed to be introduced in 2014 and add $5 per vehicle in value per year thereafter.

The value of the RSE safety applications builds as locations are equipped between 2011 and 2016; after 2016 it holds flat at $122 per vehicle. In the baseline analyses, we assume that CSW is an RSE application. Alternate scenarios can consider CSW being based on location and downloaded map data. In the later case, the value of the CSW application would be constant at $59 per vehicle, and becomes available on day one.

The Volpe study [1] uses $50 for the cost of the installed OBE. Volpe notes that they have received comments that this cost may be too low. The initial OBE in the current study is envisioned as an aftermarket unit that motorists will purchase and install. Typical of such products, the initial cost will be significantly higher than the long term cost. In the baseline scenario, we have assumed that the initial cost of the OBE will be $200, dropping over time to $50. The rate at which the cost will drop was taken in this model to be independent of annual sales in order to allow this to be a control variable. The government can influence the cost drop rate through policies that, for example, require government fleets to install OBEs, thereby insuring sales and incenting a cost drop.

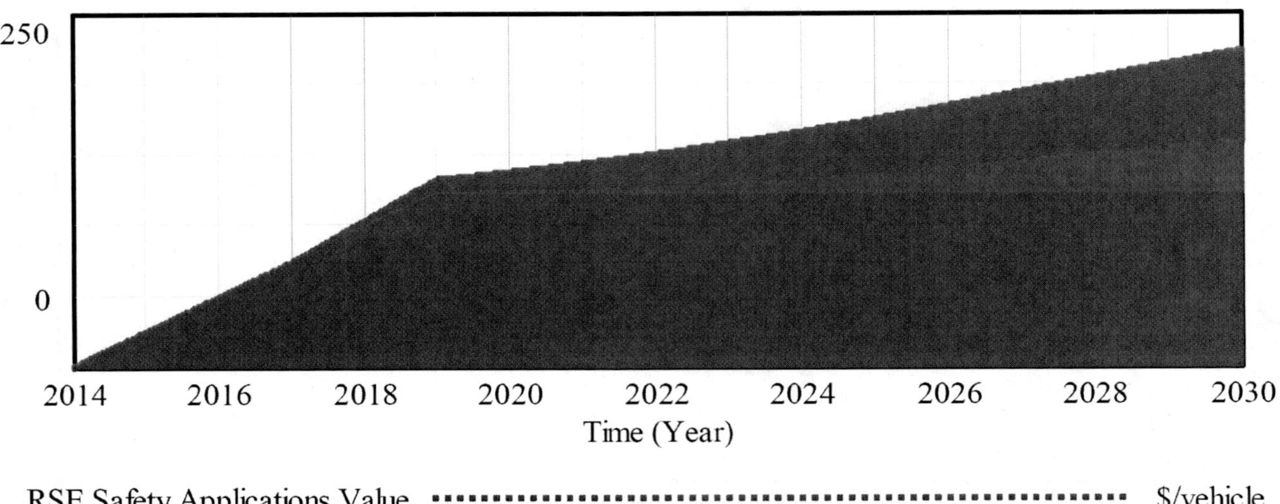

Figure 3. IntelliDrive safety applications value

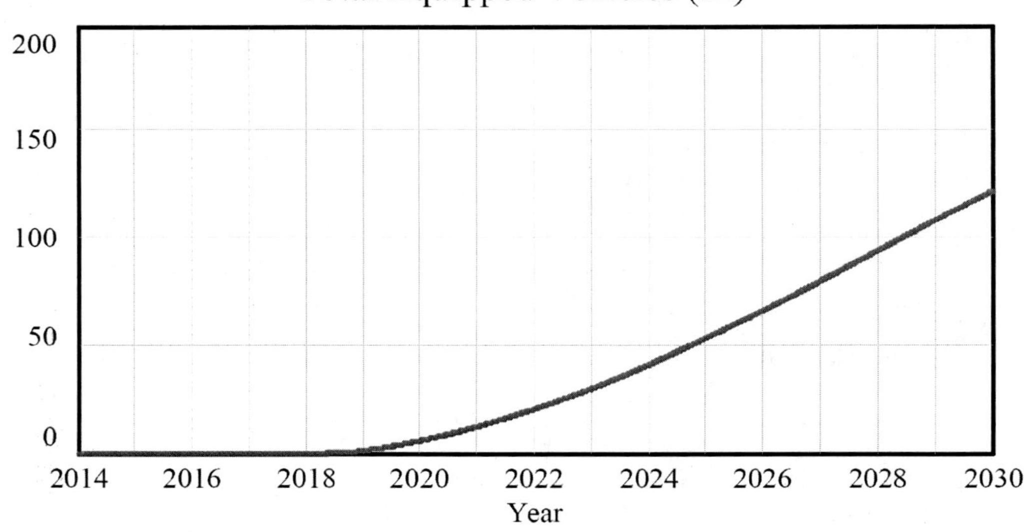

Figure 4. Equipped vehicle growth: total vehicles equipped with OBE in USA

In the current model, motorists will not adopt the system until the value exceeds their cost. Figure 4 shows the growth of vehicles with OBE after the value per vehicle surpassed the cost of the equipment in late 2018.

This equipped vehicle curve shows that the vehicle adoption rate growing through 2030, rising to 80% of all vehicles in the long run.

The costs to build-out and operate the IntelliDrive system are based on the costs in the Volpe study [1]. We assumed that the average cost to build-out an RSE is $15,000. For simplicity, we did not explicitly model the RSE replacement based on an average lifetime; rather, we set the annual maintenance cost to 20% of total imbedded cost to cover RSE replacements, operations, and maintenance. The annual costs to build, operate, and maintain the system is shown in Figure 5.

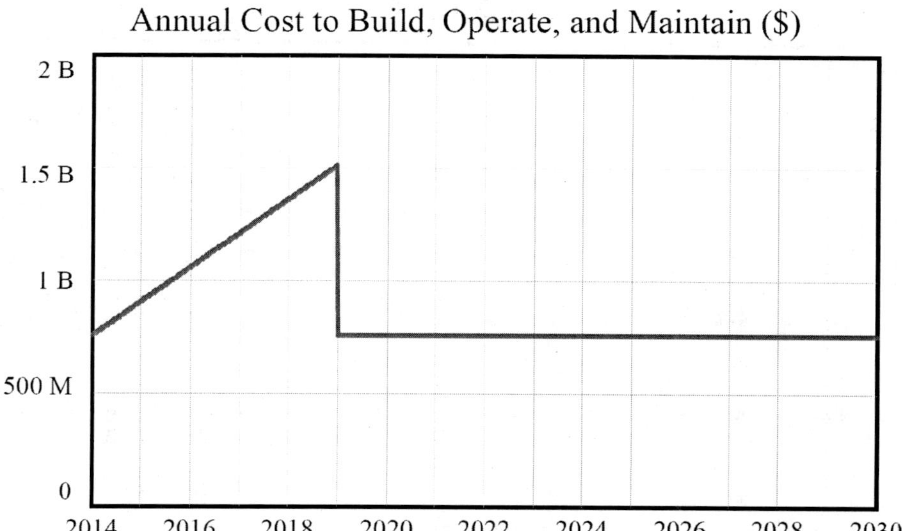

Figure 5. Expenditures to build, operate, and maintain the IntelliDrive system

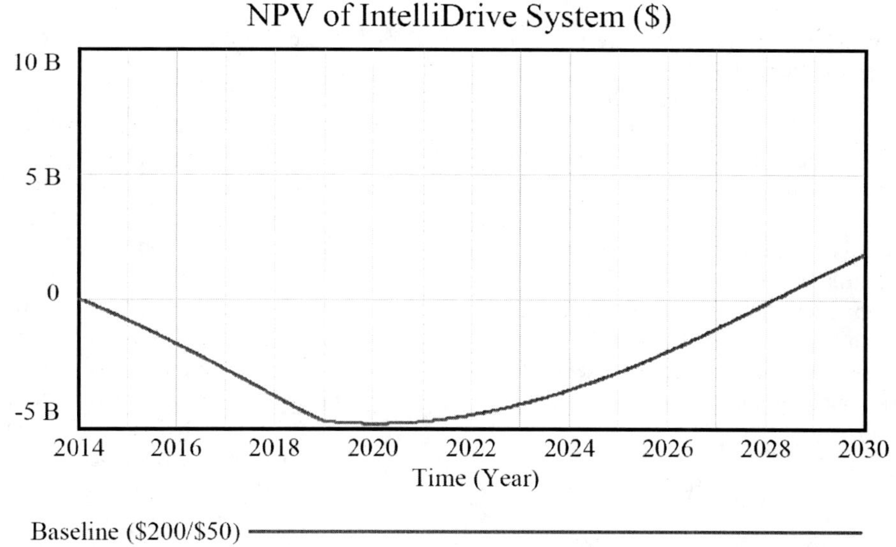

Figure 6. Net present values of IntelliDrive system with baseline scenario assumptions

The cost to build-out the RSE locations during the first five years grow as the cost of operations, maintenance, and replacements grow. In the sixth year, there is no further system build-out, so the annual costs drop to just operations, maintenance, and replacements.

As in the Volpe study [1], we can compute the net present value (NPV) to the nation as a whole by discounting the future benefits and future expenditures to the present day. In the Volpe study, all value came from benefit-producing safety applications. In that case, value was the same as benefit and could be directly compared to cost. In the current case, we are including "other" applications that increase system value, but may not have benefits to the public beyond the motorists that are using these applications, and therefore cannot be included in the NPV calculation. We set 50% of the other-applications' value as providing benefit in the baseline scenario.

Figure 6 shows the NPV of the IntelliDrive system under the baseline scenario assumptions with a 7% discount rate.

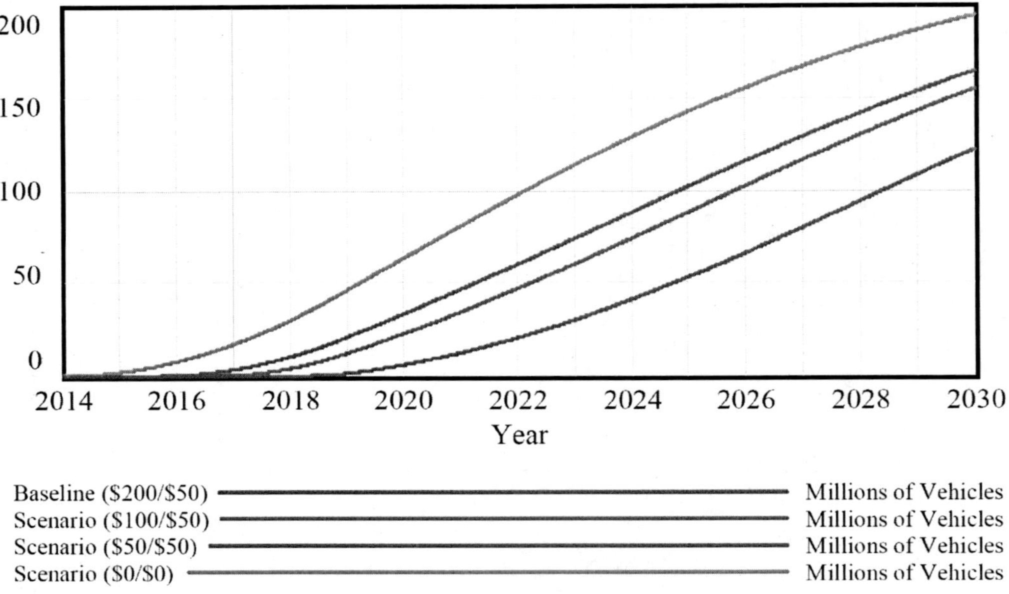

Figure 7. Effect of alternative OBE costs on IntelliDrive penetration

The NPV declines to about -$5B in mid 2019 before enough vehicles have installed OBEs to turn the NPV in a positive direction. The NPV goes positive in early 2028, and eventually reaches almost $20B. This NPV is similar to the Volpe study. This shows that under the baseline scenario, the net discounted payback period is over 14 years.

The downside of this result is the high risk: the government had to invest nearly $5B building out the system without any vehicles adopting the system until the very end of the build-out. This is untenable from a business strategy point of view; therefore, an approach is needed that brings more users into the system earlier. We examine alternative scenarios designed to do this in the next section.

ALTERNATIVE SCENARIOS

The IntelliDrive Penetration Model (IPV) can analyze many alternative scenarios involving alternative assumptions such as the speed of the RSE build-out, the value of safety and other applications, the ultimate potential system adoption level, unit costs, discount rates, and other parameters. For the purposes of this paper, we focus our attention on the cost of the OBE since it has the most impact on the IntelliDrive system NPV.

The primary reason for the delay in user adoption until 2018 in the baseline scenario is that the cost of the OBE is higher than the average system value per vehicle until then. This implies that the IntelliDrive business strategy should focus on bringing down the cost of the OBE as rapidly as possible to bring users into the system earlier, reducing risk and improving NPV.

There are many approaches that can be taken to reduce the OBE cost. For example, the government could adopt policies that reduce the effective cost of the OBE to motorists such as instituting tax credits, or providing subsidies to the manufacturers. In the E-ZPass case, the government appointed a single company, Mark IV Industries, to exclusively manufacture and supply the transponders, which are the equivalent of the OBEs in the IntelliDrive case. This allowed the government to control the OBE quality and distribution. This also allowed the government to set the up-front cost of the OBE to the motorist to $0 from day-one, resulting in immediate net positive value to the motorists, and corresponding uptake in system adoption.

In the current analysis, we do not specify or analyze specific strategies to bring the OBE cost down, although the model is capable of such analyses. Instead, we merely look at the effect of OBE cost reductions on the IntelliDrive system adoption and NPV. We looked at reducing both the initial OBE cost and the long-run OBE cost. The baseline scenario assumed the initial and long-run OBE costs were $200/$50. We now consider alternative scenarios in which these costs are reduced to $100/$50, $50/$50, and $0/$0. The resulting impacts on system adoption are shown in Figure 7.

The effects of lower OBE costs on bringing users into the system earlier are significant. Lower OBE costs mean that the time it takes for the value of the system to exceed these thresholds is reduced so that more vehicles are equipped with an OBE sooner. This leads to improved NPV and reduced payback period as seen in the following chart.

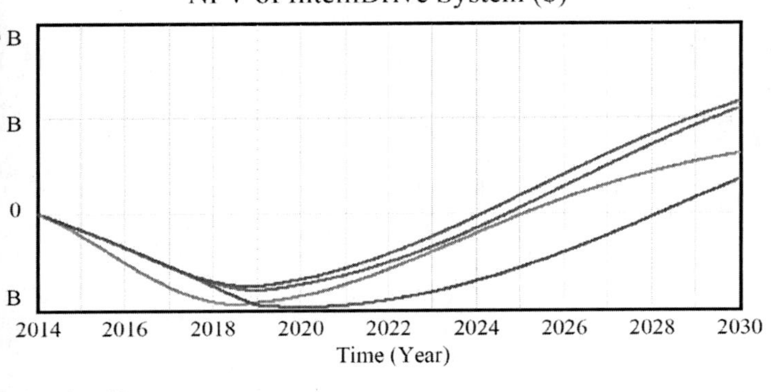

Figure 8. Effect of reduced OBE cost on IntelliDrive system NPV

This result shows an improving business case as the OBE cost is reduced, with the $50/$50 case providing the highest NPV and fastest payback. The $0/$0 OBE case, where the OBE is free to motorists on day-one, shows an immediate uptake in system adoption, however the added costs of paying for the OBEs is too much to overcome and this case is ultimately the worst of all considered scenarios. These results are central findings of the modeling, and suggest that the focus for developing an IntelliDrive business model be on finding a way to launch the system with a low cost OBE.

We also looked at the possibility of slowing the deployment of RSE locations to reduce annual construction budgets, and thereby reducing risk. Unfortunately, this leads to much slower system adoption, markedly extended payback periods, and reduced NPV. Therefore, this strategy is not recommended. This finding leads to considering a geographic deployment strategy of placing the system in localized but sufficiently large areas so that the value to motorists in these areas builds quickly, leading to rapid uptake in system adoption in these areas. More study, modeling, and analysis are needed to explore business strategies along these lines.

SUMMARY/CONCLUSIONS

IntelliDrive is a US DOT initiative to develop and demonstrate technologies for using wireless communications technologies to improve transportation safety, mobility, and sustainability. It envisions vehicles to communicate with each other and with road-side and infrastructure servers to achieve situational awareness in real time to detect and warn drivers of imminent dangers of collisions, and to provide traveler information to the drivers.

As IntelliDrive technologies mature, a crucial issue becomes how to develop a viable business case for widespread deployment of IntelliDrive systems and applications. It is widely recognized that some important IntelliDrive applications, such as vehicle safety applications based on V2V communications, will provide benefits to drivers only after a high percentage of all the vehicles are equipped with the same applications.

This paper studied the viability of a commercial business model for the IntelliDrive system with a dynamic simulation model. This model was calibrated against the market adoption of E-ZPass, which followed a similar commercial business model.

The baseline results are based on assumptions closely aligned with the Volpe study of IntelliDrive safety applications. The IntelliDrive penetration model predicts that a commercial business approach is viable only if a significant reduction in the cost to the motorist of the OBE is achieved. Specifically, the model predicts that a low cost OBE to motorists is required upon system launch to incent a strong enough adoption uptake to sufficiently reduce business risks to an acceptable level.

How can an affordable OBE be provided? One approach is to launch on luxury vehicle first, the "trickle down" adoption model. Luxury vehicles are less cost sensitive and value safety features as important to the brand. The luxury brands make up about 10% of the US fleet and could lead with V2V safety related features. This was the case for other safety related features such as airbags and ABS, with "encouragement" from NHTSA.

Another opportunity to reduce the cost of OBE equipment is through an aftermarket fitment program. At this time there is not a compelling reason for the driving public to purchase an aftermarket device. A program of the magnitude of the federal initiative related to driver distraction is needed to educate drivers about IntelliDrive safety applications and why an aftermarket device is useful. At this point, the driving public is not aware of why they need an aftermarket IntelliDrive device even if it is essentially given away. The US DOT JPO has appropriately reached out to the Consumer Electronics Industry to jointly develop this strategy. There is much work ahead.

The stakeholders who benefit are those who should take the upfront risk of "subsidizing" OBE costs by the strategies discussed. Our modeling tool helps estimate the size of the investment and determines the payback period. We envision public and private stakeholders joining forces to make this investment. Our model indicates that without government incenting significantly lower OBE costs coupled with cultivating driver awareness, it is difficult to predict if, how and when IntelliDrive will become a reality.

REFERENCES

1. Vehicle-Infrastructure Integration (VII) Initiative Benefit-Cost Analysis Version 2.3 (Draft), May 8, 2008, Prepared by: Economic and Industry Analysis Division, RTV-3A, John A. Volpe National Transportation Systems Center, United States Department of Transportation, Cambridge, Massachusetts

2. Sterman, J.D., Business Dynamics: Systems Thinking and Modeling for a Complex World, New York, NY: McGraw-Hill; 2000

3. Vensim, The Vensim Simulation environment, Vensim Professional 32 Version 5.4

4. Evaluating EZPass - Using conjoint analysis to assess consumer response to a new tollway technology, Vavra, Terry C., Green, Paul E., and Krieger, Abba M., Summer 1999,

5. E-ZPass Evaluation Report, Vollmer Associates, LLP, August 2000

6. Operational and Traffic Benefits of E-ZPass to the New Jersey Turnpike (August 2001)

7. Achieving the Vision: From VII to IntelliDrive Policy White Paper, RITA Intelligent Transportation Systems, April 30, 2010

8. Jones, W.S., "VII Life Cycle Cost Estimate," December 2006, updated April 2007, and "A VII Deployment Scenario," December 2005, US DOT ITS Joint Program Office

CONTACT INFORMATION

Robert White
Telcordia Technologies
rwhite@telcordia.com

Tao Zhang
Telcordia Technologies
tao@research.telcordia.com

Paul Tukey
Telcordia Technologies
paul@tukey.org

Kevin Lu
Telcordia Technologies
klu@telcordia.com

David A. McNamara
MTS LLC
coachdavemc@gmail.com

DEFINITIONS/ABBREVIATIONS

CSW
Curve speed warning

DSRC
Dedicated Short-Range Communications

EBL
Electronic brake lights

I2V
Infrastructure to vehicle

IPM
IntelliDrive Penetration Model

JPO
Joint Program Office

OBE
On-board equipment

OEM
Original equipment manufacturer

NPV
Net present value

RSE
Road-side equipment

SPAT
Traffic signal phase and timing

US
DOT United States Department of Transportation

V2V
Vehicle to vehicle

VII
Vehicle Infrastructure Integration

APPENDIX 1

MODEL CALIBRATION

The question of setting the parameters in the IntelliDrive model is difficult because the system has not yet begun to be deployed and decisions by motorists have not yet been made. Similar systems for which we have a full history of deployment and customer adoption can be used to provide approximate parameters if care is taken in correctly mapping the coefficients.

One such system is the E-ZPass electronic tolling system that was originally deployed in New York and New Jersey in the 1990's. Since this time, the E-ZPass system has expanded considerably to the Mid-Atlantic, Mid-West, and New England; it currently is in use in 25 agencies spread across 14 states. We focus attention on the initial E-ZPass deployment because it has been studied extensively and data are readily available.

The initial E-ZPass build-out was started in 1993 and completed in 1997. To reflect this build-out schedule, we set the budget for E-ZPass build-out so that 25% of the construction was completed in each year. E-ZPass was opened for sales in 1995. 80% was taken to be the asymptote of the E-ZPass penetration of rush-hour vehicles in the model.

Figure 9 shows that the E-ZPass Penetration model fits the tracking data very well.

There are clear similarities between the E-ZPass model and the IntelliDrive model. Both are driven by the value to the motorist and adoption is throttled by the construction of the system. There is an externality with E-ZPass as with IntelliDrive, but in the E-ZPass case, this is a negative reinforcement. This is because as drivers adopt E-ZPass, the E-ZPass electronic toll lanes become more congested while the cash toll lanes become less congested; this means that as more motorists adopt E-ZPass, its value is reduced somewhat. This negative externality effect was detected in the model fitting.

The functional form in the E-ZPass penetration growth model is identical to that in the IntelliDrive penetration growth model. In each case, the penetration growth is proportional to the average value of the system per vehicle. This common structure allows us to use the calibrated E-ZPass coefficients in the IntelliDrive penetration model. The implication of transferring these coefficients is the following: we are assuming that motorists will react to the average value per vehicle the same in both systems.

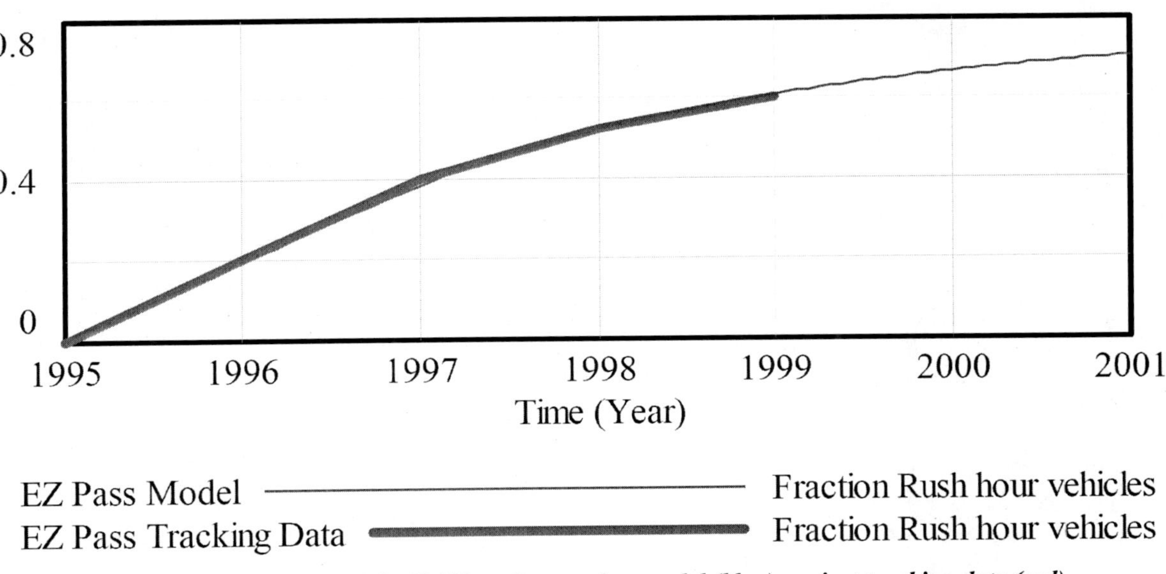

Figure 9. Resulting fit of the E-ZPass Penetration model (blue) against tracking data (red)

Performance of Aftermarket (DSRC) Antennas Inside a Passenger Vehicle

2011-01-1031
Published
04/12/2011

Radovan Miucic and Sue Bai
Honda R&D Americas Inc.

doi:10.4271/2011-01-1031

ABSTRACT

A vehicle's safety system capability can be enhanced by a cooperative Vehicle-to-Vehicle (V2V) system in which vehicles communicate their driving status data, such as location and speed, using a common Dedicated Short Range Communication (DSRC) protocol. The effectiveness of the V2V applications will depend on the number of the vehicles equipped. Market penetration significantly influences the effectiveness of V2V safety applications. Previous research indicated that it could take decades to reach 95% DSRC safety device penetration in the market if only the new vehicles are equipped with the DSRC transponders during manufacturing. In order to raise the market penetration of such technology in the foreseeable future and provide a safety benefit to the early adopters, a scenario that involves retrofit and aftermarket DSRC devices is suggested by U.S. Department of Transportation (USDOT). A typical retrofit for a passenger vehicle may mean that a DSRC antenna will be installed on the roof of the vehicle and the cable will be routed to the transponder somewhere inside the vehicle. An aftermarket device installation typically means that the antenna is placed inside a vehicle on the dashboard. However, the RF signal may be impeded by glass windows, metal pillars, seats, and even the passengers. Also, there is no built-in ground plane as in the case of a roof top mount. This paper examines the performance of commercially-available DSRC antennas mounted inside of a vehicle, shows results of field trials using 0 dBi and 9 dBi horizontal gain antennas, identifies issues, and suggests possible placements of the antenna in a vehicle.

INTRODUCTION

After 2013, the U.S. government may mandate or in other ways encourage passenger and commercial vehicles to be equipped with 5.9 GHz DSRC transponders [1]. If only new vehicles are equipped with transponders, it will take up to 40 years to reach significant market penetration [2]. For successful cooperative safety applications, high market penetration is critical. To increase the number of transponders on the road, the Vehicle Infrastructure Integration Consortium is considering several retrofit installation scenarios [3]. Some scenarios include retrofitting installations that include a roof-mounted antenna. Other scenarios include stand-alone DSRC devices that can reside on a dashboard. In-vehicle retrofit devices are easier to install since they do not need routing antenna cables from the roof to the vehicle's interior. However, communication performance of the antenna residing inside a vehicle is degraded compared to the antenna being on the roof of the vehicle. The purpose of this research was to compare DSRC communication performance with antennas placed inside a vehicle versus on the roof.

Similar research involved DSRC performance measurements with the antenna placed only on the roof of the vehicle [4, 5, 6, 7], while this research focused on in-vehicle DSRC antenna performance.

IN-VEHICLE DSRC ANTENNA PERFORMANCE

EXPERIMENT SETTINGS

This section describes the experimental equipment used for testing. We converted two ordinary passenger sedans into experimental platforms by equipping them with additional hardware. Each of the two experimental vehicles is equipped with a Denso Wireless Safety Unit (WSU) with integrated Atheros 802.11p-based DSRC radio, Novatel Global Positioning System (GPS) receiver and antenna, and a software application running on the WSU processor. The two vehicles are sending a Basic Safety Message (BSM) in wireless short message type, as defined in the Institute of Electrical and Electronics Engineers (IEEE) 802.1p draft standard [8]. Each vehicle sends messages periodically every 100 ms. The data content of the BSM is defined in Society of Automotive Engineers (SAE) J2735 standard [9]. The over-the-air BSM size includes 51 bytes of overhead, 222 bytes of security information, and 105 bytes of data, for a total of 378 bytes. Messages are sent at 20 dBm with ~3 dB cable loss yielding ~17 dBm total over-the-air power. BSMs are sent with data rate of 6 Mbps on a continuous dedicated safety channel CH172 (5.855-5.865 GHz).

We used two commercially-available antennas: a long pole antenna with 9 dBi at horizon with embedded ground plane and elevated radiating element (Antenna A), and a ground plane dependent "hockey puck" mono-pole antenna with 0 dBi at horizon (Antenna B). Both antennas are omnidirectional at horizon. As shown in Figure 1, the experiments used the following four antenna setups:

a). Antenna A mounted inside the vehicle below the rear mirror

b). Antenna B placed on the dashboard

c). Antenna B mounted on a metal sheet placed on the dashboard

d). Antenna B mounted on the roof of the vehicle

a)

Figure 1. Antenna setups: a) Antenna A mounted inside the vehicle and below the rear mirror, b) Antenna B placed on the dashboard, c) Antenna B mounted on a metal sheet placed on the dashboard, and d) Antenna B mounted on the roof of the vehicle.

b)

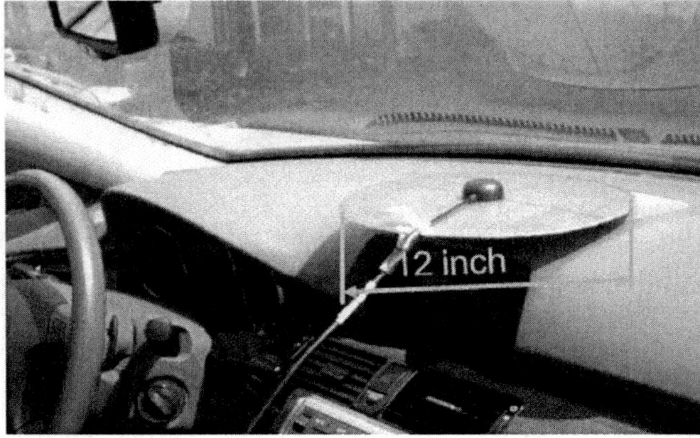

c)

d)

Figure 1 (cont.). Antenna setups: a) Antenna A mounted inside the vehicle and below the rear mirror, b) Antenna B placed on the dashboard, c) Antenna B mounted on a metal sheet placed on the dashboard, and d) Antenna B mounted on the roof of the vehicle.

TEST LOCATIONS

One goal was to test the communication range between the two vehicles and examine the omnidirectional performance with different antenna setups. For the purpose of evaluating in-vehicle antenna performance, two locations were chosen. The first location, mainly for range measurement, was a two-lane high speed open road, and the second location was an open parking lot.

Open Road

The communication range test involved multiple runs of the two vehicles approaching each other at 55 mph on a two-lane open road. The selected road has no sizable buildings, trees, or telephone poles nearby. In some of our results we differentiate incoming direction (when the vehicles are approaching each other) and outgoing direction (vehicles passed and moving away from each other).

Parking Lot

As shown in Figure 2, a large empty parking lot was used in evaluating antenna performance with respect to an angle of incidence wave on a horizontal plane (directionality test). Vehicle 1 is stationary and positioned at midpoint of the edge of the parking lot facing north. Movement of Vehicle 2 is bounded by marked semicircle. The semicircle is centered at Vehicle 1 and the radius is 61 m. Vehicle 2 travels in both directions clockwise and counterclockwise. Data is recorded for 6 runs. In our results we filtered out some points and used only data points marked as "Area of Interest" in Figure 2. The tests are repeated for Vehicle 1 facing south, east and west. These four tests were executed for all four antenna setups.

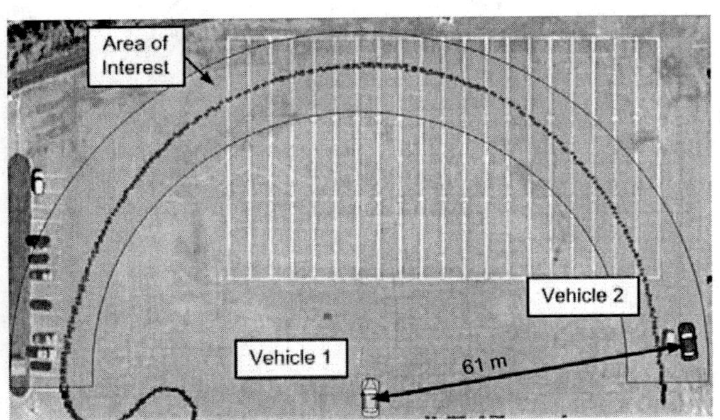

Figure 2. Large empty parking lot.

PERFORMANCE METRICS

This section describes the metrics for evaluating DSRC antenna performance.

Received Signal Strength Indicator

A Received Signal Strength Indicator (RSSI) is a measurement of the power present in a successfully-received message. The unit of received power is in dBm. For our purposes, RSSI values were read from the lower layer of the radio chipset. The basic safety message contains location information. For the range testing, the results are RSSI points versus distance between the vehicles. Directionality results show RSSI points versus angles of incidence waves.

Packet Success Ratio

A Packet Success Ratio (PSR) is a ratio between number of successfully received messages at the receiver and number of sent messages from transmitter. To compute PSR for a given test, we compared logged file from the receiver against a log file from the transmitter. For the range testing, PSR values are associated with a specific distance. The result shows an average PSR per distance bin for multiple runs. Similarly, for the directionality testing, PSR values are associated with angle bins. The result shows average PSR values per angle bins.

RESULTS

Range Test: Open road

Antenna range performance is shown in Figure 3 with each line representing an average PSR for 10 runs of a particular test. In the plot, the X axis represents the distance of separation (in meters) and the Y axis represents PSR. Since antennas are reciprocal and the plots were similar for both vehicles, only results for one vehicle are shown.

In a 9 dBi gain antenna test case, when vehicles are incoming, the first BSM was received at more than 1500 m and with more robust communication (e.g., 70% PSR occurred when the vehicles were 0-350 m apart). When the vehicles were outgoing, the range performance was similar. There was almost no performance difference in the 0-350 m range in either direction. However, this was not the case for the 0 dBi gain antenna. In the incoming direction, the first BSM was received at more than 600 m, and 70% PSR occurred when the vehicles were 0-151 m apart with a no ground plane antenna mount, and 0-195 m apart for an antenna with the ground plane. Here, communication performance differed with the direction. When vehicles were outgoing, 70% PSR was achieved when the vehicles were 0-35 m apart with an antenna with the ground plane, while the mount with no ground plane did not reach a 70% PSR level at all.

Figure 4 shows a plot of signal strengths of received data packets relative to the distance of separation between the transmitting and receiving vehicles from all measurements recorded in an open road environment. In the plot, the X axis represents distance of separation (in meters) and the Y axis represents signal strengths (in dBm). Each point represent signal strength values for successfully decoded packets at the receiver end. Unsuccessful packets are not shown. As expected, for the 9 dBi antenna, RSSI values are stretching from 0 to 1600 m and the values are typically higher than that of the 0 dBi antenna setups. There is a marginal improvement with the 0 dBi antenna with the ground plane mount compared to the 0 dBi antenna without ground plane. Figure 4 shows a plot of the RSSI points. The points are almost overlapping in the two cases.

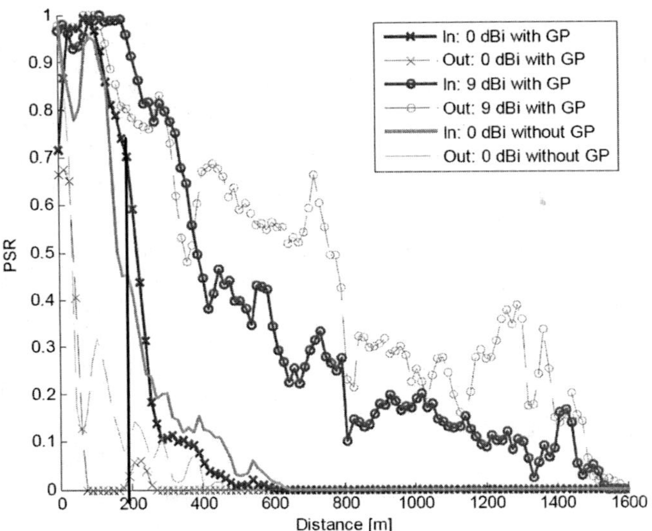

Figure 3. Open road PSR range test. In = incoming direction (vehicles are approaching each other), Out = outgoing direction (vehicles are moving away from each other), GP = ground plane

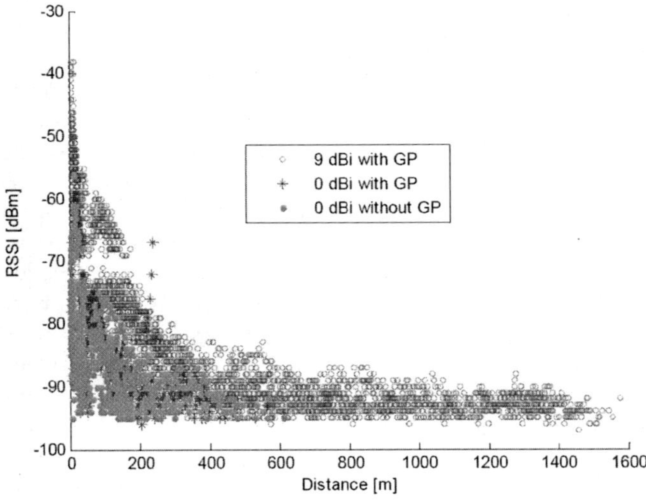

Figure 4. Open road RSSI range test

Directionality test: Parking lot

Figure 5 is a top-down view polar diagram showing PSR directionality for the open parking lot test. The center represents 0% PSR, while the most outer circle is 100% PSR. All resultant lines are made out of superimposed results from the four side tests (static vehicle facing north = 90°, south = 270°, west = 180°, and east = 0°). The outermost line (black line with triangular data markers) with almost 100% PSR for all angles is a resultant line of the roof top mount the 0 dBi antenna. This rooftop mount is the benchmark and shows a complete omnidirectional performance. The red line with circle data markers is the result of the 9 dBi in-vehicle antenna. The PSR is 80-100% for all angles of incidence except for the southwest region where the PSR is a bit lower. This is probably because the driver of the static Vehicle 1 contributes to the physical obstacle on the wave path. The blue and green lines are results of the 0 dBi antennas with and without ground plane respectively. The setup with the ground plane shows little advantage over the results with the setup without ground plane. Both results show a moderate PSR around 70% in a narrow angle span 75-115° for no ground plane setup, and 75-125° for ground plane setup. PSR in the east, west, and south direction is very low and scattered.

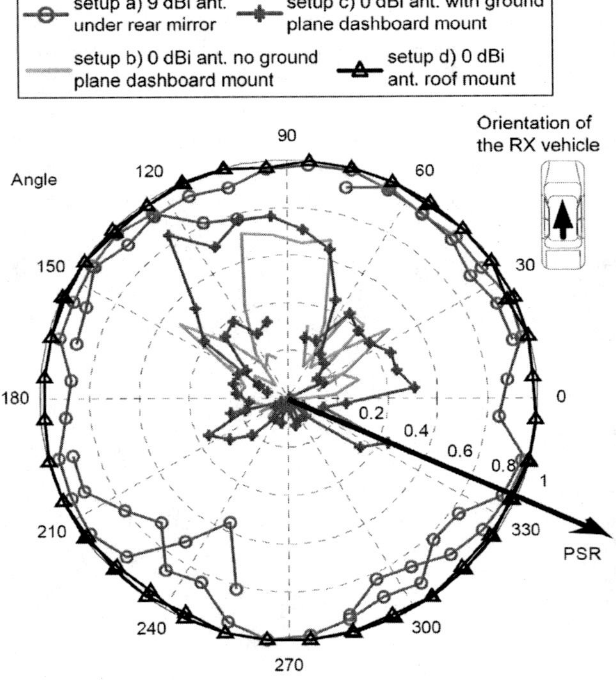

Figure 5. Parking lot directionality PSR versus incident angle for four antenna setups: a) Antenna A mounted inside a vehicle below rear mirror, b) Antenna B placed on the dashboard, c) Antenna B mounted on a metal sheet placed on the dashboard, and d) Antenna B mounted on the roof of the vehicle.

Figures 6 and 7 are top-down view polar diagrams showing RSSI directionality for the open parking lot test. Each point represents a signal strength value for a successfully-decoded packet at the receiver end. The center represents lower packet reception power threshold (−96 dBm), while the most outer circle is normalized for maximum recorded packet power (−57 dBm). The figures show that RSSI values of the 0 dBi antenna with ground plane mount and 0 dBi without ground plane almost overlap. There is only a marginal improvement using the ground plane setup.

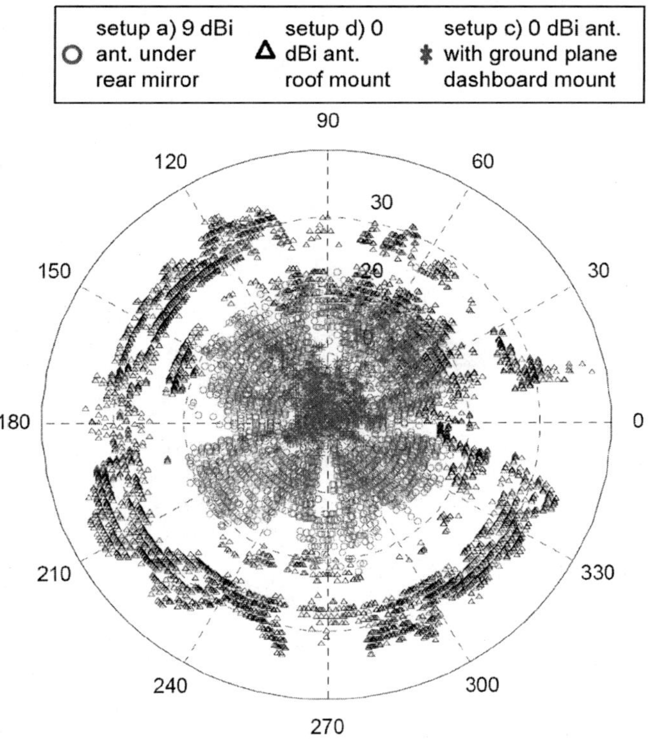

Figure 6. Directionality test: RSSI versus incident angle for antenna setups a), c), and d).

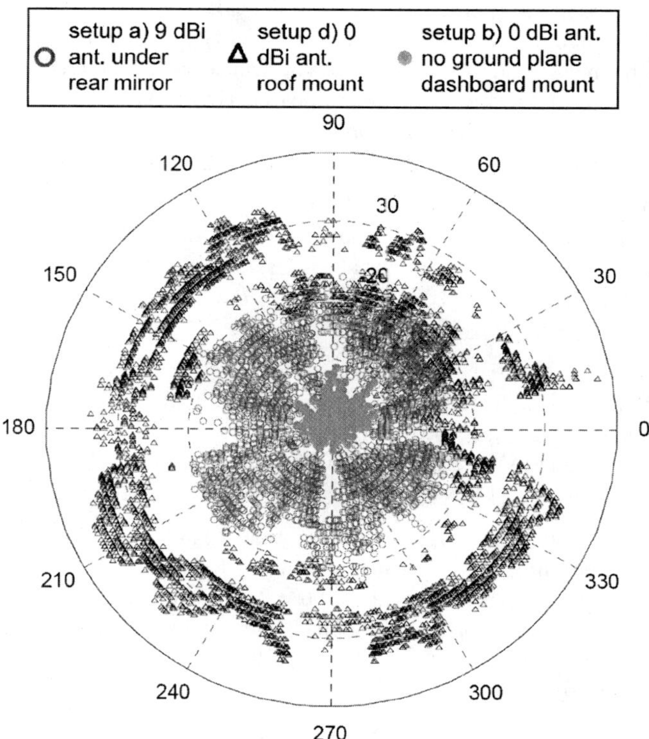

Figure 7. Directionality test: RSSI versus incident angle for antenna setups a), b), and d).

SUMMARY/CONCLUSIONS

For the purpose of accelerating DSRC market penetration for passenger vehicles, this research shows that a stand-alone aftermarket DSRC device design with an in-vehicle antenna is a feasible option. The performance of several in-vehicle antenna setups were investigated. Transmission range testing was done in a typical open road environment and antenna directionality was tested out on a large empty parking lot. The best in-vehicle antenna performance was achieved with an embedded ground plane antenna and a raised radiating element. This setup achieved a sufficient large gain that is close to roof mount performance.

REFERENCES

1. IntelliDrive, "IntelliDrive Workshop Summary: Enabling Devices with DSRC," http://www.intellidriveusa.org/documents/2010/06/ITS%20JPO%20DSRC-Enabled%20Device%20Workshop%20Summary%20June%201%202010.pdf, July, 2010.

2. Carter, A., Chang, J., "Market Penetration Study Results for V2V Safety Applications," presented at ITS America 20th Annual Meeting, Houston, Texas, USA, May 4, 2010.

3. Vehicle Infrastructure Integration Consortium, "Meeting of VIIC Cooperative Work Plan 2009 - 2013," http://www.itsmichigan.org/AnnualConf2010/presentations/VIIC%20ITS%20Michigan%204-26%20R1.pdf, July 2010.

4. Cheng, L., Henty, B.E., Stancil, D.D., Bai, F., and Mudalige, P., "Mobile Vehicle-to-Vehicle Narrow-Band Channel Measurement and Characterization of the 5.9 GHz Dedicated Short Range Communication (DSRC) Frequency Band," *Selected Areas in Communications, IEEE* **25**(8): 1501-1516, 2007, doi:10.1109/JSAC.2007.071002.

5. Guillermo, A., Kathleen, T., and Ingram, M.A., "Measured Joint Doppler-delay Power Profiles for Vehicle-to-vehicle Communications at 2.4 GHz," presented at IEEE GLOBECOM 2004, *IEEE* **6**:3813-3817, 2004, doi:10.1109/GLOCOM.2004.1379082.

6. Yin, J., ElBatt, T., Yeung, G., Ryu, B., Habermas, S., Krishnan, H., Talty, T., "Performance evaluation of safety applications over DSRC vehicular ad hoc networks," presented at VANET 2004, Philadelphia, PA, USA, Oct. 1, 2004, doi:10.1145/1023875.1023877.

7. Cheng, L., Henty, B.E., Bai, F., Stancil, D.D., "Highway and rural propagation channel modeling for vehicle-to-vehicle communications at 5.9 GHz," presented at 2008 IEEE Antennas and Propagation Society International Symposium, San Diego, CA, USA, July 5-11, 2008, doi:10.1109/APS.2008.4619037.

8. IEEE Computer Society, *"IEEE Standard for Information technology-Telecommunications and information exchange between systems-Local and metropolitan area networks-Specific requirements Part 11: Wireless LAN Medium Access Control (MAC) and Physical Layer (PHY) Specifications Amendment 6: Wireless Access in Vehicular Environments,"* IEEE Std. 802.11p- 2010, doi:10.1109/IEEESTD.2010.5514475.

9. SAE International Surface Vehicle Standard, "Dedicated Short Range Communications (DSRC) Message Set Dictionary," SAE Standard J2735, Rev. Nov. 2009.

CONTACT INFORMATION

Dr. Radovan Miucic is a communication research engineer at Honda R&D Americas, Inc. Automobile Technology Research Division.

Sue Bai is a senior research engineer at Honda R&D Americas, Inc. Automobile Technology Research Division.

Cybercars for Sustainable Urban Mobility - A European Collaborative Approach

2010-01-2345
Published
10/19/2010

Michel Parent
Inria/Imara

Abstract

We know that the existing urban transport systems based on the private vehicle (necessarily relying mostly on fossil fuels) are not sustainable in terms of energy and land needs. On the other hand, public transportation systems are also not very efficient and do not provide a good service anywhere and anytime.

Over the last twenty years, a new concept has emerged through strong cooperation between researchers, automotive companies, suppliers and transit operators. It is the concept of a co-modal systems. This means well-designed systems that will combine the use of various transportation modes and in particular the individual vehicles and the mass transit systems. A key element of such a system is the Cybernetic Tansportation Systems (CTS), which are based on fully automated urban vehicles. This paper will present these CTS and how they have emerged through a European collaborative approach.

These environmental friendly novel systems offer far-reaching solutions that will drastically mitigate or solve the problems that we encounter in current urban transportation systems. They will yield much more effective organisation of the urban mobility, with a more rational use of motorised traffic; less congestion, pollution noise and CO_2 emissions and better accessibility and safety. The result will be a higher quality of living, an enhanced integration with the spatial and also societal development and advancement towards sustainability.

Cybernetic Transportation Systems (CTS)

The Cybernetic Transportation Systems (CTS) concept has emerged in the early 1990's as a researcher dream: to bring fully automated clean urban vehicles as an attractive alternative to the use of private cars (1). These vehicles were meant to be operated by a public transport supplier and fully integrated in an information system that would give the users the best choices in their travel demands through a mix of soft modes (walking, biking), mass transport and individual transport.

The first step towards this goal was the development of a carsharing systems based on electric cars and information management. This was the French Praxitele project (2) which took place between 1993 and 1999 with a large scale demonstration in the city of Saint Quentin-en-Yvelines, near Paris with 50 electric cars and 13 stations. During this programme, a technical solution for the relocation of empty vehicles was developed with the platooning concept (3) which was a first step towards full automation (4) and introduced the concept of dual-mode vehicle with manual or fully automated modes.

In 1997, a bold system with four fully automated electric vans was put in place at Schiphol airport by the company Frog Navigation Technologies and operated without any accident for several years. At about the same time, the University of Bristol developed the ULTra, a PRT (Personnal Rapid Transit) based on fully automated electric road vehicles and INRIA in France developed the CyCab, an innovative electric vehicle for manual (joystick) or fully automated operation (5).

At that time, it became clear to many that a collaborative approach was needed to bring the concepts of the CTS to reality. An informal consortium was therefore formed to present research projects for financing to the European Commission. These projects that introduced the concept of the cybercars and CTS started in 2000 and have been since then constantly supported by various Directorates of the European Commission (EC).

National and European Research Projects

The first cooperative projects financed by the Commission on the CTS where the CyberCars and CyberMove projects. They lasted from June 2000 until December 2004. These 2 projects received about 2.5 millions euros of funding each, one from the INFSO Directorate (Information Society), one from the Research Directorate. Both programmes had roughly the same partners that included research organizations (INRIA, Bristol University, Technion Israel, La Sapienza in Rome, TNO), industrial companies (Yamaha Europe, Robosoft, Frog, Ligier), an urbanist (GEA) and a transport operator (Veolia). The CyberCars project focused on the technology and demonstrated various techniques for navigation and collision avoidance. The CyberMove focused on the implementation of CTS in cities and the impacts that could be expected.

The technology developments continued with Cybercars-2 (2006-2008) which included cooperative behavior of the vehicles through communications and dual-mode vehicles (where driving could be switched between manual, assisted or fully automatic driving. The research center of Fiat (CRF) was involved in the development of such vehicles.

In parallel, the large scale project (Integrated Project in the EC language) CityMobil (2006-2011) looked at a bigger picture of automated urban transport and had as an objective the presentation of these technologies to potential cities. Three large scale demonstrations where therefore defined plus a number of "showcases". The big demonstrations were the PRT system (in fact a CTS on dedicated tracks) of Heathrow which is now in operation, the BRT system of Castellion in Spain with guided hybrid buses on a dedicated infrasructure and a CTS for the exhibition center of Rome. The showcases presented cybercars and dual mode vehicles to a number of cities throughout Europe: Daventry in UK, La Rochelle in France, Vantaa in Finland, Trondheim in Norway, Clermont-Ferrand in France and Formello and Orta in Italy.

The work of the consortium in CityMobil was to study the difficulties and the advantages on implementing these systems in real environments and to issue recommendations for further deployment. One key issue was the certification process that is still a big hurdle for the introduction of automated road transport. Progress is being made with the work of TNO to help the cities and their suppliers introduce the systems.

Due to the success of the showcases and the interest of the European cities in knowing more about the potential of these technologies, a further funding was granted by the Commission with the CityNetMobil (2009-2011). Its main goal besides further showcases was to set a network of cities interested in the implementation of CTS and dual mode carsharing systems.

The next European project that is supporting the development of CTS is the PICAV project (2009-2011), also financed by the Research Directorate. The goal of this project is to develop a specific tiny urban vehicle dedicated to ancient cities of Europe. This vehicle is developed to be used either as a CTS or a dual-mode vehicle. The city for experimentation is the city of Genoa in Italy.

The last project, which started in 2010, was financed first by the French state (in 2008), then by the Research Directorate of the EC. It is the CATS project. This project is also developing a new urban vehicle but of larger size (up to 6 passengers). This vehicle could be used as a regular electric carsharing vehicle but it is also designed to be used as a variable length bus with a professional driver at peak times. Two techniques to implement this function are considered: a mechanical connection or an electronic one. The company developing the vehicle is the French Lohr company, an enterprise already well know in the transportation field for its tramway on rubber wheels and for the new generation of the automated metro VAL (in cooperation with Siemens).

European Cooperation on CTS

All these projects have brought in Europe a rich field of lasting cooperation between a large numbers of entities over the years. Many of the key players have known each other for more than ten years. This has allowed strong cooperation for the dissemination of knowledge among the different partners but also has fostered dissemination among the potential users. Several new industrial players have also emerged from research centers through these projects and we have now in Europe quite a number of striving companies ready to enter the field of CTS.

Among the key industrial players that can propose CTS, we should mention (starting from the oldest):

- 2GetThere (NL) issued from Frog Navigation
- ATS Ltd (UK) issued from Bristol University
- Robosoft (FR) issued from INRIA
- Induct (FR) also issued from a cooperation with INRIA
- CriticalMove issued from Coimbra University

But we should also mention the research lab of Fiat (CRF) that early on thought about the concept of the dual-mode vehicle for advanced carsharing and CTS and the Lohr (FR) company who is now developing an innovative vehicle for advanced carsharing.

We should mention also some small and large companies in the field of services that have become interested through these projects to operate such systems as part of their transportation system:

- Veolia (FR) one of the largest public transport operator in the world,
- BAA who is operating the ULTra system at Heathrow,
- Connexxion (NL) who is operating the Rotterdam CTS (from 2GetThere),
- ATAC IT) who is in charge of the CTS now in implementation in Rome,
- GEA (CH), a consultant company in charge of the design and implementation of CTS systems,
- VuLog (FR) an INRIA start-up who is involved in advanced carsharing.

And finally a set of public and private research organization that have been involved (and often leading the way) from the very beginning in many of these projects:

- INRIA (FR) a public research institute in Information and communication technologies (ICT) that set up a whole research team devoted to CTS in 1991,
- Bristol University that developed the concept of PRT based on cybercars in the early 1990's,
- TNO (NL), one of the top institute in Europe which cooperated early with Frog for the certification of their CTS,
- La Sapienza (IT) of Rome with its Transportation Research Department (DITS) that has been leading the way in the evaluation of such systems,
- Coimbra University that has contributed in the development of many technologies,
- Southampton University and its Transportation department that has been deeply involved in the design and evaluation of CTS and PRT.

Finally we must mention all the cities that have cooperated strongly with the partners of the projects to give inputs for studies and for doing experiments. In these cities, technicians and politics are now fully aware of the potentialities of CTS and they are in contact with the suppliers and the consultants. Here are some of the cities which are now considering the use of CTS or advanced carsharing: Coimbra (PT), Antibes-Sophia Antipolis (FR), Nancy (FR), Brussels (BE), Daventry (GB), Vantaa (FI), Trondheim (NO), La Rochelle (FR), Formello (IT), Orta (IT),...

Conclusions

Through a number of closely related research projects financed by the European Commission over the last 10 years, a large number of transport specialists has emerged in Europe. These specialists come from research organizations, large vehicle manufacturers, specialist manufacturers, consulting agencies, transport operators, administrations and they have frequent exchanges. This large group is now pushing forward for a fast introduction of new urban transportation schemes in many cities in Europe and worldwide. The next ten years will indeed be very exciting.

ParkShuttle (2GetThere)

ULTra (ATS Ltd)

CyberCab (2GetThere)

Critical-Move

References

1. Michel, Parent, Pierre-Yves, Texier. *A Public Transport System Based on Light Electric Cars.* Fourth International Conference on Automated People Movers. Irving, USA. March 1993.

2. Daniel, Augello, Evelyne, Benéjam, Jean-Pierre, Nerrière and Michel, Parent. « Complementarity between Public Transport and a Car Sharing Service ». First World Congress on Applications of Transport Telematics & Intelligent Vehicle-Highway Systems. Paris, France. Nov. 1994.

3. Pascal, Daviet, Michel, Parent. « Platooning for Small Public Urban Vehicles ». Fourth International Symposium on Experimental Robotics, ISER'95 Stanford, California, June 30-July 2, 1995.

4. Michel, Parent, Jean-Marc, Blosseville. *Automated Vehicles in Cities : A First Step Towards the Automated Highway.* SAE Future Transportation Technology Conference. Costa Mesa, USA. August 11-13, 1998.

5. Michel, Parent, Sylvain, Fauconnier. « Design of an Electric Vehicle Specific for Urban Transport ». Congrès EVT'95. Paris, Nov. 1995.

Important web sites

- www.cybercars.org the portal to many projects about cybercars, CTS and advanced carsharing
- www.citymobil-project.eu, the portal of the CityMobil project

SAE 2010 Commercial Vehicle Engineering Congress
55th Annual L. Ray Buckendale Lecture (*Session Code: CV801)*
October 5-6, 2010 Rosemont, Illinois

2010-01-2053

Merge Ahead: Integrating Heavy Duty Vehicle Networks with Wide Area Network Services

Mark Zachos (mark@dgtech.com)
DG Technologies President & CEO
SAE Board of Directors &
SAE J1939-84 HD-OBD Task Force
Chairman

ABSTRACT

Commercial vehicle operators have many options available to them for managing their assets. Whether in an on-highway fleet, agricultural / off-road, construction, or military, available real-time vehicle information is growing. While accessing this data via applicable Wide Area Networks (WANs) is commonplace, new technologies are just beginning to develop to take advantage of all of the connectivity possibilities to further aid in delivery of goods and services. As an enabler to expanding these fleet management applications, vehicle on-board networks (commonly referred to as "in-vehicle" or simply "vehicle networks") are expected to support a growing number of vehicle related technological solutions.

This paper provides background on vehicle networks, including key terminology, an introduction to standards based protocols, and critical SAE vehicle network related standards. While an historical view of vehicle network topologies and a rationale for the very first vehicle networks is summarized, growth applications such as fleet management system use of vehicle network data generated is emphasized.

To provide an understanding of the importance of a vehicle network backbone, a comparison to modern local area networks (LANs) is provided, along with the structure of the data packets or Protocol Data Units (PDUs). Next, standards based vehicle networks supporting Heavy Duty vehicles are described to explain information is conveyed. These standard protocols include SAE J1708, J1587, J1939, J2534, CAN (ISO 11898), and ATA/TMC RP1210. In order to illustrate how these standard protocols work, this paper provides a detailed overview of SAE J1939 including the J1939-7 and J1939-73 standards, as well as Heavy-Duty On Board Diagnostics (HD OB) standard Messaging and Diagnostics that use J1939. The paper describes how these protocols and those related with LAN and WAN networks complement each other to provide end-to-end connectivity to support a variety of fleet management applications.

Next, the promise of leveraging integration of vehicle networks with LANs and Wide Area Networks (WANs) is discussed. While the industry has recently begun the implementation of the aforementioned internetworking, future inter-vehicle networking scenarios will be described, along with proposed standards required for implementation.

Finally, the future for vehicle networking is outlined including opportunities for standards development in the area of security, bandwidth allocation, application-specific vehicle network protocols, and emerging WANs.

1. INTRODUCTION

The goal of this paper is to provide solid reference material for those involved in vehicle networking and, in particular, to provide basic tutorial material for young engineers just entering the commercial vehicle profession. By merging high technology on-board vehicle networks with off-board functions linked via local and wide area computer networks, numerous exciting new commercial and military vehicle communications applications have been, and will continue to be developed.

Modern Heavy-Duty (HD) vehicles are equipped with multiple Electronic Control Units (ECUs) that operate components on the truck such as engines, transmissions, brake systems, instrument clusters and lamps. The term "truck" refers not only to the tractor (cab) and trailer, but also to military and off-highway vehicles including those in the construction and agricultural industries. We will often use "HD" or "truck" to signify all HD vehicles.

ECUs need to share information in order to implement the complex control mechanisms in a powertrain system. The ECUs are also capable of performing system diagnostic functions and reporting diagnostics information to tools, such as hand held diagnostic scan tools (generically, Vehicle Diagnostic Adapters (VDAs)) or PCs, for assisting technicians with vehicle maintenance. The sharing of information is accomplished via an on-board vehicle network. HD equipment often has multiple vehicle networks, which are used for communicating both standards based information (e.g. J1939 message data) and private OEM information (proprietary data). A vehicle network linking ECUs operates in a similar manner to an office building local area network that shares information between office computers.

The following sections of this paper explain the operation of vehicle networks. This paper then describes how wireless wide area network technology, such as WiMAX and LTE (often marketed as mobile 4G networks), is utilized to communicate vehicle information from HD equipment to an office computer, or network of PCs (LAN). This "merging" of vehicle data, with applications running on office computers, enables new capabilities from remote diagnostics to improved logistics and results in better fleet operating efficiencies, cost savings, emissions improvements, and related benefits.

2. VEHICLE NETWORK BACKGROUND

A. *The Need for Vehicle Networks*

Why are vehicle networks needed, and how are they utilized in modern heavy duty equipment?

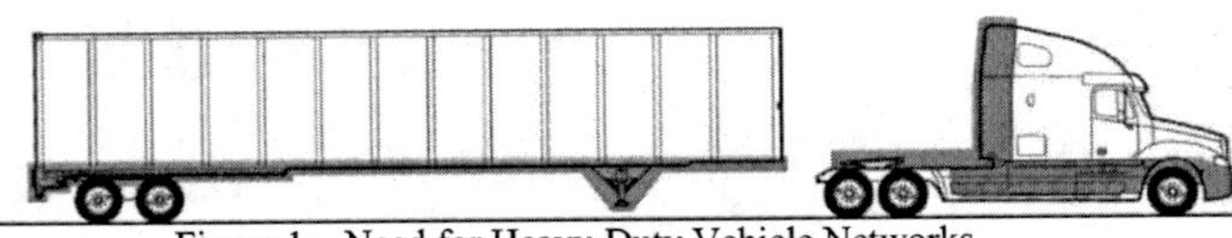

Figure 1 – Need for Heavy-Duty Vehicle Networks

As described in the previous section, although Figure 1 shows a truck cab (tractor) and trailer, we now know the HD vehicle industry includes many more elements, and industries, than this. But how does this differ from the passenger car industry?

First of all, different components (engines, transmissions, brakes, etc.) from different well-known tier one suppliers are mixed together on the same equipment. Many vehicles are customer specified, so variation abounds. Even without a custom order, a HD vehicle most often comes standard with an engine from tier one A, transmission from tier one B, and brakes from tier one C. Therefore, the HD vehicle OEM requires a communications system (i.e. vehicle network) to implement a common method of communications between all of these components In addition, OEMs must manage their own standards based instrument cluster, body controller, and related systems. The HD industry has promoted open communications that is not proprietary, although proprietary messages may still exist. Typical documented open communications messages are: vehicle speed, temperatures, engine speed, and pressures. There are also standard system commands and diagnostic information.

A standard diagnostics interface for HD vehicles is a key enabler for the industry to link on-board functions to off-board devices. Later in this paper we will provide information on vehicle diagnostics interfacing, and an overview of emerging HD-OBD (On Board Diagnostics), and the off-board interface RP1210 API.

There are distinct systems that come together to form a common vehicle diagnostics network system. Figure 2 identifies each element of the in-vehicle network system beginning with the onboard vehicle/equipment communications side and moving to Original Equipment Manufacturer (OEM) or component manufacturer diagnostic applications.

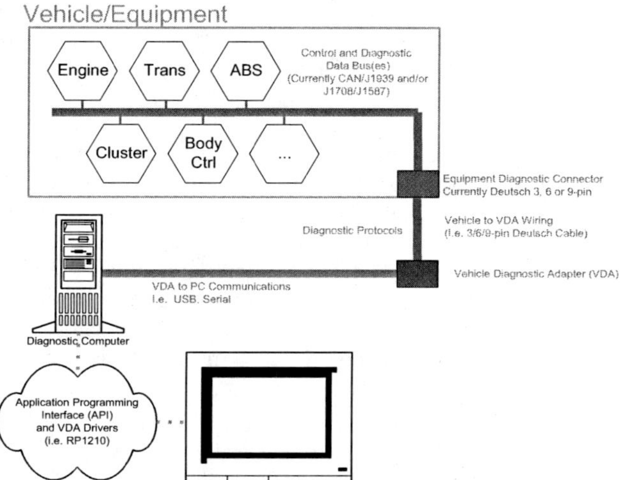

Figure 2 - Vehicle/Equipment Data Bus Architecture and Diagnostic Connectivity

B. Introduction to Standards-Based Protocols [Ref 18]

i. SAE Technical Standards Committee Overview

SAE International supports the development and publication of vehicle standards under the direction of the Technical Standards Board (TSB). SAE's TSB has organized several Councils that focus on specific industry groups. In the TSB's Truck and Bus (T&B) Council, for example, work focuses on HD On-Highway Vehicle standards. Also, under the T&B Council, many Technical Committees exist in which the actual work of standards writing is accomplished. These Technical Committees are responsible for the preparation, development and maintenance of all relevant technical reports within their scope. Technical Committees consist of technical experts from government, industry, regulatory agencies and academia.

The following is a list of some of the Technical Committees reporting to the T&B Council.

- Hybrid Communications Network for Power Management
- Hybrid Safety
- Hybrid Energy Storage
- Hydraulic Hybrids
- Electrical/Electronic Advisory Group
- Low Speed Communications Network
- Control and Communications Network
- Event Data Recorder
- Electrical Systems
- Brake and Stability Control Advisory Group
- Foundation Brake
- Brake Actuator
- Brake Systems
- Wheel
- Stability Control Systems
- Air Brake Tubing & Tube Fittings
- Aerodynamics/Fuel Economy
- Tire

The types of Technical Reports generated by these Committees are:

- **SAE Standards:** These Technical Reports are documentations of broadly accepted engineering practices or specifications for a material, product, process, and procedure or test method.
- **SAE Recommended Practices:** These Technical Reports are documentations of practice, procedures and technology that are intended as guides to standard engineering practice. Their content may be of a more general nature, or they may propound data that have not yet gained broad acceptance.
- **SAE Information Reports:** These Technical Reports are compilations of engineering reference data or educational material useful to the technical community.

ii. The ISO

Another standards organization actively working on vehicle networks is the International Organization for Standardization (termed "ISO" for its French acronym), based in Geneva. The following is an overview of the scope of work being done in the ISO/TC22/SC3/ WG1 (Working Group):

- Standardization in the field of data communication and diagnostic communication
- Observation of worldwide OBD regulation
- Harmonization of national and international OBD standards
- Related data access and transmission security issues
- Data transmission between road vehicles and off-board diagnostic devices.

Their role also includes the proposal of new standards where necessary and appropriate. This may include globalization of a national standard, whether from the automotive industry in Germany (Verband der Automobilindustrie or VDA), SAE International or other standards organizations. In fact, vehicle related standards are only one small component of ISO standards, which also include data communications, telecommunications, and many other areas of standardization.

iii. The American Trucking Association (ATA) TMC Recommended Practices

The mission of the American Trucking Associations (ATA), Inc. is to serve and represent the interests of the trucking industry. The ATA has organized several Councils to serve the industry. One of the Councils related to vehicle network technology is their Technology & Maintenance Council (TMC). Another council that is related to data communications and use of information technology is their Information Technology & Logistics Council (ITLC).

Concentrating on the TMC, their purpose is to improve transport equipment, maintenance and maintenance management. The Council develops Recommended Engineering and Maintenance Practices "RP"s that are voluntarily adopted by fleets, OEMs and component suppliers. For example, the RP1210 document is a standard diagnostics Application Programming Interface (API) RP developed by TMC's S.12 Onboard Vehicle Electronics Study Group.

This nomenclature is commonplace, specifically calling something that may be thought of as a "standard" something other than the term "standard". Here a Recommended Practice or RP is no different than how much of the Internet has been built, not based upon standards, but instead RFCs or "Request for Comment". The Internet Engineering Task Force, IETF, develops RFCs which do gather many industry comments, which are resolved over time and slowly the RFC is commonly adopted as a standard way to perform some capability.

iv. The IEEE

The Institute for Electrical and Electronics Engineers, Inc. (IEEE), a non-profit organization, is a worldwide leader in the advancement of technology. It is a leading authority in subjects ranging from aerospace systems, computers, and telecommunications to biomedical engineering, electric power, and consumer electronics. As of early 2009, IEEE had over 1000 active standards and more than 400 standards under development. Its membership includes 39 societies organized into five technical councils (Electronic Design Automation, Superconductivity, Nanotechnology, Sensors, and Systems) representing a wide range of technical interests. For example, the IEEE Sensors Council comprises 23 societies with a combined membership of 260,000 as of early 2007.

The work of one Council and one Technical Society are of particular interest to automotive engineers. The IEEE Vehicle Technology Society is concerned with land, airborne, and maritime mobile services; portable, commercial, and citizen's communications services; vehicular electro-technology, equipment and systems of the automotive industry; and traction power, signals, communications, and control systems for mass transit and railroads.

Many IEEE Standards influence the design, testing, and operation of HD systems. These include Standards related to power and energy, instrumentation and measurement, mobile and stationary batteries, nanotechnology, telecommunications, and transportation safety, especially highway communications and rail safety. One relevant standard to this paper is the IEEE 1609 standard set now under development. IEEE 1609 is a set of documents that specify vehicle Dedicated Short Range Communications (DSRC) primarily for safety systems applications. An in-vehicle network interface message set which would link to the DSRC system is defined in SAE J2735.

Such standards development implies that vehicle data is destined to travel beyond the vehicle, via standard two-way communications methods. Whether DSRC, WANs, or LANs; standards will "pave the way" for interaction from, and to, HD and other vehicles.

v. United Nations Economic Commission for Europe (UNECE)

The objective of the UNECE is to foster sustainable economic growth in, and among, its member countries. UNECE provides a forum for communication; brokers international legal instruments that address trade, transport, and the environment; and provides statistics, economic and environmental analysis.

The United Nations' Economic and Social Council set up the UNECE in 1947. It is one of the five regional commissions of the United Nations. Its 56 member countries are located in the EU, non-EU Western and Eastern European countries, southeast Europe, the Commonwealth of Independent States, and North America. Over 70 international professional organizations and other non-governmental organizations take part in UNECE activities.

UNECE has quite a history when it comes to road vehicles, as their UNECE WP.29 Working Party on Road Transport Vehicles (WP.29) is now working on issues relating to vehicle networks.

WP.29 was established in 1952 as the Working Party on the Construction of Vehicles under the Inland Transport Committee of UNECE. In March 2000, WP.29 became the "World Forum for Harmonization of Vehicle Regulations (WP.29)." The objective of WP.29 is to initiate and pursue actions leading towards worldwide harmonization and development of technical regulations for vehicle safety, environmental protection, energy efficiency, and anti-theft performance.

In 2001, WP.29 decided that there should be a Global Technical Regulation (GTR) dealing with HD vehicle emission diagnostics systems. The group working on the GTR is referred to as WWH-OBD (World Wide Harmonization of OBD). The ISO 27145 document now under development supports the WWH-OBD vehicle diagnostics communications requirements.

3. HISTORY & TOPOLOGIES – VEHICLE NETWORK PROTOCOLS AND STANDARDIZATION

Today, it is common for a fleet to custom order a truck, bus, construction, or agricultural HD vehicle of OEM "Brand X" with an engine of "Brand E", a transmission of "Brand T" and an ABS brake system of "Brand B". This even occurs though OEM "Brand X" may have an intimate business affiliation with, or even own, the engine, transmission, and/or brake manufacturer.

This is in contrast to passenger car OEMs. While their standards are open, actual implementations are not quite as open. Because of high volume and vertical integration, passenger car OEMs often feel free to deviate from industry standards for reasons of competitive advantage or convenience. In the automotive OEM world, there is no equivalent situation for vehicle systems to be assembled with competitive components as is common in the commercial vehicle world. Buyers cannot custom order engine, transmission, and braking systems from companies other than those the OEM offers. Vehicle network interfaces for automobiles, therefore, may represent simplified design and selection alternatives available to the customer.

In the 1980's, requirements for vehicle emissions reduction and cleaner air brought government involvement leading to legislated On-Board Diagnostics (OBD) standards. This required use of a common vehicle communications interface. This happened at a time when the automotive companies started development and use of networked electronic components. This allowed them to utilize the required OBD system for customer service diagnostic access. As OBD regulations evolved, standards (such as SAE J1962 Diagnostics Connector and SAE J2012 Diagnostics Trouble Codes) for interfacing test equipment to vehicles were developed by SAE International. While on the automotive front, SAE standards development has been limited to supporting OBD emissions regulations and has not been expanded to include non-emissions system component communications, the scope of J1939 for HD vehicles has expanded to include other areas of interest.

Although clean air requirements for the HD industry came several years later than those for passenger cars, the heavy-duty truck and bus industry has voluntarily moved ahead to adopt standards that ease mixing and matching of on-vehicle network components. In fact, one of the heavy-duty industry voluntary standards, SAE J1939, has been accepted by the U.S. Environmental Protection Agency (EPA) and California Air Resources Board (CARB) as an approved OBD protocol. SAE J1939 is also a standard that defines an in-vehicle (serial) network used to communicate control and status information between Electronic Control Units (ECUs) on a HD vehicle. The following sections will now explore the function of in-vehicle network serial communications.

A. Wire Harness without Serial Communications

In the "old days", wire harnesses were simple links between the various electrical components of a vehicle. Figure 3 shows a typical 1940's vehicle wire harness connector. The number of connections in the vehicle used for power and control (data) were usually less than 100.

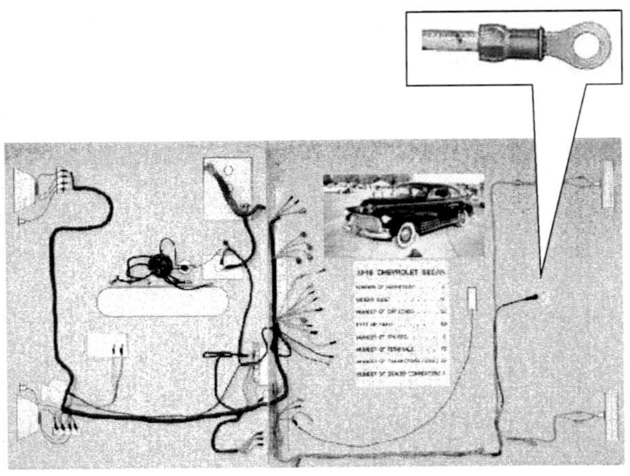

Figure 3 - 1940's Vehicle Wire Harness Connector

B. Vehicle Network Topologies for Today

Figure 4 shows an example of a modern automotive wiring harness. It is much more complex. This vehicle harness has more than 6400 terminals and up to 500 connectors. The total harness may consist of a significant length of wires and can be very heavy. These harnesses include the vehicle network's hard wire (physical layer) that carries data signals between nodes (ECUs).

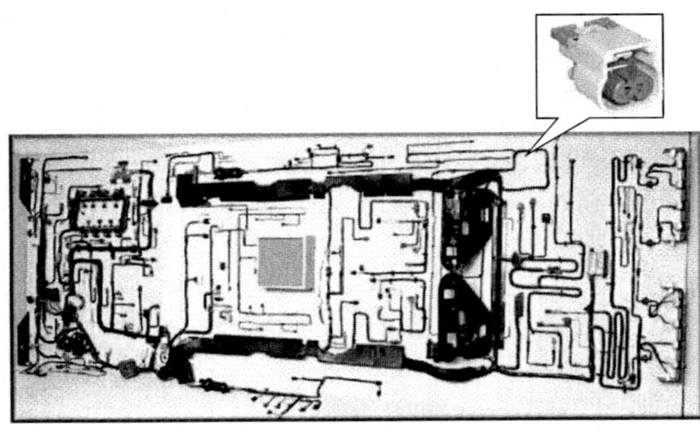

Figure 4 – Today's Typical (Automotive) Vehicle Wire Harness

The topology of a vehicle network can vary, depending on the network application. A Ring topology, Figure 5, is typically used for networks requiring redundant connections in mission critical applications.

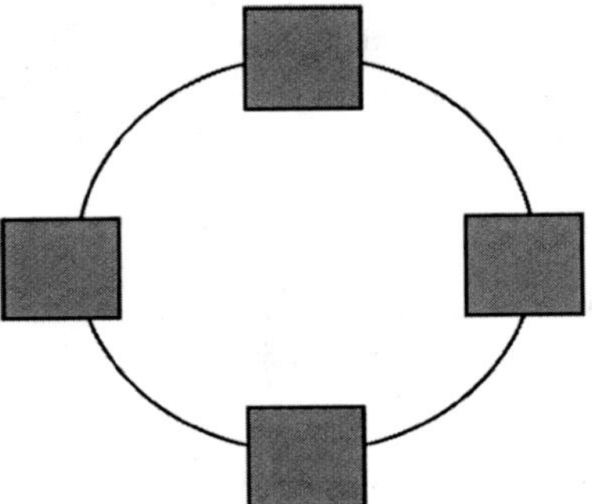

Figure 5 –Ring Network Topology

A Star topology, Figure 6, is typically used in less complex, often lower cost applications.

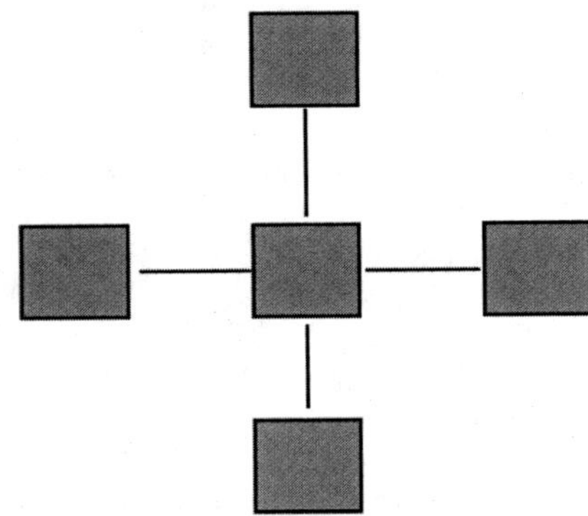

Figure 6 –Star Network Topology

In network applications with nodes operating as "peer-to-peer", any node is allowed to communicate with any other node. For this architecture, similar to the world of LANs and Ethernet, the Bus topology is commonly used. Note that the Controller Area Network or CAN protocol, a mainstay of vehicle networks, is a Bus topology design, Figure 7.

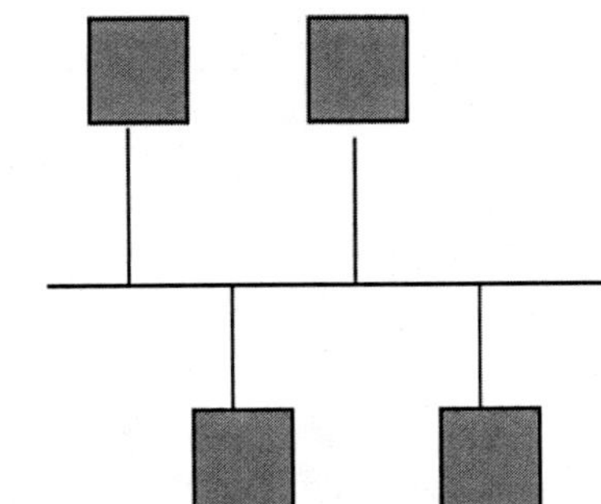

Figure 7 –Bus Network Topology used by the CAN Vehicle Network Protocol

C. Vehicle Network Protocol Features

What vehicle network features must to be supported to handle today's HD vehicle network environment, applications, and emerging uses? Although this is easily the subject of another paper on that specific subject, the following are typical features that are specified for all vehicle network applications.

- Message assignments for each node
  - Data Transmitters
  - Data Receivers
- Message error checking methods
- Dominant and Recessive States (network voltage levels)
- Message Frame Configuration
  - Header descriptor
  - Message priority
- Bus Arbitration Method
  - Destructive, Non-Destructive
- Large Message Transport Mechanism
- Parameters for Data
  - Length
  - Bit Resolution And Offset
  - Broadcast Rate
    - Periodically
    - As Necessary
    - On-request

For example, the bus arbitration method from the list above describes how a device obtains access to the network on a timely basis. This is important in determining how to architect a network. Critical vs. non-critical applications, must be prioritized. This becomes more important as data travels to and from a vehicle. As a point of reference, let's discuss something everyone is familiar with, a LAN using an Ethernet network. Whether at home or work, devices utilize a CSMA/CD (Carrier Sense Multiple Access/Collision Detect) protocol at the OSI layer 2 (more on this later) to determine which device has access to the single Ethernet network Since two or more stations share this common transmission medium (a LAN's Category 5 Cabling or Wi-Fi).

To transmit a PDU of data or information, a node must wait for an idle period on the medium when no other station is transmitting. It then transmits the frame by broadcasting such that it is "heard" by all the other stations on the network. If another device tries to send data at the same time, a "collision" occurs. After a reset frame, an algorithm called "truncated binary exponential back off" is used to generate a random time to retransmit, avoiding another collision.

Similarly, Vehicle Networks use protocols and specific methods to send information and recover from collisions, all referred to as bus arbitration.

4. HD VEHICLE NETWORK PROTOCOLS

A. J1708/J1587: Low Data Rate (Speed) HD Vehicle Network [Ref 1] [Ref2]

SAE International Technical Standards Committees have been active for over 25 years developing vehicle communications specifications for Automotive Passenger Vehicles and Heavy-Duty Commercial Vehicles. Network specifications have been developed for different types of vehicles (e.g. light duty or heavy duty) as well as for different networks applications on the vehicle (low data rate diagnostic applications, high data rate control applications, etc.). Many modern vehicles will have more than one network to support different on-board applications.

After the success that the automotive industry enjoyed with the advent of vehicle computing devices (i.e. ECUs) for increasing networked functionality and meeting emissions requirements, the HD industry followed closely behind with the "physical layer" (Figure 8, Layers 1 & 2 "Physical and Data Link Layers", SAE J1708) and a corresponding set of standardized messages defined in SAE J1587 (Figure 8, Layer 7 "Application Layer"). These two documents commonly referred to together as J1708/J1587 protocol, were developed as a joint effort between the American Trucking Association's Technology & Maintenance Council (TMC), and SAE International.

Besides solving some of the vehicle network challenges of the day, SAE J1708 and J1587 provided a very good learning experience for managing system interface issues. The development process helped engineers identify the problems that would need to be solved in future vehicle networks.

- Most Commonly Found in HD World

7	Application	Diagnostic Apps, Transfer a File, Send Messages
6	Presentation	Handles File Format Differences
5	Session	Provides Synchronization of Data Flow
4	Transport	Provides End to End Delivery
3	Network	Switches and Routes Information (E.G. Router)
2	Data Link	Delivers Information to the Next Node
1	Physical	Transmits Bit Stream on Physical Medium

Figure 8 – The Open Systems Interconnection (OSI) Model upon Which Most Network Standards Are Based

Due to its fairly low speed (9.6Kbps), SAE J1708/J1587 is not considered a preferred "control" network today, but it remains in use as an information-sharing and diagnostics network. Ten years ago it was predicted that it would be gone in ten years. It will, however, most likely be here for another ten years, as it is slowly being replaced by newer, faster general-purpose, and even slower specific-purpose network protocols.

The following are defined in the SAE J1708/J1587 protocol and are typical structures and functions that are defined for most vehicle network protocols.

- Priority - 3 Bits (0 = Highest, 7 = Lowest)
- MID (Message IDentifier)
    - Engine, Transmission, Etc
- PID (Parameter IDentifier)
    - Vehicle Speed, Engine Speed, Etc.
    - Packed PIDs
- Destructive Bus Arbitration
- Checksum
    - Mathematical Validation (Good Message / Bad Message)
- Diagnostic Fault Codes
    - Active Faults, Inactive Faults
- Active Diagnostic Codes:
    - Indicate A Fault Is Presently Occurring
    - Check Engine Light, ABS Light
        - MIL – Malfunction Indicator Light
    - No Lights Defined By J1708/J1587
        - Left Up To Instrument Cluster Software
- Inactive Codes:
    - Not Presently Occurring, But Has Been Detected;
        - The Conditions To Create The Fault Are Not Present "Right Now"
    - Clearing Inactive Faults
        - Possible Only To Clear "Inactive" Faults
    - SID (Subsystem IDentifier)
        - Problem With A Component (e.g. fuel injector)
    - FMI (Failure Mode Indicator)
        - "Shorted High", "Data Above Normal", Etc.
    - Occurrence Count
        - How Many Times Has This Fault Happened?

B. The CAN Protocol

Searching for a higher speed network to support modern HD vehicle power-train control applications, the SAE determined that the Bosch **C**ontroller **A**rea **N**etwork (CAN) protocol could be adopted for its requirements.

The International Standards Organization (ISO) has adopted the CAN protocol as a standard for high speed data communications (ISO 11898-2). A summary of the CAN protocol features follows.

- CANID description (the CAN IDentifier)
    - 11-bit CANID = "Standard CAN"
    - 29-bit CANID = "Extended CAN"
        - (J1939 Uses "Extended CAN")
- Error Checking
    - Bit Stuffing Rule
        - No More Than Five Consecutive 1s Or 0s
- Non-Destructive Bus Arbitration
    - Nodes may transmit many different messages
    - Each message has an unique priority assign to it via software
    - Bus arbitration occurs when 2 or more nodes transmit a message at exactly the same time.
    - The message with the highest priority wins the arbitration and its message transmission succeeds, the others fail and will retry.
- Outstanding Error Detection Capability
- No Master or Slave control module
    - All nodes communicate Peer to Peer.
- All nodes see all messages on network at the same time.
- 0 to 8 bytes of data per CAN frame.
    - This can be changed dynamically by software.
- Network needs two nodes to operate.
    - An off board diagnostic tool can normally act as a node.

The features noted above describe the Data Link Layer (Figure 8, Layer 2) of a vehicle network, but are not a complete protocol definition. Vehicle networks also require definitions for an Application Layer (i.e. messages, Layer 7, Figure 8) and Physical Layer (Layer 1, Figure 8). SAE therefore added those missing protocol layers by when incorporating CAN into the new SAE J1939 specification.

The CAN protocol Data Link Layer features have been implemented in an ISO 11898-2 compliant device called the CAN Controller Chip. The chips are made by a number of semiconductor suppliers.

Figure 9 below is an example CAN Node block diagram with the CAN Controller Chip. It also shows a CAN Transceiver Chip which implements a two wire differential Physical Layer transmitter-receiver circuit as specified in ISO 11898-1.

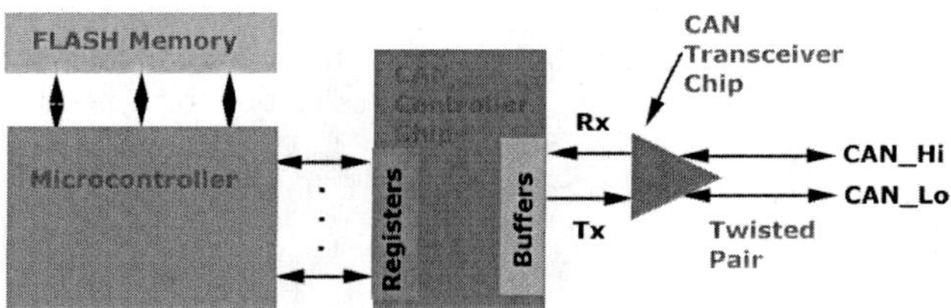

Figure 9 – CAN Node Block Diagram with CAN Controller Chip

Communication with the other CAN nodes on the Bus is done via the CAN Controller device. It streams data serially using a CAN Transceiver to produce the voltage levels on the CAN_Hi and CAN_Lo wires. The microcontroller is programmed to implement specified messages (such as defined in the SAE J1939-71 Application Layer) common to all nodes on the bus. The messages are designed to communicate commands, status information and diagnostics data between nodes.

C. J1939/CAN: High Speed HD Vehicle Network for Control and Diagnostics Overview *[Ref 3]*

i. SAE J1939 Scope

The SAE J1939 specification is a "suite" of documents that are selectively employed by system design engineers. The document J1939 (without a "dash number") is known as the "top level document". Its scope, as shown below, describes the overall operation of the protocol. The other documents in the J1939 suite are given "dash numbers" (e.g. J1939-11) that denote specified OSI Layers (e.g. J1939-<u>1</u>1; the <u>1</u> indicates this document is an OSI Layer 1 related).

The SAE J1939 documents are intended for light, medium, and heavy-duty vehicles used on or off road as well as appropriate stationary applications which use vehicle derived components (e.g. generator sets). Vehicles of interest include, but are not limited to, on- and off-highway trucks and their trailers, construction equipment, and agricultural equipment and implements.

The purpose of these documents is to provide an open interconnect system for electronic systems. It is the intention of these documents to allow Electronic Control Units to communicate with each other by providing a standard architecture.

The CAN protocol corrected many of the issues found in J1708/J1587, especially the problem of arbitrating messages on the bus, and providing the increased speed needed to serve vehicle networks and, at the time, then relevant and new or emerging applications. Again, SAE J1939 utilizes the CAN protocol and defines additional requirements such as physical layer and message layer documents. SAE J1939 is very robust and at 250Kbps runs at a speed 25 times as fast as J1708/J1587.

What J1939/CAN did was to solve most, if not all, problems encountered with SAE J1708/J1587 and introduced many new features, including:

<u>Device-to-Device Messaging</u>. There were "source addresses" defined in J1708/J1587, but there were no messages specifically designed for "destination addresses" except for proprietary messages. J1939 includes destination-specific messages, called Protocol Data Unit 1 (PDU1) types, as well as general-purpose broadcast messages, called PDU2 types. The PDU1 types allow controllers on the data bus, such as a hydraulics pump controller on a mining truck, to easily ignore messages not specifically destined for it. This allows controllers more time for their own sensor monitoring and control.

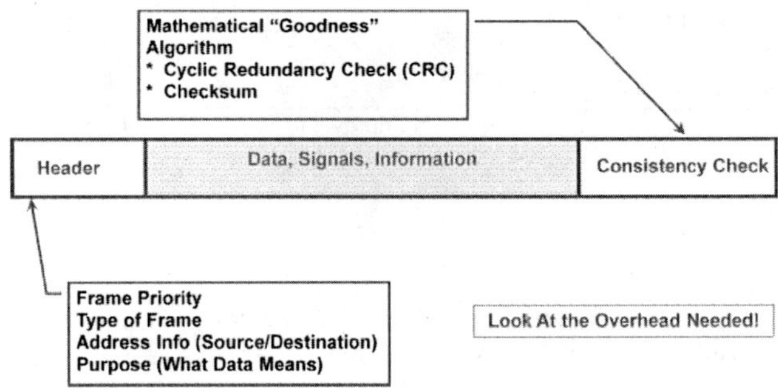

Figure 10 – A Standard Format for a Vehicle Network Protocol Data Unit (PDU). For Example, J1939 And J1587 Follow This General Format.

<u>Good "Large Message" Support</u>. Underlying CAN messages are limited to 8-bytes, whereas J1708/J1587 messages are 21-bytes. However, the J1939 committee

uses those 8-bytes to their maximum potential, breaking them into "Parameter Group Numbers" or PGNs. PGNs are simply groupings of related types of messages, such as engine messages. When messages greater than 8-bytes are needed, there is a mechanism for getting those messages where they needed to go. This method, called the "Transport Layer" (Layer 4, Figure 8), offers both flow-controlled messaging with handshaking (called Ready-To-Send/Clear-To-Send or RTS/CTS) as well as a faster approach without handshaking, called Broadcast Announce Message or BAM.

Industry-Specific Messages. Not only were problems solved, and new "protocol" features added, but J1939/CAN opened up a new arena in messaging, as the construction, agriculture and industrial-stationary OEMs have even added their own messages and the process of dynamically figuring out which address to claim on the bus.

Figure 11 is a summary of some of the J1939 network features vs. J1708/J1587.

Feature	*J1939*	*J1708/J1587*
Command messages	Yes	Yes
Data messages	Yes	Yes
Data rate	250K bits/sec.	9600 bits/sec.
Physical layer	Differential twisted pair wires	Differential twisted pair wires
Bus voltage spec.	ISO 11898-1	RS-485
Network topology	Bus	Bus
Max. length of Bus	40 m	40 m
Highest priority message always gets transmitted first	Yes	No

Figure 11 – J1939/CAN Improvements over J1708/J1587

ii. Elements of the SAE J1939 Standard

Some specific elements of the SAE J1939 standard are described for the following 4 Areas:

- Physical Layer,
- Data Link Layer,
- Application Layer, and
- Use of J1939 Family of Standards:

Physical Layer

SAE J1939 has defined several Physical Layer requirements documents. The bus wiring documents are J1939-11 and J1939-15, either of which may be used on a vehicle. The document J1939-13 defines the HD-OBD connector requirements.

SAE J1939-11

"This document defines a physical median of shielded twisted pair. These 2 wires have a characteristic impedance of 120 Ω and are symmetrically driven with respect to the electrical currents. The designations of the individual wires are CAN_H and CAN_L. The names of the corresponding pins of the ECUs are also denoted by CAN_H and CAN_L, respectively. The third connection for the termination of the shield is denoted by CAN_SHLD." [Ref4]

This document is commonly referred to as the Truck and Bus Physical Layer, as:

- The bus data rate is fixed at 250K bits/sec.
- The communication media is a shielded, twisted-pair cable with a drain requiring a termination resistor at each end.
- Network connections are made using a three-pin, unshielded connector. The three pins are defined as *CAN_H*, *CAN_L* and *Shield*.
- Maximum 30 nodes per segment.

SAE J1939-15

"This document describes a physical layer utilizing Unshielded Twisted Pair (UTP) cable." [Ref5]

This document is commonly referred to as the "Lite" or "Reduced" Physical Layer, as:

- The bus data rate is fixed at 250K bits/sec.
- Based on unshielded, twisted-pair (UTP) cable with a drain requiring a termination resistor at each end.
- Maximum 10 nodes per segment.

SAE J1939-13

Newer vehicles typically have the diagnostics connector installed near the driver's side. The connector (Figure 12) has 9 pins including CAN_Hi and CAN_Lo for the J1939 signals (the earlier vintage 6 pin connector does not include J1939 signals).

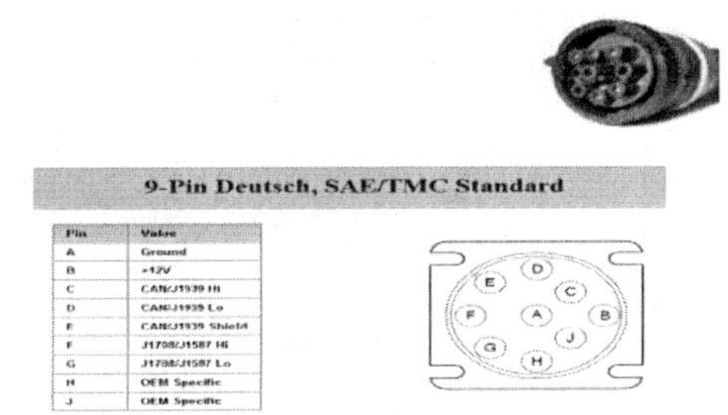

Pin	Value
A	Ground
B	+12V
C	CAN/J1939 Hi
D	CAN/J1939 Lo
E	CAN/J1939 Shield
F	J1708/J1587 Hi
G	J1708/J1587 Lo
H	OEM Specific
J	OEM Specific

Figure 12 – J1939-13 Diagnostics Connector

Data Link Layer

J1939-21 specifies the Data Link Layer requirements for the network including the Protocol Data Unit (PDU) definitions for use of the 29 CAN Identifier bits as message types (see PGN description below), and the rules used when more than 8 bytes of data needs to transferred by a message (i.e. transport protocol, Figure 8, Layer 4).

SAE J1939-21

"This document describes the data link layer using the CAN protocol with 29-bit Identifiers. SAE J1939 only allows the use of the described data link layer." [Ref6]

Specifically, J1939-21 specifies the bit definitions of the PDUs (Figure 13), the five message types supported (command, request, broadcasts/response, acknowledgement, group function), and the transport protocol used to transfer large data blocks.

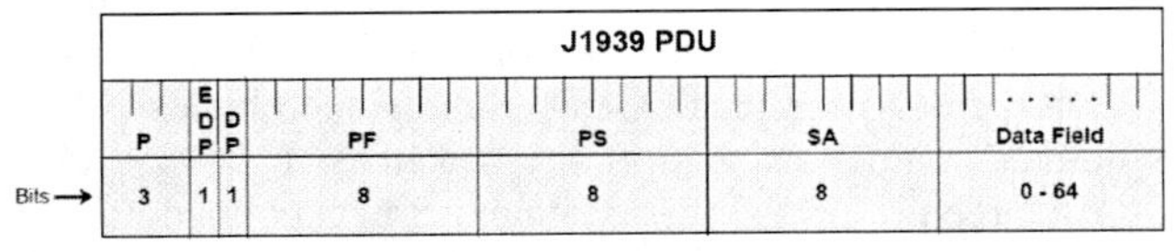

Definitions: P is Priority, EDP is Extended Data Page, DP is Data Page, PF is PDU Format, PS is PDU Specific, and SA is Source Address

Figure 13 – J1939-21 Data Link Layer Bit Definitions

Application Layer

SAE J1939 has defined several Application Layer requirements documents, including J1939-71 and J1939-73. These documents specify how an entire message consisting of the PDU and the data bytes are encoded.

SAE J1939-71

As a J1939 application standard, SAE J1939-71 specifies common *control and status messages* for HD vehicles. These control and status messages are continuously sent between ECUs on a vehicle to operate the vehicle's systems.

For example, Figure 14 below shows the format of message PGN65215 which is defined as EBC2 (Electronic Brake Control message 2 - wheel speed information). The ECB2 message has been assigned a Parameter Group Number (PGN) which defines what parameters are included in the data field of the message.

pgn65215 - Wheel Speed Information - EBC2 -

Transmission Repetition Rate:		100 ms	
Data Length:		8 bytes	
Data Page:		0	
PDU Format:		254	
PDU Specific:		191	
Default Priority:		6	
Parameter Group Number:		65215 ($00FEBF_{16}$)	
Bit Start Position /Bytes	Length	SPN Description	SPN
1-2	2 bytes	Front Axle Speed	904
3	1 byte	Relative Speed; Front Axle, Left Wheel	905
4	1 byte	Relative Speed; Front Axle, Right Wheel	906
5	1 byte	Relative Speed; Rear Axle #1, Left Wheel	907
6	1 byte	Relative Speed; Rear Axle #1, Right Wheel	908
7	1 byte	Relative Speed; Rear Axle #2, Left Wheel	909
8	1 byte	Relative Speed; Rear Axle #2, Right Wheel	910

Figure 14 – Real World Example of SAE J1939-71 Information

SAE J1939-73

Another J1939 application standard, SAE J1939-73 specifies *messages for HD-OBD* functions. For example, Figure 15 below shows the format of message DM1 (Diagnostics Message 1). This message contains data indicating an ECU's Diagnostic Trouble Code (DTC), Failure Mode Indicator (FMI), and other information useful for the vehicle's maintenance.

"This document identifies the diagnostic connector to be used for the vehicle service tool interface and defines messages to accomplish diagnostic services. California-regulated OBD II requirements are satisfied with a subset of the specified connector and the defined messages. Diagnostic messages (DMs) provide the utility needed when the vehicle is being repaired. Diagnostic messages are also used during vehicle operation by the networked electronic control modules to allow them to report diagnostic information and self-compensate as appropriate, based on information received. Diagnostic messages include services such as periodically broadcasting active diagnostic trouble codes, identifying operator diagnostic lamp status, reading or clearing diagnostic trouble codes, reading or writing control module memory, providing a security function, stopping/starting message broadcasts, reporting diagnostic readiness, monitoring engine parametric data, etc." [Ref 8]

HD diagnostics are typically done by a technician using a VDA connected between the SAE J1939-13 vehicle diagnostics port and a computer running an application. The DM1 (Diagnostics Message 1) message is the only diagnostics message which is transmitted continuously by an ECU. All of the other approximately 50 Diagnostics Messages (DM) defined are only transmitted upon receiving a Request message from another node (e.g. from a VDA). The DM1 message has been assigned PGN65226 (Fig 15).

Transmission Rate:	A DM1 message shall be transmitted, regardless of the presence or absence of any DTC, once every second and on state change. To prevent a high message rate due to intermittent faults that have a very high frequency, it is recommended that no more than one state change per DTC per second be transmitted. For example, if a fault has been active for 1 second or longer, and then becomes inactive, a DM1 message shall be transmitted to reflect this state change. If a different DTC changes state within the 1 second update period, a new DM1 message is transmitted to reflect this new DTC.
Data Length:	Variable (When there is only one DTC to report then unused bytes 7 and 8 of the CAN frame shall be set to 255 (as per J1939-21)).
Extended Data Page:	0
Data page:	0
PDU Format:	254
PDU Specific:	202
Default Priority:	6
Parameter Group Number:	65226 (00FECA$_{16}$)

Byte	Bits	Description
Byte: 1	bits 8-7	Malfunction Indicator Lamp Status
	bits 6-5	Red Stop Lamp Status
	bits 4-3	Amber Warning Lamp Status
	bits 2-1	Protect Lamp Status
Byte: 2	bits 8-7	Flash Malfunction Indicator Lamp
	bits 6-5	Flash Red Stop Lamp
	bits 4-3	Flash Amber Warning Lamp
	bits 2-1	Flash Protect Lamp
Byte: 3	bits 8-1	SPN, 8 least significant bits of SPN (most significant at bit 8)
Byte: 4	bits 8-1	SPN, second byte of SPN (most significant at bit 8)
Byte: 5	bits 8-6	SPN, 3 most significant bits (most significant at bit 8)
	bits 5-1	FMI (most significant at bit 5)
Byte: 6	bit 8	SPN Conversion Method
	bits 7-1	Occurrence Count

A Suspect Parameter Number (SPN) is a 2 byte value which designates the data parameter used with the messages.
A Diagnostic Trouble Code (DTC) is made up of four (4) independent fields, as follows:

Suspect Parameter Number	(SPN)	19	bits
Failure Mode Identifier	(FMI)	5	bits
Occurrence Count	(OC)	7	bits
SPN Conversion Method	(CM)	1	bit

Figure 15 – Messages for HD-OBD functions, Diagnostic Message 1 (DM1)

A Diagnostic Trouble Code (DTC) (Figure 16) is made up of four (4) independent fields. A Suspect Parameter Number (SPN) is a 2 byte value which designates the data parameter used with the messages.

<table>
<tr><td></td><td colspan="32">DTC</td></tr>
<tr><td></td><td colspan="8">Byte 3
8 least significant bits of SPN
(bit 8 most significant)</td><td colspan="8">Byte 4
second byte of SPN
(bit 8 most significant)</td><td colspan="8">Byte 5
3 most significant bits of SPN and the FMI (bit 8 SPN msb and bit 5 FMI msb)</td><td colspan="8">Byte 6</td></tr>
<tr><td></td><td colspan="19">SPN</td><td colspan="5">FMI</td><td>CM</td><td colspan="7">OC</td></tr>
<tr><td>J1939 Frame Format</td><td>8</td><td>7</td><td>6</td><td>5</td><td>4</td><td>3</td><td>2</td><td>1</td><td>8</td><td>7</td><td>6</td><td>5</td><td>4</td><td>3</td><td>2</td><td>1</td><td>8</td><td>7</td><td>6</td><td>5</td><td>4</td><td>3</td><td>2</td><td>1</td><td>8</td><td>7</td><td>6</td><td>5</td><td>4</td><td>3</td><td>2</td><td>1</td></tr>
<tr><td></td><td>1</td><td>0</td><td>1</td><td>1</td><td>1</td><td>0</td><td>0</td><td>0</td><td>0</td><td>0</td><td>0</td><td>0</td><td>0</td><td>1</td><td>0</td><td>0</td><td>0</td><td>0</td><td>0</td><td>0</td><td>0</td><td>0</td><td>1</td><td>1</td><td>0</td><td>0</td><td>0</td><td>0</td><td>1</td><td>0</td><td>1</td><td>0</td></tr>
</table>

Figure 16 – Diagnostic Trouble Code (DTC) J1939 Frame Format

SAE J1939-84

"The purpose of this Recommended Practice is to verify that vehicles and/or components are capable of communicating a minimum subset of information, in accordance with the diagnostic test modes specified in SAE J1939-73: Diagnostic Services. This document describes the tests, test methods, and results for verifying diagnostics communication from an off board device to a vehicle and/or component." [Ref 9]

Use of J1939 Family of Standards

Incredibly enough, there are even standards for how to *use* standards. SAE J1939 has these types of documents as well, for example:

SAE J1939-03

"SAE J1939-03 provides requirements and guidelines for the implementation of On Board Diagnostics (OBD) on heavy-duty vehicles (HDV) using the SAE J1939 family of standards. The guidelines identify where the necessary information to meet OBD regulations may be found among the SAE J1939 document set. Key requirements are identified here to insure the interoperability of OBD scan tools across individual OBD compliant vehicles. Market-defined regulations permit the use of SAE J1939 to meet OBD requirements. Implementers are cautioned to obtain and review the specific regulations for the markets where their products are sold. This document is focused on guidelines and requirements to satisfy the State of California Air Resources Board (ARB), the authors of 13 CCR 1971.1, United States Environmental Protection Agency, Euro IV and V requirements from European Commission directives, and UN/ECE WP 29 GRPE WWH OBD Global Technical Regulation (GTR)." Ref [7]

5. DIAGNOSTICS AND REPROGRAMMING STANDARDS

A. RP1210 API

This section provides an overview of the HD vehicle Diagnostics Interface and the RP1210 API.

How vehicles perform diagnostics (and prognostics) is facilitated by connecting a tool as a node to the in-vehicle network. The tool is usually and off board device (such as a VDA) temporarily attached to the network. But the tool could also be an on board device permanently connected to the vehicle's network. The tool is designed to transmit diagnostic request messages to other ECUs on the vehicle and to process the responses received from the ECUs.

While vehicle network standards help to provide a common framework for exchanging information on the vehicle, it is the Application Programming Interface (API) that allows for the off-board diagnosis and reprogramming of vehicle components via vehicle network interface devices, including personal computers. For the vehicle networks which apply to heavy-duty industry, the API of common interest and standardization is TMC's Recommended Practice (RP) 1210. RP1210 supports standards J1708/J1587, CAN, J1939, as well as these other HD vehicle network standards: ISO 15765, J1850, J2497 (PLC4TRUCKS), and ISO 9141.

Simply stated, RP1210 allows one diagnostic / reprogramming protocol adapter to be used for many OEM/component applications, and is a parallel to the SAE J2534 standard the automotive industry is using (discussed next). In a poll taken in 2007, there were over 150 commercially used RP1210 compliant software applications in the field. Some OEMs initially resisted adopting the standard (mostly those having a vertically-integrated component vehicle design), but due to fleet pressure, today all HD vehicle OEMs, engine, transmission and ABS component suppliers support the RP1210 standard.

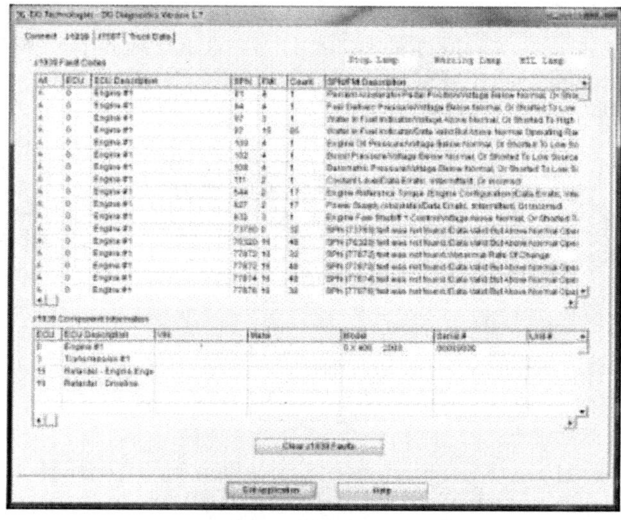

Figure 17 – Technician Performs Vehicle Network Diagnostics via a PC and Protocol Adapter

The history of RP1210 was born from truck fleet frustrations of the late 1980's and early 1990's. As electronic vehicle components came on the scene, every engine, transmission, and ABS manufacturer came out with their own PC diagnostics programs. Not only did they write software, but also they created or sourced their own J1708/J1587 adapters and cabling. The problem was that with no standards in place, one supplier's software would seldom work with other supplier's adapters. To add complexity, writing device drivers for a fledgling operating system (Windows 3.1/95) was very difficult, tedious, and often-turned up operating system bugs and forced "work-arounds".

Although they were not really electronically complicated, the adapters were overly complex and expensive. In visiting a HD service bay, one would be amazed at the spider web of cables and adapters that filled the tool crib wall. To service 3 truck OEMs, 4 brands of engines, 3 brands of transmissions and 2 brands of ABS brake units in a service shop having 10 bays a company would have to purchase 120 adapter/cable sets, and having to pay upwards of $500 for each, easily spending over $100,000 just to equip the shop. Adding to the excessive cost was the irritation produced from the physical time a technician needed to switch from one application to another; he had to unhook the old adapter and cable and he might have to reboot the PC. In addition there were also costs for software updates and new adapters that were designed to support the "new" protocol on the block (i.e. J1939) as well as supporting the older legacy protocols.

Fleet owners approached the TMC S.12 Offboard Vehicle Electronics Study Group with the problem. Soon, the TMC Recommended Practice RP1210 and "Coopetition" in Diagnostics was born.

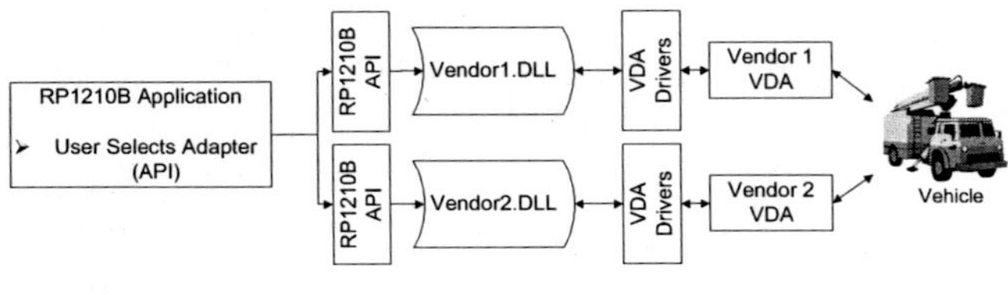

Figure 18 – The TMC's RP1210 functional diagram. A single vendor's Vehicle Datalink Adapter (VDA) can support all RP1210 compliant applications.

RP1210 is a cooperative effort written by HD OEM and component suppliers who volunteer their time to further the industry as a whole. RP1210 allows one diagnostic / reprogramming protocol adapter to be used for many OEM/component applications. RP1210 is currently in the field as "RP1210A" and is being slowly migrated to the new "RP1210B" version, which plugged a lot of "holes" and features, all the while staying backwards compatible with RP1210A. RP1210C, which will include the legacy protocol ISO9141, is scheduled for release in the 2011 timeframe.

B. SAE J2534 API

"This SAE Recommended Practice provides the framework to allow reprogramming software applications from all vehicle manufacturers the flexibility to work with multiple vehicle data link interface tools from multiple tool suppliers. This system enables each vehicle manufacturer to control the programming sequence for electronic control units (ECU's) in their vehicles, but allows a single set of programming hardware and vehicle interface to be used to program modules for all vehicle manufacturers." [Ref 10]

This API is used for Automotive, Light Duty vehicle applications. It does, however, also have optional support for J1939 so may be used in future to support HD vehicles.

6. ADDITIONAL HD/OBD RELATED STANDARDS

The following section is an overview of several other standards related to HD-OBD and vehicle networks.

A. SAE J1930

"This document focuses on diagnostic terms applicable to electrical/electronic systems, and therefore also contains related mechanical terms, definitions, abbreviations, and acronyms." [Ref 11]

B. SAE J2497

"This SAE Recommended Practice defines a method for implementing a bidirectional, serial communications link over the vehicle power supply line among modules containing microcomputers. This document defines those parameters of the serial link that relate primarily to hardware and software compatibility such as interface requirements, system protocol, and message format that pertain to Power Line Communications (PLC) between Tractors and Trailers. This document defines a method of activating the trailer ABS Indicator Lamp that is located in the tractor." [Ref 12]

C. SAE J1979

"This document is intended to satisfy the data reporting requirements of On-Board Diagnostic (OBD) regulations in the United States and Europe, and any other region that may adopt similar requirements in the future." [Ref 13]

D. SAE J2012

"This document is intended to define the standardized Diagnostic Trouble Codes (DTCs) that On-Board Diagnostic (OBD) systems in vehicles are required to report when malfunctions are detected. This document includes: a) Diagnostic Trouble Code format. b) A standardized set of Diagnostic Trouble Codes and descriptions and c) A standardized set of Diagnostic Trouble Codes subtypes known as Failure Types." [Ref 14]

E. ISO 9141

"This document specifies a low speed vehicle diagnostics protocol using UART type devices. It defines the requirements for setting up the interchange of digital information between on-board Electronic Control Units (ECUs) of road vehicles and suitable diagnostic testers." [Ref 15]

F. ISO 15765-4

"This document specifies requirements for the emissions-related systems of legislated OBD-compliant controller area networks (CAN), such communications networks consisting of a road vehicle equipped with a single or multiple emissions-related ECUs and external test equipment. It is based on the specifications of ISO 15765-2, ISO 11898-1 and ISO 11898-2, while placing restrictions on those standards for legislated-OBD purposes." [Ref 16]

7. ON-BOARD DIAGNOSTICS

"On-Board Diagnostic systems are self-diagnostic systems incorporated into the computers of new vehicles. All 1996 and newer vehicles less than 14,000 lbs. (e.g., passenger cars, pickup trucks, sport utility vehicles) are equipped with OBD II (or OBD-2) systems, which are California's second generation of OBD requirements.

The OBD II system monitors virtually every component that can affect the emission performance of the vehicle to ensure that the vehicle remains as clean as possible over its entire life, and assists repair technicians in diagnosing and fixing problems with the computerized engine controls. If a problem is detected, the OBD II system illuminates a warning lamp on the vehicle instrument panel to alert the driver. This warning lamp typically contains the phrase "Check Engine" or "Service Engine Soon". The system will also store important information about the detected malfunction so that a repair technician can accurately find and fix the problem." [Ref 17]

Government regulators (such as the California ARB) established goals for the OBD-2 system to implement:

- A stringent new vehicle certification process
- Monitor vehicle performance based on emission regulations
- Detect all emission related faults
- Assist vehicle technicians with fixing faults
- Faster diagnostics and repair of system issues
- Tamper resistant to discourage cheating
- Constant improvement for new requirements

Figure 19 shows the standard OBD-2 automotive diagnostic connector used to interface scan tools or VDAs to a light duty vehicle.

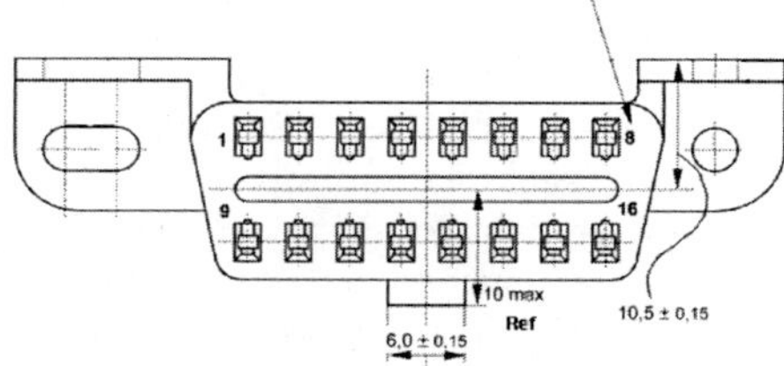

Figure 19 – The OBD-2 Light Duty Vehicle Diagnostics Connector (SAE J1962)

The goals of a vehicle OEM for the OBD-2 system are:

- To place the customer first
- Fix the vehicle right,
- Conveniently, quickly
- Reduce repair costs
- (Potentially a revenue source?)

Evolution of OBD Regulations and Service Support

Regulation	Circa	I/M Vehicle Test Equipment	OEM Vehicle Service Support	SAE Standards
OBD-1 (first generation OBD system)	1980's	Tailpipe ASM	Flashing code lamp	None
OBD-2 (second generation)	1996 - current (evolution continues)	Scan tools and PCs	Trouble codes, Monitors, Function tests, ECU reflash	J1979, J2012, J1962, J2534...
OBD-3 ??? (suggested next generation)	TBD ????	Wireless and Internet	Constant connection with vehicle	New J-doc for open vehicle communication and functional interoperation?

Figure 20 – OBD Through the Years

On Board Diagnostics systems continue to evolve (Figure 20). There are new regulations coming from California Air Resources Board (CARB), the US EPA, and Europe.

Much work is currently being done by the ISO to develop standards to support the new World Wide Harmonization of On Board Diagnostics (WWH-OBD) regulation which is intended to be adopted by European countries in 2016. The new ISO 27145 document is under development which provides the technical requirements for supporting WWH-OBD. SAE is cooperating with ISO in the development of these new standards.

There will be new vehicle service requirements for more complex emission control and safety systems. Worldwide cooperation for standards supporting these systems is increasing with a view towards lowering vehicle development and maintenance cost. Also, new technology is being adopted, such as the draft standard ISO 13400 Diagnostics over Internet Protocol (DoIP) and wireless diagnostics communications which will need further standards and HD vehicle support.

8. FLEET MANAGEMENT SYSTEMS (FMS)

A fleet management application is an application used by vehicle fleet managers to track and maintain the health of all the vehicles in their fleet. The fleet management software closely integrates with the vehicle diagnostics data and proves additional features including:

- Vehicle fleet health monitoring
- Vehicle maintenance scheduling and tracking
- Prognostics and Condition Based Maintenance (CBM+)
- Fleet reports
- Interface to parts order processing applications
- Geo-fencing & delivery documentation

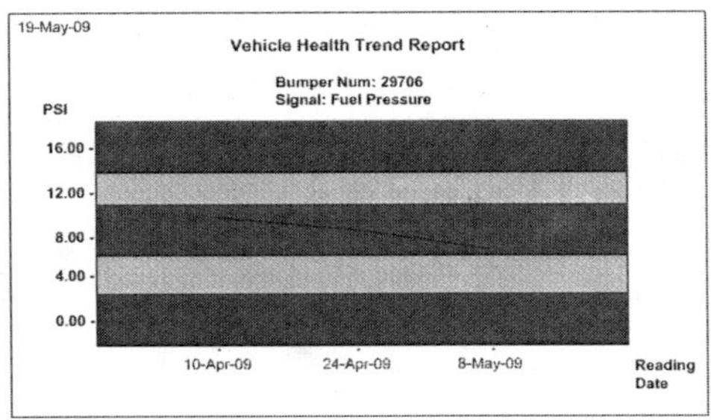

Figure 21 – Fuel Pressure Vehicle Health Trend Report

Shown in Figure 21 is an example vehicle health trend report from a FMS application that can be used for prognostics. Another type of application of a FMS may be to monitor compliance to safety regulations.

"Relating FMS to safety, the U.S. Department of Transportation (USDOT) Federal Motor Carrier Safety Administration (FMCSA) commissioned the Wireless Roadside Inspection (WRI) Program to validate technologies and methodologies that can improve safety through inspections using wireless technologies that convey real-time identification of commercial motor vehicles (CMVs), drivers, and carriers, as well as information about the condition of the vehicles and their drivers. It is hypothesized that these inspections will:

- Increase safety—Decrease the number of unsafe commercial vehicles on the road.
- Increase efficiency—Speed up the inspection process, enabling more inspections to occur, at least on a par with the number of weight inspections.
- Improve effectiveness—Reduce the probability of drivers bypassing CMV inspection stations and increase the likelihood that fleets will attempt to meet the safety regulations.
- Benefit industry—Reduce fleet costs, provide good return on investment, minimize wait times, and enable uniform roadside safety compliance checking of all motor carrier operations regardless of type and size of operations." [Ref 19]

A. Example of FMS and Backend Diagnostics Output

The FMS is designed to collect pre-configured data and provide it to a Backend Diagnostic System for off-line or on-line data analysis, data presentation and data archiving. This section outlines some of the main features.

- Vehicle Maintenance
    - Manage service intervals
    - Read, transmit, and reset trouble codes locally or remotely
    - Local (motor pool) and remote (in field) diagnostics

- Define, set, read triggers
  - Examples:
    - Over-speed
    - Over-temperature
    - Over-pressure
- Track parts used, manage service parts inventory
- Access to online diagnostic procedures, standards, and specifications
- Motor pool check in / check out
- Maintenance reminders

- Preventative Maintenance
  - Historical and statistical data analysis
  - Histograms
  - Data mining and trending

Vehicle data access, vehicle maintenance, and preventative maintenance are the key elements of a FMS. The list above illustrates functions that appropriate application and analysis software can perform with access to pertinent vehicle data.

The following Figures 22-24 demonstrate the presentation of recorded data, the use of diagnostics software using a VDA, and capturing of data, so called "data logging" to show trends and use for prognostics or CBM+. Different vehicle parameters are displayed in time-synchronized format, to enable the mechanic to better diagnose a problem by correlating various vehicle parameters.

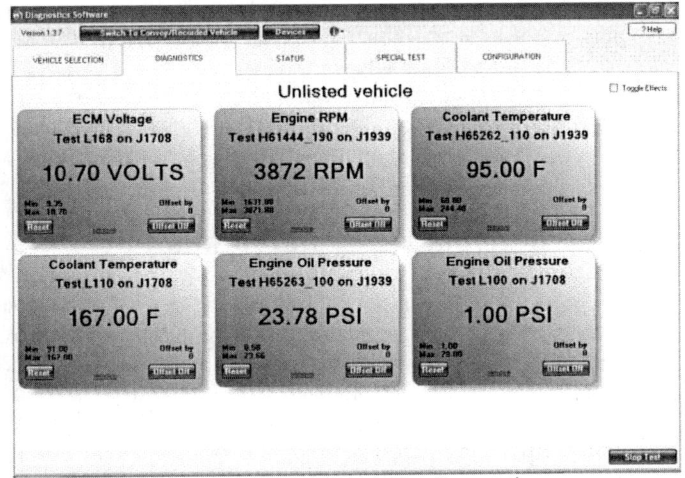

Figure 22 – FMS Vehicle Data Presentation

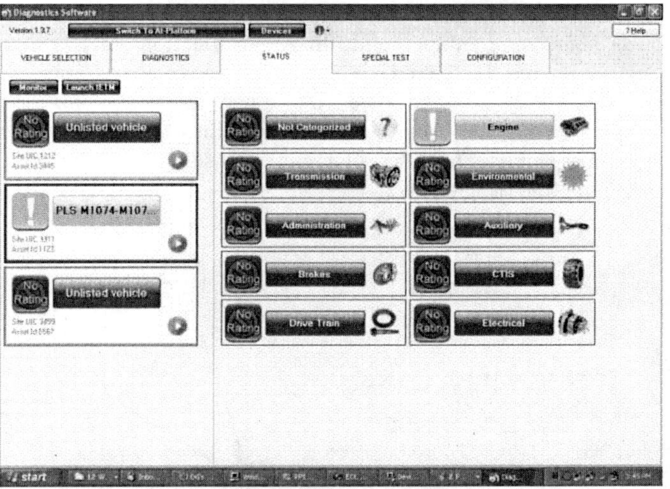

Figure 23 – FMS Diagnostics Software

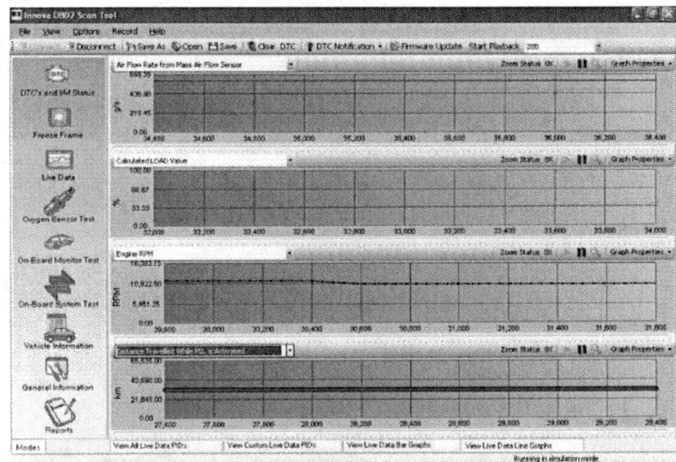

Figure 24 – FMS Benefits from Data Capture, Data logging

B. Leveraged Integration of FMS

i. Beyond VDAs, Smart Wireless Diagnostic Sensor Device

Fleet Management can now involve Intra-Vehicle Networks, using something called a Smart Wireless Diagnostic Sensor (SWDS) device. Think of a VDA that is semi-permanently mounted (e.g. to the Deutch connector shown in section 5.c.i J1939-13 connector) in a vehicle that can report information using various wireless technologies. The goal is to is to improve vehicle health monitoring and associated diagnostic systems, in effect, Fleet Management System goals.

Almost all of today's military vehicles have sophisticated on-board computers controlling the functions of various systems on the vehicle. These include systems such as engine, transmission, brakes, and other electronic controlled specialty devices. There is a wealth of information available on the vehicle network that can be used for diagnostics, CBM+ and prognostics.

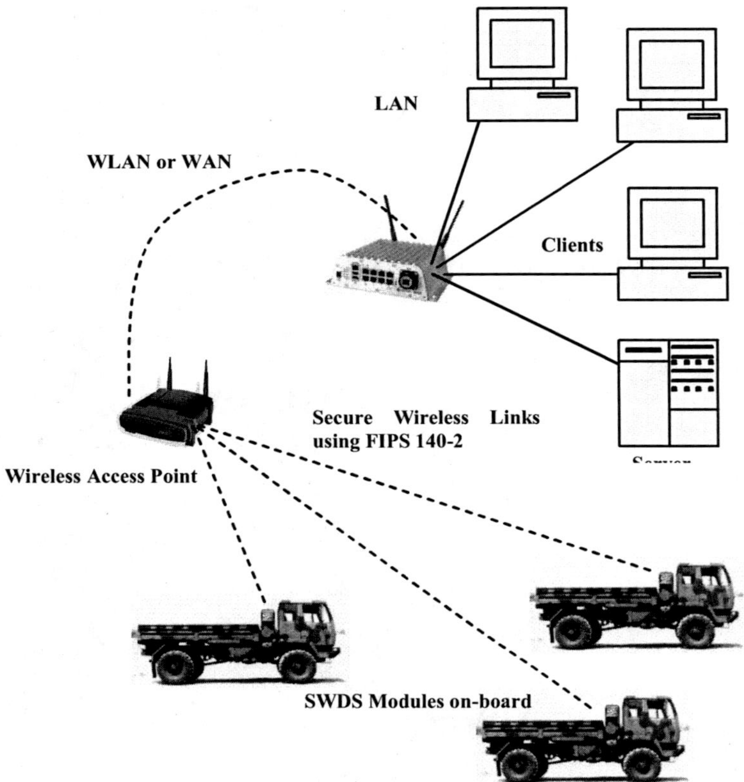

Figure 25 – FMS Topology: Merging Vehicle Networks & WANs

ii. Inter-Vehicle Networks: Convoy Sharing FMS

Military vehicles, each equipped with an inexpensive SWDS, can share information among themselves, or inter-vehicle, with the SWDS components semi-permanently installed on the vehicles. Convoy Mode (Figure 26) allows a computer, mounted in a lead vehicle, to monitor the health of all vehicles in a convoy via wireless technology as the convoy is deployed in the field.

The convoy will continuously and securely transmit the health of electronic systems via the embedded wireless technology, reporting to a lead unit that has more sophisticated equipment and applications available and running.

This economical leverage of a single vehicle's capability is only eclipsed by each vehicle reporting in real time over a cost-effective and always available WAN. In short, intra-vehicle communications can be replaced by inexpensive WAN services allowing each HD vehicle can talk to the "WAN cloud" instead of each other.

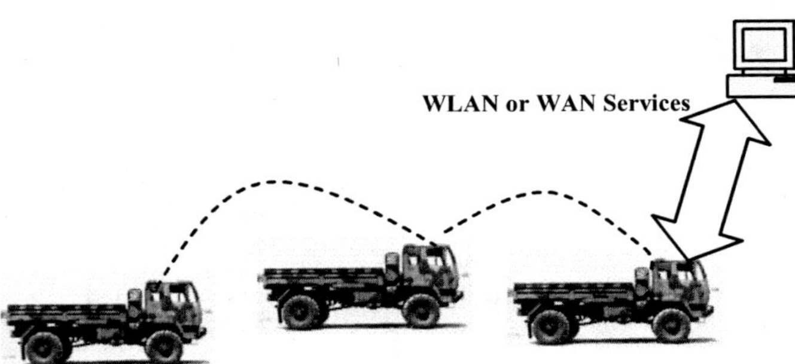

Figure 26 – HD Inter-Vehicle Info Sharing: WAN Service Beginnings

iii. Fleet Prognostics and Condition Based Maintenance (CBM+)

In the commercial fleet, CBM+ is accomplished by downloading vehicle data (hours, miles, faults, min/max temps and pressures, etc) either by somewhat expensive electronic GPS/cellular type systems (near-real-time-data) or by waiting for that vehicle to reach one of the fleet terminal locations so that a technician can physically connect a vehicle diagnostic adapter to it and download the data. Over hundreds or thousands of vehicles, and over time, the CBM+ database can be "mined" to determine specific trends and patterns as to vehicle and individual component life. This process is sometimes called "Pattern and Trend Analysis (PTA)", and saves fleets thousands of dollars every year from expensive on-highway downtime incidents and even more importantly, preemptive parts replacement based mostly on general human observations.

9. NETWORKING OF NETWORKS

The power and leverage of networking is to interwork networks. While vehicle networks have a quarter century of history, a parallel effort in the area of local and wide area networks has emerged over even a longer and even more storied past. Data circuits, T1's, T3s, CCSA Common Control Switching Arrangement networks, PSTNs Public Switched Telephone Networks, VPNs Virtual Private Networks, ARPAnet and the Internet and its associated leverages e.g. running voice apps using VoIP, cellular data networks 1G, 2G, 2.5G, 3G, and 4G all are just a few of these parallel efforts in data networking.

The bottom line is that in this regard, we are just starting HD vehicle as well as other vehicle communications…when we take into account LANs and WANs.

A. WANs - Wide Area (or Wireless Area) Networks

There are so many types of networks that often times there is a "doubling up" on acronyms. WANs, once just Wide Area Network, now can be Wireless Area Networks (or WLANs). VANs, once only Value Added Networks, are now also Vehicle Area Networks. PANS, Personal Area Networks, VPNs (Virtual Private Networks) and more all exemplify that there are so many networks that we are running out of acronyms. It also is evidence that there are many effective ways to transmit data.

The following will present an overview of some technologies that might be expected to "merge" together with in-vehicle networks to provide remote application connections with on-board vehicle systems.

B. LANs - Local Area Networks

Vehicle Networks are, in-effect Local Area Networks, or LANs, for vehicles. They are interconnected via "gateways", which are becoming increasingly complex and becoming "bridges or routers". There is a distinct similarity between this environment and that of an office and home office: they both communicate data in a similar manner.

Common communications protocols developed by ISO, IEEE, ANSI, and other standards bodies enable the operation of office and home LANs, Figure 27. They are built upon the OSI reference model, discussed earlier, and also are based on standards. One of these standards, Ethernet, or IEEE 802.3, is used by most of the world as the basis for today's office and home LANs. Wi-Fi is effectively wireless Ethernet.

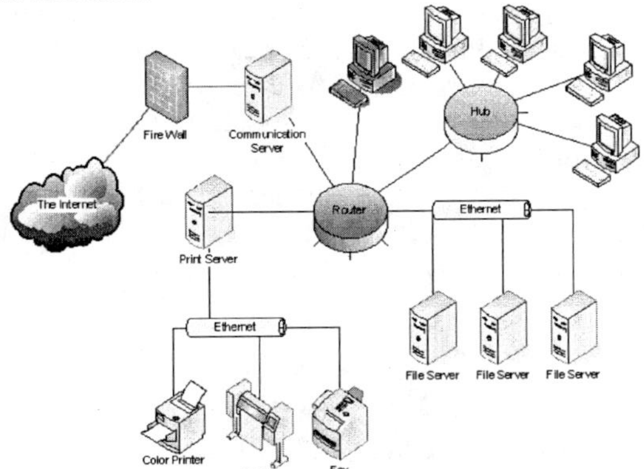

Figure 27 – Typical Local Area Network with Internet and/or WAN Access

Because of this wide usage, two questions come to mind. First, is there any overlap of technology that can be shared among these diverse environments? Next question: is there any reason to interconnect, or interwork (short for internetwork), office and home LANs with vehicle networks? The answer to both questions is: yes. Here's how and why.

Ethernet is used in LANs everywhere, and it is inexpensive. There have been many studies, especially on heavy-duty and military vehicle platforms, which have looked at GbE, or Gigabit Ethernet, as an alternative to some of the vehicle networks previously described. It's not that GbE would need to replace them, but it could be interworked via some gateway functionality on a vehicle. A possible solution would be with the use of a bridge, hub, or router. There are so many routers on the market today, and they would just need to have some of the vehicle network protocol standards to leverage all of the wireless and wired Ethernet interoperability.

Looking to the future, when dealing with capabilities past the generic "plugging-in" or "connecting wirelessly" into the vehicle diagnostic port and performing diagnostics, possibilities at the top of the list include: (a) near real-time vehicle information transfer via wide area networks, and (b) Prognostics (Proactive Diagnostics).

Commercial providers of near real-time vehicle information for fleets (for example, Qualcomm, PeopleNet, and Xata) offer Prognostics or Condition Based Maintenance (CBM) applications which require a very large infrastructure to be in place within a commercial or military fleet. This infrastructure includes computer servers with large memory capacity, high speed database systems, and the software and reports to mine the database for patterns and trends. One of the toughest aspects is to get that information, in an understandable form, back to fleet maintenance managers.

Significant effort is involved in the planning and implementation of a CBM+ back-end system, as well as installing a hardware computer platform into the vehicle that is able to collect and process the on-vehicle data. Besides the proper hardware platform, the best solution will include an API (Application Programming Interface) to support functions of the prognostics client "plug-in" module and integrated support for the fleet logistic operating implementation

C. Vehicle Networks to LANs

There are hundreds of millions of vehicle networks in use today, but also a similar number of LANs with investment in wired or wireless technologies. As these technologies are local, they are both short range at this time. This investment in technology develops into a practical approach to interwork LANs with vehicle networks, which results in optimizing applications and storage. Of course, we should be able to "plug" our vehicle, either wired or wireless, into our LAN to share apps and data. Whether it's downloading information for entertainment or other data, it can be accomplished efficiently over these short range networks. At issue is, how much time does it take the vehicle owner to select and transfer information that may already be "owned" vs. virtually having access to all information. Which is better?

For example, a home or office LAN today does not need much information from a vehicle; it is the other way around. The LAN has access to inexpensive data archival and historical data, but depending upon the type of data, not everything. If a collection of MP3s are resident on the LAN, a portion of them, or all of them, could be made available to the vehicle network in real-time, but is not cost effective to transmit to the vehicle for storage in terabyte numbers.

More effectively, a vehicle network should be accessing a WAN, or Wide Area Network service, to virtually have access to all possible MP3s. The issue is the low cost of LAN information that has already been paid for vs. the convenience of the WAN and access "on demand". In short, a reason to interwork with a LAN is cost effective data access.

Figure 28 – HD Ford Sync and Some Supported Devices

An example of merging these environments, vehicle networks and LANs, lies in Ford's SYNC, Figure 28. SYNC is a factory-installed, "in-car communications and entertainment" system (commonly referred to as ICE) jointly developed by Ford and Microsoft. It is based on the Microsoft Auto platform, formerly known as Windows CE for Automotive. Microsoft Auto Platform, and therefore SYNC, is an embedded operating system based on Windows CE for use on computer systems in automobiles. It utilizes a Motorola ARM 11 processor and 256MB of DRAM, flash. memory/USB port and speech technology from Nuance Communications (Figure 29).

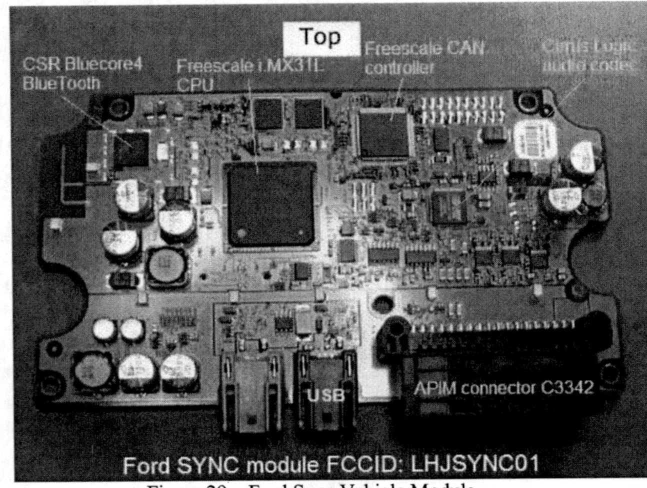

Figure 29 – Ford Sync Vehicle Module

Ford SYNC allows drivers to bring nearly any mobile phone and some digital media players into their vehicle and operate them using voice commands, the vehicle's steering wheel, or radio controls. This architecture may effectively circumvent the need to attach to an existing Local Area Network.

D. Vehicle Networks to WANs

How integrated does SYNC need to be with the vehicle network? How should this architecture be effectively interworked with Wide Area Network services and, once selected, how should the networks be secured? We cannot separate these questions, because of their cause and effect.

A vehicle needs a good WAN connection, but ideally it should be just one, serving the vehicle networks' diagnostic needs, information updates, along with entertainment, or ICE for an automobile. The reason for a single WAN connection is simply cost. There have been many proposals of what is economical for heavy-duty truck & bus vehicles, autos, along with common military vehicles: Satellite services (e.g. On Star, Hughes Telematics), Emerging WANs (WiMAX (section i. follows), LTE (section ii. follows), MANs (metropolitan area networks)) and many more.

There is, however, one solution that seems to stand out, cellular data networks. Companies that are offering services using so-called 3G cellular GSM (Global System for Mobile Communications) and CDMA (Code Division Multiple

Access) networks had issues, but so called 4G networks (actually still technically 3G) based on requirements from the ITU (International Telecommunication Union) from cell companies are either using WiMAX (Worldwide Interoperability for Microwave Access) as their backbone, or emerging and competing LTE (Long Term Evolution) networks for high speed broadband capabilities hold promise for being true cable and DSL competitors for fixed and mobile users.

i. Wi-Fi to the MAX

WiMAX, an IEEE standard, is a telecommunications protocol that originally was a high bandwidth fixed wireless access technology but met with success for mobile networking when its mobile variant, backed by Intel, was released as a standard. So, WiMAX provides both fixed and fully mobile internet access.

The current WiMAX revision, developed by the Broadband Wireless Access Working Group within the LAN/MAN subcommittee, provides up to 40Mbps with the IEEE 802.16m update expected offer up to 1 Gbps fixed speeds. In terms of performance, although the name implies Wi-Fi to the Max (Figure 30), this technology is not plagued with problems users have come to know as synonymous with Wi-Fi. Since WiMAX is a telecom service provider solution, the consumer is not responsible as the network administrator for security. Instead, network security professionals are responsible for security.

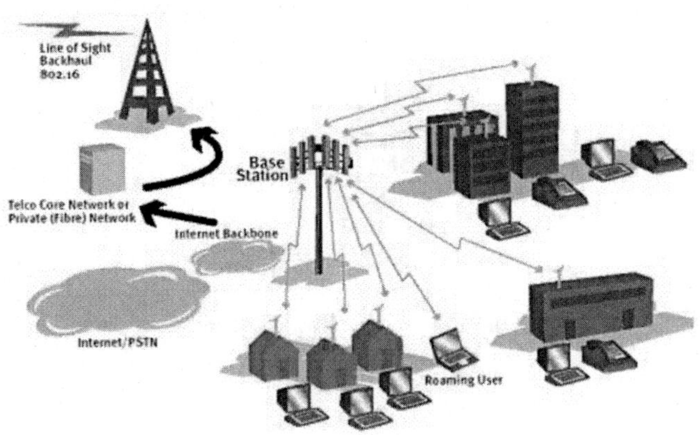

Figure 30 – WiMAX Use & Architecture

With regard to performance and reliability, WiMAX uses Multiple Input Multiple Output (MIMO) technology (Figure 31), provided on the latest publicly available Wi-Fi standard 802.11n, which provides multiple antennas to either serve as backup data transmission sources or to regularly share in this responsibility. These WiMAX MIMO antennae have another advantage, as they use allocated frequencies leased from governments, like the FCC in the US, to avoid interference issues with cordless phones, microwave ovens, Wi-Fi routers, etc. that effected Wi-Fi. WiMAX is implemented by Sprint, for example, in the US.

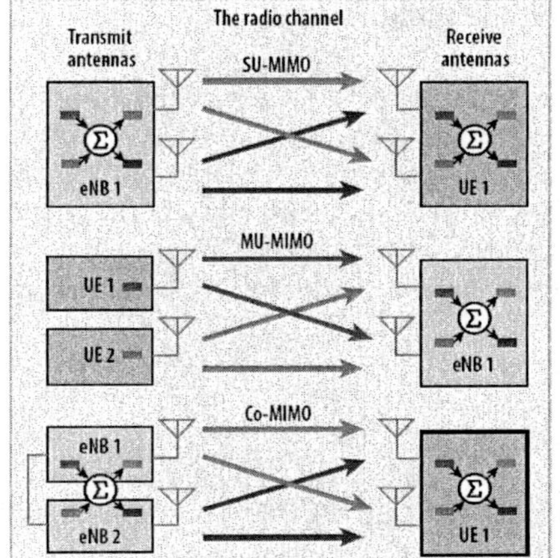

Figure 31 – WiMax Multiple In Multiple Out (MIMO Antenna
SU is Single User, MU Multiple User, Co Cooperative Users

ii. Long Term Evolution (LTE)

LTE (Long Term Evolution) was developed in 2005 by the 3rd Generation Partnership Project, a partnership of standards organizations responsible for GSM standards.

What does LTE do that WiMAX doesn't, as they both implement MIMO? The big difference is that LTE implements an all-IP (Internet Protocol) architecture and is no longer using circuit switching as its base technology. LTE is being implemented by Eriksson in Europe, AT&T and Verizon in the US.

Which of these technologies comes out in front? Or, what if both do? Or, what if something else better emerges in the future? It doesn't matter which wins as the stage is set:

computer networking if moving toward telephony with WiMAX and telephony is moving away from circuit switching to IP networks with LTE. These technologies can either coexist, or morph into a new need for standards to meld them together. The point is, the cellular data network is emerging as the mobile network, and either a phone/smart phone or the same internal components that provide you with "cellular data dial tone" can be how the network is accessed by HD and other vehicles. There is, however, one additional issue that should be mentioned, and it is very important. Is there a need for speed?

iii. WANs on the Cheap

The first vehicle network to WAN apps, possibly accessed via a LAN along the way, may not need the most robust 4G network speeds. Remember, J1708 data is nowhere near the speed of a 6 to 10Mbps "4G" network, and neither is J1939 at 250Kbps. So, what about WANs on the cheap, the older services from cellular and other providers? Where did they go?

The answer may be surprising. Nowhere. Inexpensive 1G, 2G, 2.5G networks are still available, accessible, and running for price points that may be much more in line with what is needed to begin the interoperability of HD and other vehicle networks with WANs.

iv. The Most Important WAN

Even though we access the robust cellular data network for mobile applications using WiMAX or LTE, the backbone of these networks may be the Internet, especially if service providers are using economically viable WAN architectures. Thus, we will potentially flood the Internet with data. The Internet's architecture is not ready for all of these mobile apps to traverse its backbone. Thankfully, this is also being worked on.

INTERNET2®

First is the Internet 2 project (www.internet2.edu), a not-for-profit advanced networking consortium comprising more than 200 U.S. universities in cooperation with 70 leading corporations and 45 government agencies which since 1996 has been jointly run in Ann Arbor, Michigan and Washington, D.C. This project leads to increased backbone technologies (Figure 32) and support of fixed high speed applications, middleware, security, network research and performance measurement capabilities which are critical to the progress of the Internet. This is important to the success of mobile networking, and thereby critical to vehicle networks growing in both application and economical sharing of data.

Figure 32 – Internet 2 Backbone Overlay Network

Next, the National Science Foundation (NSF) has launched the Global Environment for Network Innovations (GENI) (www.geni.net) which also holds the promise to aid in building the next generation internet.

Currently a $367M project, GENI is a virtual laboratory for exploring future internets at scale, creating major opportunities to understand, innovate and transform global networks and their interactions with society. This is a longer term solution, which over time like the Internet 2 project can provide collaborative and exploratory environments for academia, industry and the public to catalyze groundbreaking discoveries and innovation.

10. CONCLUSIONS

There are many emerging heavy-duty vehicle network application needs (Figure 33), and an ever-increasing number of active vehicle and trailer components are being added with each new architectural design.

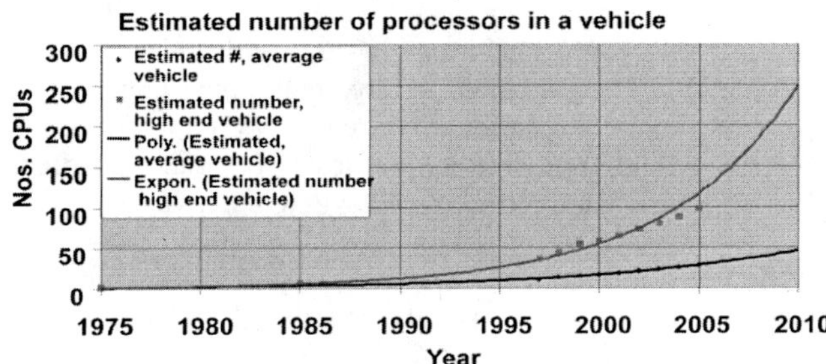

Figure 33 – Ever-increasing CPUs on a Vehicle

With the help of the many standards volunteers from industry, SAE, ISO, IEEE, TMC, and other standards bodies will keep vehicle network standards continuously updated to meet industry demands, and lower overall operating expenses.

A. *Future of Vehicle Networking*

Some new HD vehicles have heavy vehicle network traffic on the J1939 network that is slowing communication and negatively affecting performance. But, there are some novel network design solutions that can be used on these vehicles to adapt the network so that we can keep using J1939 for quite some time in the future.

For example, several OEMs are architecting or splitting the data bus into separate networks, for example, one for control, and one for non-critical items such as air conditioning controls. A "gateway" device connects these two buses and passes information back and forth as needed. Another option is to "pump-up-the-speed" by sending network traffic at a higher bit rate, 500K bits/sec., on the bus (this is now being worked on by the SAE J1939-14 Task Force).

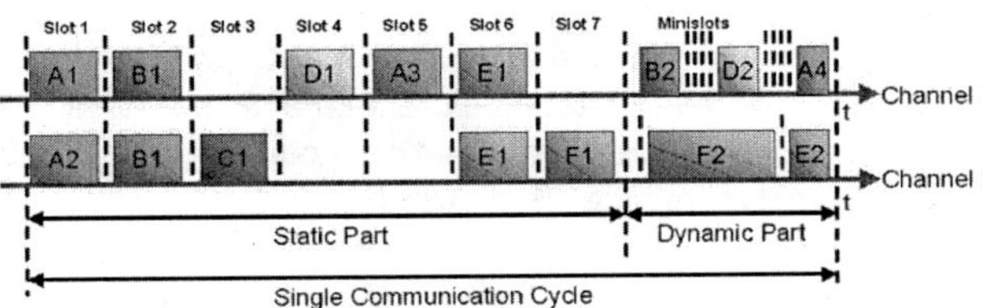

Figure 34 - Example of an Emerging Network Standard – FlexRay, which the Heavy Duty Industry is likely to adopt in the future.

While J1939/CAN works well for vehicle control, it is not fast enough for future developments such as brake-by-wire, steer-by-wire, and infotainment applications. The bottom line is that J1939/CAN is serving our heavy-duty industries exceptionally well, and it will be used for many years to come. However, looking to the future and working with the automotive industry, emerging is the need for even higher-speed protocols like Ethernet and FlexRay (Figure 34), along with fiber optic networks like MOST.

Figure 34 above shows the FlexRay protocol's time slot synchronized communication cycle. FlexRay runs up to 10 times faster than CAN based protocols. Some high end automobile OEMS are now using FlexRay networks for controlling vehicle stability systems. SAE International is currently developing a new standard for FlexRay systems (J2813). On the other hand, there is also a need for lower-cost/lower-speed application specific networks like SAE J2602 which incorporates the Local Interconnect Network (LIN) protocol. SAE J2602 can reduce vehicle wire count and offers simple messages for items such as heated mirrors, adjustable seats and pumps. Figure 35 shows possible high level vehicle architecture with multiple networks, while Figure 36 provides an overview of some common and emerging vehicle network protocols and their applications.

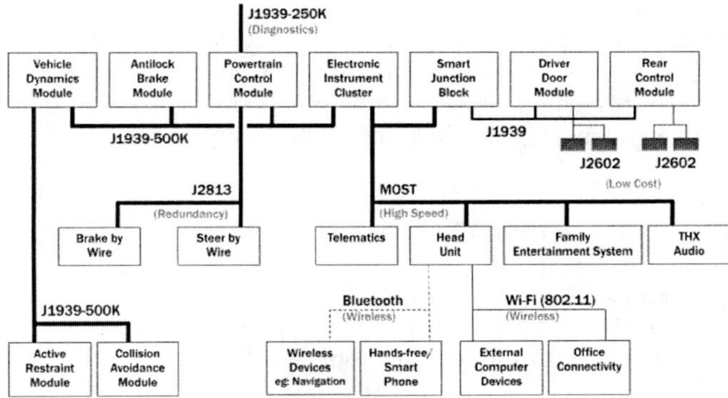

Figure 35 – Ever-increasing number of Vehicle Networks to Handle Emerging Heavy-Duty Applications

As you can see, a modern vehicle has several different networks, each designed for different applications, and future vehicles will have even more. Some OEMs are producing vehicles today with 6 or more in-vehicle networks.

Vehicle-Network Protocol	Base Technology	Applications
SAE J1939	CAN	Control and Diagnostics
SAE J2602	LIN	Input/Output Functions
SAE J2813	FlexRay	Control
MOST	MOST	Multimedia data
IEEE 802.11	Wi-Fi	WAN Connection
IEEE 802.16	WiMax	High speed wireless voice and data
IEEE 1609	Wi-Fi	Vehicle to Vehicle data, Vehicle to Road Side data

Figure 36 – Common & Emerging Vehicle Network Protocols

One thing that is consistent is change. One should note that HD OEMs are interested in eliminating the SAE J1708 network, possibly migrating its functions to the J1939 network. Also, we mentioned above that SAE is developing a faster 500K bits/sec. J1939 network to handle the increased communications traffic (that is J1939-14) to be introduced in the next few years.

Figure 37 – Vehicle Networks used for more than Vehicles

Heavy Duty vehicle networks are no longer confined to vehicle applications. For example, stand-alone compressor in Figure 37 utilizes standards-based network messaging for control and information flow, while some medical equipment for heart and lung functionality utilizes FlexRay.

B. *Getting IT Right*

Getting IT, or Information Technology, right, is crucial to Vehicle Networks and their ability to evolve to both integrate with and offer new applications.

i. *Adaptive Bit Rate Technology*

One example of getting IT right, relying upon cellular data networks, and supporting the evolution of the Internet to make applications run efficiently over this environment. For example, a new algorithm was found to be all it took, e.g. adaptive bit-rate technology, to allow information to efficiently stream data rather than broadcast for next generation of live, multi-channel television over the Internet.

ii. *Security*

Finally, as all of these vehicle networks and associated and developed "services" are interconnected via cellular data network services, the need for security will arise. Perhaps the most relevant way to describe why is just to ask what is different between this network and a PC that needs protection on a LAN. One could argue that since the vehicle is in motion, even more is at stake with virus, DOS (Denial of Service) attacks, Malware and the like.

In a recent paper by university researchers [Ref 20], several experiments were performed where vehicles were actually "hacked". Performance, safety, and other elements of a typical vehicle's functionality are easily compromised, all of which will be exacerbated, or at least potentially enabled, by WAN access via interface to HD vehicle networks.

"Modern automobiles are no longer mere mechanical devices; they are pervasively monitored and controlled by dozens of digital computers coordinated via internal vehicular networks. While this transformation has driven major advancements in efficiency and safety, it has also introduced a range of new potential risks...demonstrate the fragility of the underlying system structure. We demonstrate that an attacker who is able to infiltrate virtually any Electronic Control Unit (ECU) can leverage this ability to completely circumvent a broad array of safety-critical systems." [Ref 20]

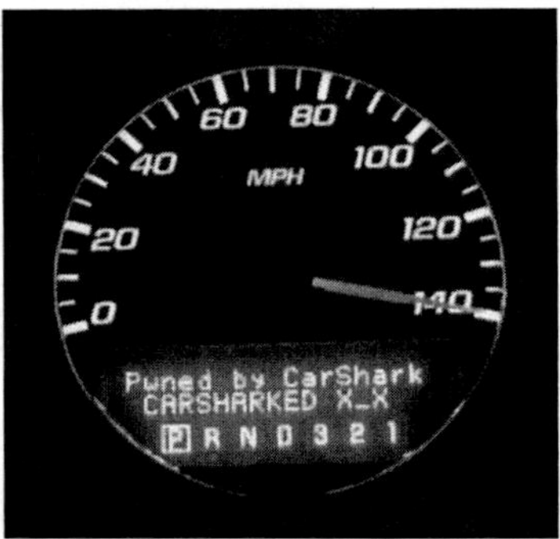

Figure 37 - Burning Rubber or Saved by Zero? 140MPH in Park

Security concerns become even more important, as the vulnerabilities get even worse as they show by experiments that, "We find that it is possible to bypass rudimentary network security protections within the car, such as maliciously bridging between our car's two internal subnets. We also present composite attacks that leverage individual weaknesses, including an attack that embeds malicious code in a car's telematics unit and that will completely erase any evidence of its presence after a crash. [Ref 20]

The experimenters easily managed to disable communications to and from all the ECUs, placed the Engine Control Module (ECM) and Transmission Control Module (TCM) into reflashing mode while the vehicle was moving, and took advantage of OEM noncompliant Access Control in that both vehicle network related firmware and memory (Figure 37). This last item allowed them to access ECUs with emissions, anti-theft, and safety functionality which was supposed to be protected by a challenge/response access control protocol.

A self-destruct scenario was also created, with only 200 lines of code. Combining control over various components, a 60-second count-down was displayed on the Driver Information Center (the dash), accompanied by clicks at an increasing rate and horn honks in the last few seconds. This sequence culminated with killing the engine and activating the door lock relay, rendering the electronic door unlock button useless.

Clearly, security is gaining importance and will become critical to the proper operation of vehicle networks as the merging of these networks with LANs and WANs takes place.

C. Merge Ahead

As this paper describes, "Merge Ahead" means looking ahead to interconnect Heavy-Duty and all vehicle networks with Local Area and/or Wide Area Networks. In-vehicle networks are expected to enable further growth in HD applications for fleet management, prognostics, safety and many other areas. New applications will certainly emerge supported, and at times enabled, by Wide Area Network technology, including access via LANs, and integrated with SAE J1939 and other in-vehicle networks. These applications should not only raise new Vehicle Network questions which are yet to be explored, but actually are the nexus of the bold new beginning vehicle networking. It all started about a quarter century ago, but it's far from over.

Then again, maybe vehicle networking all started long before then…with a single wire.

11. ACKNOWLEDGEMENTS

The author wishes to thank Karl E. Schohl and Ken DeGrant for their assistance with providing input and background information for this paper.

12. AUTHOR'S BIOGRAPHY

Mr. Mark P. Zachos is the founder and CEO of DG Technologies (DG), a company specializing in Vehicle Diagnostics and Communications Network Technology. Mr. Zachos was elected to SAE International's Board of Directors in 2010 for a three year term, responsible for redefining SAE's strategic direction, providing leadership that links SAE's members to the operating organization and ensuring organizational performance on a macro level.

At DG Technologies, Mr. Zachos is responsible for the company's overall operating business as well as promoting, selecting, developing and integrating advanced enabling technologies that support all of DG Technologies' programs and strategic initiatives. He is also responsible for new technology "mining" and commercialization, and for the technical direction of all DG products and services.

Mark has been very active in SAE Technical Committee work for over 20 years. He is Vice-Chairman of both the Vehicle E/E Diagnostics Committee and the Vehicle Systems Network Architecture Committee, and is a contributing member of the SAE Truck and Bus Communications Sub-Committee. He is also Chairman of the J2602, J2561, J2411, J2178, J1699-3, J1939-82 and J1939-84 Task Forces.

Mark has had direct involvement with the development of over 20 technical standards relating to vehicle networking; Organized, authored, and co-authored dozens of technical papers and led technical sessions at SAE World Congress and other SAE and industry events; Participated with SAE and Industry in the development of OBD-2 technology providing the US EPA and the California Air Resource Board (CARB) with new technical standards for diagnostics of vehicle emission systems; Organized an SAE Toptech on Future Technology for Vehicle Communications; Actively recruited new SAE members; and Contributed on behalf of SAE to other technical organizations including many ISO Working Groups and IEEE.

A past recipient of SAE's Forest McFarland Award and the Outstanding Contribution Award from SAE's Technical Standards Board, Mark is extremely honored to receive the 2010 L. Ray Buckendale Lecture award. To this end, Mark's goal is to use this opportunity to detail the importance of vehicle networks not only to the commercial vehicle industry that DG Technologies is an integral part of, but also to engineers and students to emphasize both practical and emerging application of vehicle networks.

13. REFERENCES

[REF1] SAE J1708 Physical Layer
http://standards.sae.org/wip/j1708

[REF2] SAE J1587 Electronic Data Interchange Between Microcomputer Systems in HD Vehicle Applications
http://standards.sae.org/j1587_200807

[REF3] SAE J1939 Recommended Practice for a Serial Control and Communications Vehicle Network
http://www.sae.org/standardsdev/groundvehicle/j1939.htm

[REF4] SAE J1939-11 Physical Layer, 250K bits/s, Twisted Shielded Pair
http://standards.sae.org/j1939/11_200609

[REF5] SAE J1939-15 Reduced Physical Layer, 250K bits/s, Unshielded Twisted Pair
http://standards.sae.org/j1939/15_200808

[REF6] SAE J1939-21 Data Link Layer
http://standards.sae.org/wip/j1939/21

[REF7] SAE J1939-03 On Board Diagnostics Implementation Guide
http://standards.sae.org/j1939/03_200812

[REF8] SAE J1939-73 Application Layer - Diagnostics
http://standards.sae.org/j1939/73_201002

[REF9] SAE J1939-84 OBD Compliance Test Cases
http://standards.sae.org/j1939/84_200812

[REF10] SAE J2534 Recommended Practice for Vehicle Pass-Thru Programming
http://standards.sae.org/j2534/1_200412

[REF11] SAE 1930 E/E System Diagnostics Term
http://standards.sae.org/j1930_200810

[REF12] SAE J2497 Power Line Carrier Communications for Commercial Vehicles
http://standards.sae.org/j2497_200706

[REF13] SAE J1979 E/E Diagnostic Test Modes
http://standards.sae.org/j1979_200705

[REF14] SAE J2012 Diagnostic Trouble Code Definitions
http://standards.sae.org/j2012_200712

[REF15] ISO 9141 Diagnostic Systems – Requirements for interchange of digital information
http://www.iso.org/iso/catalogue_detail?csnumber=16737

[REF16] ISO 15765-4 Diagnostics on CAN
http://www.iso.org/iso/iso_catalogue/catalogue_tc/catalogue_detail.htm?csnumber=33619

[REF17] California Air Resources Board (CARB) On Board Diagnostics (OBD)
http://www.arb.ca.gov/msprog/obdprog/obdprog.htm

[REF 18] Freund, Deborah M.,
Foundations of Commercial Vehicle Safety: Laws, Regulations, and Standards, SAE 2007-01-4298.

[REF 19] Wireless Roadside Inspection Proof of Concept,
Federal Motor Carrier Safety Administration, US Department of Transportation
September 2009 Report, FMCSA-RRA-09-007_WRI-POC

[REF 20] IEEE Computer Society 2010 IEEE Symposium on Security and Privacy
Experimental Security Analysis of a Modern Automobile
Karl Koscher, Alexei Czeskis, Franziska Roesner, Shwetak Patel, and Tadayoshi Kohno
Department of Computer Science and Engineering, University of Washington
Stephen Checkoway, Damon McCoy, Brian Kantor, Danny Anderson, Hovav Shacham, and Stefan Savage
Department of Computer Science and Engineering, University of California San Diego

SAE J1939-13 Off-Board Diagnostic Connector
http://standards.sae.org/j1939/13_200403

SAE J1939-71 Vehicle Application Layer
http://standards.sae.org/j1939/71_201002

SAE J1962 Diagnostic Connector
http://standards.sae.org/j1962_200204

TMC RP1210
http://www.atabusinesssolutions.com/p-271-atas-tmc-2010-2011-recommended-practices-manual.aspx

ISO 27145 WWH-OBD Requirements
http://www.iso.org/iso/iso_catalogue/catalogue_tc/catalogue_detail.htm?csnumber=44022

SAE J2813 FlexRay for Vehicle Applications
http://standards.sae.org/wip/j2813

SAE J2602 LIN Network for Vehicle Applications
http://standards.sae.org/j2602/1_200509

Microwaves & RF, Analyze Antenna Approaches for LTE Wireless Systems
Moray Rumney, Janine Whitacre | ED Online ID #19534 | August 2008
http://www.mwrf.com

PointRed Telecom
http://www.pointredtech.com/mobile_wimax_technology.htm

3GPP The Mobile Broadband Standard
http://www.3gpp.org/

Internet 2
http://www.internet2.org

Global Environment for Network Innovations (GENI)
http://www.geni.net

14. KEY TERMINOLOGY & NOMENCLATURE

Acronym	Term	Definition
API	Application Programmers Interface	A documented methodology for writing software programs so that they operate cooperatively with other programs. In this document, the API refers to a DLL containing a set of functions that the diagnostic application can call to connect and send/read messages to/from a vehicle data bus using a VDA and associated VDA device drivers. Also see DLL.
bits/sec	Bits per Second	Speed of the bit data being transferred on the network.
CAN	Controller Area Network	An equipment standard designed to allow microcontrollers and devices to communicate with each other. In this document, the term CAN refers to the Bosch CAN 2.0B specification as specifically called out in J1939-21 (250k bits/sec., 29-bit identifiers, etc) with no deviations from the J1939-21 specification.
CARB	California Air Resources Board	Part of the California state government responsible for developing vehicle emissions requirements.
CBM	Condition Based Maintenance	Vehicle maintenance function which repairs or replaces components before a fault occurs.
Comm	Communication	The Transfer Of Information Between Parties
DLL	Dynamic Link Library	A DLL is an implementation of an API. It is a mechanism to link applications to libraries at run-time instead of at compile-time. The libraries are separate files and are not copied into an application's executable as with static linking. The RP1210 API is specifically called out by RP1210 to be in a DLL form.
DTC	Diagnostic Trouble Code	See "Fault Code".
ECM	Electronic Control Module	Synonymous with ECU and sometimes referred to as Engine Control Module (which is also categorized as an ECM/ECU).
ECU	Electronic Control Unit	Synonymous with ECM, the ECU represents an electronic component or vehicle control module (i.e. engine, brakes, and transmission). This term will be used throughout this document. You will see the term TCM/TCU, BCM/BCU for transmission and body controllers as well as other flavors of xCM/xCU. They are all categorized as "generic" ECU/ECM's.

Fault Code	Fault, Fault Code, Diagnostic Trouble Code (DTC)	A generic term representing a vehicle/equipment component that detects a situation/problem/issue that is not part of normal operation and needs to notify other components or a diagnostic application that there is a problem. Fault codes are defined in the J1587 and J1939-73 documents. Along with fault codes, come definitions like MID, PID, SID, SPN, FMI.
FMI	J1708 and J1939 Failure Mode Indicator	An FMI is a means to describe how a component failed causing it to report a fault code. For example, "shorted high".
Frame	Frame	Single smallest information possible to transmit on a network (Data Bus).
Header	Header	Part of the message frame used to convey information about what data is contained. The Header normally does not contain vehicle data (payload) and is considered message overhead.
HD	Heavy Duty	In this document, a generic term that refers to medium and large diesel driven equipment ranging from generators, ground support equipment and trucks.
HD-OBD	Heavy Duty On-Board Diagnostics	Diagnostics functions providing information for monitoring the proper function of vehicle systems.
ISO	International Standards Organization	Standards organization with headquarters in Europe.
ISO 11898	ISO 11898	A set of standards related to the CAN protocol.
ISO14229	ISO14229	For this document, ISO14229 (UDS) is a message layer protocol running on top of the CAN protocol (as defined by J1939-21) and is being used by some OEMs to troubleshoot and diagnose their 2007+ engines.
ISO15765-2	ISO15765-2	A document relating to a serial control and communications vehicle network developed by the ISO body permitting the transmission of any length message from 0 to 4095 bytes. Messages that use the ISO15765-2 messaging format are commonly referred to as "Segmented Messages" or "Multi-Frame Messages", although single frames are also permitted by this format. In this document, the ISO15765-2 transport mechanism is used by Detroit Diesel to send ISO14229 messages.
J1587	J1587	SAE document entitled "Joint SAE/TMC Electronic Data Interchange Between Microcomputer Systems in Heavy-Duty

		Vehicle Applications." In this document, J1587 is the message layer riding on the J1708 physical layer. All references of J1587 herein refer to J1587 messages running on the J1708 physical layer.
J1708	J1708	SAE document entitled "Serial Data Communications Between Microcomputer Systems in Heavy-Duty Vehicle Applications". In this document, J1708 is the physical layer for the J1587 message layer.
J1939	J1939	A series of documents relating to a serial control and communications vehicle network developed by SAE. In this document, J1939-71 and J1939-73 are a message layer riding on the physical layer.
J1939-11	J1939-11	Document that is part of the J1939 specification dealing with electrical characteristics of the standard J1939 data bus. Note - there is another document called J1939-15 (sometimes referred to as "J1939 light") which deals with the same topic but does not call out for a shield over the J1939 data bus wires.
J1939-21	J1939-21	Document that is part of the J1939 specification dealing with how J1939 messages are formatted, how they map into the 29-bit CAN 2.0B identifier, as well as how large messages are transmitted. This is probably the most important J1939 document along with J1939-71 and J1939-73.
J1939-71	J1939-71	Document that is part of the J1939 specification dealing with the definition of the J1939 messages. These definitions get to the bits and bytes of how data parameters shall appear on the data bus (i.e. "engine speed", "vehicle speed").
J1939-73	J1939-73	Document that is part of the J1939 specification dealing with all things diagnostic message related (i.e. "Fault Codes", "OBDII Readiness").
J2497	J2497	Document that defines a communications link between a tractor and a trailer using a Power Line Carrier (PLC) physical layer for transferring data between nodes.
LAN	Local Area Network	A communications medium, often based on Ethernet, allowing information sharing among computers and group access to a Wide Area Network service.
Message	Message	A communications transaction at the "Application Level". It may require several

		Frames to transfer all the message data.
Message Layer	Message Layer	The message layer (sometimes referred to as the "Application Layer") defines how messages on a physical layer are sent, received and interpreted by ECUs. There are three message layers defined in this document; ❑ J1939 uses the CAN physical layer ❑ J1587 uses the SAE J1708 physical layer ❑ ISO14229 (UDS) uses the CAN physical layer with CAN parameters as defined in J1939-21. This is OEM specific. ❑ ISO15765-2 uses the CAN physical layer with CAN parameters as defined in J1939-21. This is OEM specific.
MID	J1708/J1587 Message Identifier	Defined in J1708 and in J1587 as the source address of the controller that is sending a message. For example, MID 128 is the Engine and MID 130 is the transmission. MID is also part of a J1587 fault code.
Network	Network	A system to communicate information between nodes. Sometimes also referred to as a Data Bus.
Node	Node	A Module (or ECU) connected to a network
OBD	On Board Diagnostics	See HD-OBD
OBD-2 (or OBD-II)	OBD-2	On Board Diagnostics second generation requirements as described by CARB regulations.
OSI	Open Systems Interconnect	The ISO OSI (Open Systems Interconnect, ISO IEC 7498) 7-layer model was created to describe all networks.
Physical Layer	Physical Layer	The physical electronics responsible for component to component, and vehicle/component to PC communications. (The physical layers outlined in this document are SAE J1939-11, J1939-15 and J1708.)
PID	J1708/J1587 Parameter Identifier	Defined in J1587 as the label that identifies the next piece of data in a J1587 message. It is also used to identify a fault code where there is a problem with a "parameter" (i.e. "oil pressure low") as opposed to a specific component (SID) on a vehicle (i.e. "injector cylinder 1").
PLC	Power Line Carrier	Network using the vehicle's Power Line as the carrier for message communications.

		(Also see J2497).
Prognostics	Prognostics	"Predictive-Diagnostics" function which predicts vehicle component failures before they occur.
Protocol	Protocol	Rules that nodes must follow to communicate
Random Access	Random Access	A protocol that has no time synchronization between nodes, operating with a message arbitration method.
RP1210	RP1210	A TMC document worked on by the S.12 Onboard Vehicle Electronics Study Group entitled "S.12 – RP1210x – VMRS 053 WINDOWS™ COMMUNICATION API" (x = revision, currently "B"). This document defines a standard API that VDA vendors provide so that vehicle diagnostic applications can communicate with their components. The current version of RP1210 is RP1210B; however RP1210A is the most commonly implemented standard by VDAs and applications at this time. RP1210B mainly closed holes in the RP1210A specification.
RP1210 Compliant Application	RP1210 Compliant Application	Any application that is written to be RP1210 compliant and allows the user to select any RP1210 adapter that meets minimum specifications as defined in this document. The application should work for any VDA that is RP1210 compliant and supports the protocols and operating system needed by that application.
RP1210 Compliant VDA	RP1210 Compliant VDA	Any VDA that is written so that their API is RP1210A compliant and meets the minimum specifications as defined in this document.
RS-485	RS-485	A term used in this document as a reference to the SAE J1708 physical layer (which is a slightly modified version of RS-485). RS-485 is formally defined by the American National Standard Institution (ANSI) standards body.
SAE	SAE International	SAE is a standards body producing standards associated with mobility industries including Commercial Vehicle.
Scan Tool	Scan Tool	Hand held device used for diagnostics communications to a vehicle.
Serial Communication	Serial Communication	Transferring data one bit at a time (0 or 1)
SID	J1708/J1587 - Subsystem Identifier	Defined in J1587 for use with faults to identify a problem with a specific replaceable component on the vehicle/equipment (i.e. "injector cylinder 1")

		as opposed to a problem with a generic parameter (PID) (i.e. "oil pressure low").
SPN	J1939 Suspect Parameter Number	SPN for this document refers to a list of fault code to text translations that are part of the J1939 base document. These text translations are defined so that generic faults can be transmitted in a manner that speeds vehicle/equipment diagnosis.
Time Triggered	Time Triggered	A protocol that does have time synchronization between nodes, allowing messages to be sent only when it's designated time.
TMC	Technology and Maintenance Council	TMC is a part of the American Trucking Associations and is a standards body producing maintenance standards for medium and heavy duty engine driven vehicles. The RP1210 document is maintained by the TMC S.12 subcommittee.
UDS	Unified Diagnostic Services	See ISO 14229.
VDA	Vehicle Datalink Adapter	The physical device, when connected to the vehicle data bus, provides translation between the data bus and a diagnostics application.
WAN	Wide Area Network	One of many data networking services provided to Local Area Networks and potentially Vehicle Networks, allowing for information sharing in stored and real-time.

Editor's and Special Contributors' Biographies

About the Editor

Dr. Andrew Brown, Jr.
Executive Director & Chief Technologist, Delphi Corporation

Dr. Andrew Brown, Jr. is Executive Director & Chief Technologist for Delphi Corporation, and as such he provides leadership on corporate innovation and technology issues to help achieve profitable competitive advantage. Dr. Brown also represents Delphi globally in outside forums on matters of innovation and technology including government and regulatory agencies, customers, alliance partners, vendors, contracting agencies, academia, etc. Prior to this assignment, Dr. Brown had responsibility for common policies, practices, processes and performance across Delphi's 17,000 member technical community globally and its budget of $2.0 billion, including establishing Delphi's global engineering footprint with new centers in Poland, India, China, and Mexico, among others.

In April of 2009, SAE International's Executive Nominating Committee named Dr. Andrew Brown Jr., as its candidate for 2010 SAE International President. He was elected as 2010 SAE International President and Chairman in November of 2009, and was sworn into office in January of 2010.

As an NAE member, Dr. Brown was appointed by the National Research Council (NRC) to serve as chair of the Committee on Fuel Economy of Medium and Heavy Duty Vehicles. The report developed by this group was recently referenced by President Obama in his enhanced efforts on fuel economy improvement.

Dr. Brown joined Delphi coming from the GM Research and Development Center in Warren, Michigan, where he was Director - Research, Administration & Strategic Futures. He also served as a Manager of Saturn Car Facilities from 1985 to 1987. At Saturn, he was on the Site Selection Team and responsible for the conceptual design and engineering of this innovative manufacturing facility.

Dr. Brown began his GM career as a Project Engineer at Manufacturing Development in 1973. He progressed in the engineering field as a Senior Project Engineer, Staff Development Engineer, and Manager of R&D for the Manufacturing Staff. During this period, he worked on manufacturing processes and systems with an emphasis on energy systems, productivity improvement and environmental efficiency. Before joining GM, he supervised process development at Allied-Signal Corporation, now Honeywell, Incorporated in Morristown, New Jersey.

Dr. Brown earned a Bachelor of Science Degree in Chemical Engineering from Wayne State University in 1971. He received a Master of Business Administration in Finance and Marketing from Wayne State in 1975 and Master of Science Degree in Mechanical Engineering focused on energy and environmental engineering from the University of Detroit-Mercy in 1978. He completed the Penn State Executive Management Course in 1979. A registered Professional Engineer, Dr. Brown earned a Doctorate of Engineering in September 1992.

Special Contributors

Steven H. Bayless

Steven H. Bayless is the Director of Telecommunications and Telematics at the Intelligent Transportation Society of America (ITS America). He is responsible for providing guidance to ITS America's Board of Directors and senior staff on matters involving wireless and the automotive industry.

Steven previously served as staff advisor and Presidential Management Fellow in the Secretary of Transportation's policy office at the headquarters of the U.S. Department of Transportation. Steven had cabinet-level lead related to research and development, spectrum management, and telecommunications policies.

As a detailee to the White House, he assisted in formulation of several presidential directives on aerospace policy, focusing in particular on satellite navigation. He also advised the State Department and the Federal Aviation Administration in negotiations on space and aviation cooperation with the European Union, the Russian Federation, and Japan. In the surface transportation domain, Steven supported secretarial policy initiatives regarding the reorientation of federal research and development, highway safety, and transportation infrastructure finance reform.

Steven's career also includes tenure as a management consultant with American Management Systems (now CGI) in Washington, D.C. and as a project coordinator for DuPont Europe in Budapest, Hungary. Steven holds a specialized master's degree in International Security Studies and Business from the Fletcher School at Tufts University. His bachelor's degree is in Economics and Foreign Affairs from the University of Virginia. He has also completed graduate work at Harvard Business School and MIT Sloan School of Management in Cambridge Massachusetts, and at Eötvös Loránd University in Budapest, Hungary.

Scott Belcher

Scott F. Belcher became the President and CEO of the Intelligent Transportation Society of America (ITS America) in Washington, D.C. in September 2007 after a successful legal and nonprofit management career.

Scott brings more than 20 years of private and public sector experience to his current position. Prior to joining ITS America, Scott served as executive vice president and general counsel at the National Academy of Public Administration (NAPA) in Washington, D.C. Previously, he held senior management positions at a number of prominent trade associations. Scott also worked at the Environmental Protection Agency and in private practice at the law firm of Beveridge & Diamond, PC.

Scott's vision for moving ITS America to the next level includes raising awareness of the value of the organization among consumers, legislators, and the media, and seeking increased federal funding of ITS America initiatives. This vision will help guide the U.S. transportation network to a level of enhanced safety, reduced traffic congestion, decreased fuel consumption and emissions, and lower economic burden on our society.
Scott holds a juris doctor degree from the University of Virginia, a master of public policy degree from Georgetown University, and a bachelor of arts degree from the University of Redlands.

Doug Welk

Doug Welk is the Global Chief Engineer of the advanced engineering group supporting Delphi's Infotainment and Driver Interface Product Business Unit. He is responsible for defining a technology strategy for automotive cockpit electronics. Doug leads teams of engineers in creating an array of next generation products for future connected vehicles. His current projects focus on the areas of open computing platforms, automotive apps, portable device connectivity, software defined receivers, intelligent transportation systems, and in-vehicle HMI. Doug is also chairman of the GENIVI Alliance where he helps lead the definition and implementation of a common, industry-standard Linux software platform for in-vehicle infotainment products. Doug graduated from Purdue University and holds both a bachelor's and a master's degree in Electrical Engineering. He began his career at the Delco Electronics division of General Motors, which later became Delphi Automotive. He has been actively engaged in the development of key technology including radio data system (RDS) receivers, navigation systems, multiplexed data busses, and broadband data connectivity. Doug has several patents, intellectual property disclosures, and technical papers in these areas.

Tim Bolduc

Tim Bolduc is a staff engineer in the advanced engineering group supporting Delphi's Infotainment and Driver Interface Product Business Unit. In this role, Tim is responsible for leading project teams in developing an array of next generation features for future connected vehicles. Tim holds a master's degree in Electrical Engineering from Stanford University. He began his career at the Delco Electronics division of General Motors, which later became Delphi Automotive. He has been actively engaged in the development of key technologies. Tim has several patents, intellectual property disclosures, and technical papers related to this work.

Gerald J. Witt

Gerald Witt manages advanced driver interface development within Delphi's Advanced Infotainment and Driver Interface group. He has a diverse systems engineering and management background with more than 20 years of experience managing advanced HMI technology development. Gerald's areas of focus include displays, controls, driver monitoring, workload management, distraction mitigation, driver assistance, and the associated human factors research spanning both the active safety and infotainment domains. Gerald holds an Automotive Technology degree and a bachelor's degree in Electrical Engineering from Southern Illinois University. His technical achievements include numerous patents, corporate awards, and publications.

Keenan Estese

Keenan Estese is an Engineering Group Manager at Delphi who is responsible for connected vehicle technologies and applications. Keenan has led advanced development activities in the areas of open computing platforms, portable device connectivity, intelligent transportation systems, and bridging active safety with infotainment (i.e., "Connecting with Safety"). Keenan began his career at Delco Electronics, which later became Delphi Automotive. He has worked as an engineer and team leader in various areas such as integrated circuit design, design automation, and advanced infotainment systems. Keenan holds a master's degree in Electrical Engineering from Purdue University and a bachelor's degree in Electrical and Computer Engineering from the University of Cincinnati.

Dr. In-Soo Suh

In-Soo Suh is currently serving as an associate professor in the Cho Chun Shik Graduate School for Green Transportation, KAIST, Daejeon, Korea. His areas of expertise and research fields are the electric vehicle system with wireless and conductive charging infrastructure strategy, the vehicle system integration focused on electrical power train, NVH and structural acoustics, and the green transportation technology for future urban application related to the smart grid and communication technology.

In addition to Dr. Suh's 15-year industrial experience in global automotive OEM's with Chrysler Corporation and GM Korea, he continues to expand his career in academia with emphasis on applied engineering research and education in green mobility. As a vehicle group leader in the KAIST Wireless Power Transfer Project Group, one of his important contributions was leading the systematic efforts on the first public launch of the OLEV system in Seoul Grand Park. This has been recognized as one of the best 50 inventions of 2010 by Time magazine. He also received a design innovation award from the Society of Design and Process Science in 2011. Dr. Suh holds a PhD in Mechanical Engineering from Massachusetts Institute of Technology in Cambridge, MA, USA, where he majored in structural acoustics. He obtained his bachelor's and master's degrees in Mechanical Engineering from Seoul National University in Korea. He has been a general member of SAE since 1991 and an organizing committee member of SAE Noise and Vibration Conference since 2001. He is also a member of the committee for SAE TIR J2954 standard, Wireless Charging of Electric and Plug-in Hybrid. Dr. Suh can be reached by email at insoo.suh@kaist.ac.kr.